The Only Game In Town

The LOCOTROL Story

Fergus Moffat

Copyright © Fergus Moffat
December 2021
All rights reserved.

ISBN: 978-0-6485290-1-9

IngramSpark, La Vergne, TN, USA
A self-publishing print-on-demand platform, through
Ingram Content Group (Lightning Source), Melbourne, Australia

Cover image: Deep in the Rocky Mountain Trench, and having cruised along Lake Windermere for its entire 14.5 km (9 mile) length, a loaded CPR Locotrol train approaches the lake's northern end. The Fairmont Hot Springs Resort lies on the distant forested slopes of Fairmont Ridge above the chalk-coloured escarpment ahead of the Lead locomotive. Snow-capped Indian Head Mountain (aka Chisel peak, 2687 m/8850 ft) looms above.
Image: From a 1970s Harris product brochure.

For Milt

FOREWORD

At a time when there is much discussion about the development of autonomous vehicles and robots, and when remotely-controlled machines such as drones are becoming relatively commonplace, it is refreshing to look back several decades to a time when the rail industry—often criticised for the slow uptake of technology—had already developed and implemented state-of-the-art remote control technologies for some of its largest-scale operations.

For years, railways in North America had been working to operate progressively longer and heavier freight trains, mainly for the movement of high-volume freight such as minerals. In the years following World War II, the mass introduction of diesel-electric locomotives enabled coupled multiple units to be worked together by one driver with two, three or more at the head of each train. However, if there was a need to have additional locomotives pushing at the rear of the train or placed elsewhere within it, these had to be separately crewed. As a solution to one problem, this led to another: asynchronous operation of the motive power groups, and the potential promotion of unintended in-train forces.

At the same time, locomotives were becoming more powerful in both traction and dynamic braking. The combination of high tractive effort and the potent retardation properties of dynamic braking could produce enormous longitudinal forces within train consists. This made safe handling of very long and heavy trains challenging at best and created a derailment potential in some circumstances. Further, conventional air braking systems were becoming increasingly challenged on very long trains due to the propagation delays for the application and release of brakes—in turn creating the potential of further significant in-train forces.

In this incredibly detailed work, Fergus Moffat describes how these conditions produced a search for better solutions involving the use of additional locomotives dispersed to various positions within the train and under the control of the driver at the head-end. The eventual solution hinged on the development of secure space radio communications and increasingly sophisticated electronic control of locomotive control systems and train braking. This was the genesis of LOCOTROL, and the distributed motive power capability it provided on long, heavy trains in a variety of countries. Today, this technology has been further extended by the inclusion of wired ECP braking and a concomitant diminution of the need for RF communication.

This book is for both rail professionals and historians who might be interested in understanding the origins and technical development of these systems, as the author incorporates a valuable personal connection to these developments. These include a graphic part-diarised account of a 4-month installation project in India (which, although satisfactorily completed, ultimately fell victim to the turgid internal politics and massive bureaucracy of Indian Railways) and a highly personalised portrayal of the man most responsible for the introduction of LOCOTROL to the Canadian Pacific Railway—the first large-scale global customer for the technology. The CPR experience with LOCOTROL is vital since it paved the way for global acceptance of the product.

In my case, as a former senior rail executive in Queensland where that network's electrified coal-train operations have been extensively supported by LOCOTROL since the mid-1970s, this book struck a particular chord and reminded me of how dependent are such operations on the use of distributed power.

It is important that the history of technological development be accurately recorded and perhaps more so in the case of a niche application such as distributed power, that has so assuredly contributed to the safe operation of long, heavy freight trains. LOCOTROL was the progenitor of the technology, and Moffat's account—being the only known detailed history of these developments—is a remarkable contribution to such literature.

John Hearsch
President, Rail Futures Institute
Heathcote Junction, VIC
November 2019

PREFACE

This is the story of the development of remote-controlled railroad distributed motive power. Distributing locomotive power throughout the length of a long, heavy train and controlling those dispersed locomotives remotely, shares the motive force throughout the train rather than concentrating it all conventionally at the head-end. The technology provides a railroad with the tools to multiply the carrying capacity of trains without adding expensive trackage. With synchronous traction and braking control, the system provides the locomotive engineer with opportunities to more safely and effectively handle trains and is most beneficial in mountainous territory. Many railroads have found that it is the only solution to increasing network capacity other than to install duplicate or triplicate trackage.

This book is not intended as an instruction manual on how to operate LOCOTROL[1]. The reader should not expect that, and for that reason, a lot of available material that might have been included, has not been. Despite that, I have included some descriptive narrative and graphic substance around how LOCOTROL operates and how it is operated, the better to describe and explain it. The LOCOTROL product continues to evolve, and—not intending to pursue this incremental development as far as the current day—I have limited descriptions of it to the iteration known as LOCOTROL III. Since taking over Harris Controls, GE Transportation have greatly advanced the technology, integrating it with the electronic control of locomotives and train air braking.

Prior to the advent of ECP braking, distributed power had arguably been the single greatest technological advance for railroading since the introduction of the automatic coupler and the air brake triple valve. Originally known by various names before receiving the proprietary name LOCOTROL, this was the first distributed power scheme to be proven in regular service, and for 40 years was the only practical application of this technology. Nowadays, with wireline ECP braking, control of distributed power can be carried through the braking cable rather than by RF[2] transmission, and the use of the latter as a medium for achieving remote control in this context will probably become obsolete.

⊱⊰

Your author knew little about LOCOTROL prior to arriving in the Pilbara iron ore region of Western Australia from New Zealand in January 1981, and he had never heard the term 'distributed power'. On the narrow-gauge (1067 mm) New Zealand Railways, a 'bank' was our term for a significant hill or grade. When we needed temporary extra power for the grade we double-headed or 'banked', in some instances, from the rear. 'Banker engines' were how we referred to Helpers and Pushers. Whether they coupled onto the head-end or the hind-end of the train, bankers were crewed units; however, when they were attached at the head-end, we 'double-headed' (and with diesel-electric power, the MU jumper cable would be connected and the train crew would get a brief rest).

---

1 LOCOTROL is a registered trademark and product name of GE Transportation, a Wabtec company.
2 See Glossary.

Upon my arrival at the Mount Newman Mining Company in Port Hedland, Western Australia, I was confronted not only with American railroading (a heavy-haul standard-gauge permanent way built by a US construction company; 3600 hp Alco Century-Series locomotives; a rule book based solidly on what was in those days known as the North American Uniform Code of Operating Rules) but was also introduced to LOCOTROL. Not that it was being used much at the time, but in 1975 and 1976, Charlie Parker—then Assistant Chief of Motive Power & Rollingstock with the Canadian Pacific Railway—had visited MNM on a consultancy basis and had overseen the installation of two sets of LOCOTROL 105SS[3] equipment. Parker had been 'grabbed' for this while he was in Australia helping the Queensland Government Railways get a LOCOTROL operation up and running for their coal haulage task. Even then, MNM had an eye to the future and an expected increase in iron ore production.

I duly received training on the operation of this early LOCOTROL technology and was able to experience operating it on numerous, if not frequent, occasions. I continued to encounter these two LOCOTROL Lead units during my work and developed a curiosity about the equipment—including an ore car that had been adapted for use as a 'Robot' control car—and how it functioned. Over the ensuing couple of years, I hounded our operations supervisors as well as the supervisors of the Ore Car Repair, Locomotive Servicing, and Locomotive Maintenance Shops for extra information, and learned what I could about this fascinating technology. In the process, I acquired some very useful technical documentation including maintenance manuals.

During the 1980s, the company determined that distributed power had a necessary future in its train operations and moved to acquire the newer LOCOTROL II technology, eventually instituting a full DP train schedule: seven trains each way, every day over the 430 km between the mines and the port. In 2020—29 years after the author left the company—it has moved beyond LOCOTROL III and continues to operate LOCOTROL trains with communications piggy-backing off a Knorr-Bremse wireline ECP braking system.

It was during the installation process for the aforementioned expansion of the LOCOTROL operation that I met Milt Deno, manager of the Railroad Product Line for (then) Harris Corporation's *Controls & Composition Division*. Milt was an 'old-school' Canadian railroader with a solid mechanical trades, railroad operations, and management background with CPR and over a decade of intimate involvement with LOCOTROL.

Along with his trusty sidekick, electronics engineer Gene Smith (and sometimes others, such as Steve Heneka or Lewis Cox), Milt had been travelling throughout North America and across the globe during that decade, bringing the LOCOTROL product to many countries whose railroads had a requirement to commence long-train operations (usually heavy-haul, bulk mineral transport) or to facilitate long trains becoming even longer. Through collaboration with Milt—as a locomotive engineer, crew supervisor and instructor, and eventually as a member of an overseas project team—I came to know and admire the man, and we established a warm and mutually respectful relationship. I was gratified to learn, during 2014, that my 2-year-old nomination of Milt for induction into the Canadian Railway Hall of Fame had been accepted (in the *Industry Trailblazers* category). I had intended to keep the nomination a secret to surprise him, but a curious granddaughter—for whom Milt's iMac was no challenge—found the details in a Google search and I had to come clean. Milt was not able to travel to Ottawa for the ceremony and—living in retirement in Florida—passed away in April 2019, aged 86.

I had often thought about how seminal to railroad operations was the concept of remotely-controlled distributed power and felt that it deserved to be better-acknowledged throughout the global railway environment, including among interested lay-people. For this reason, at some time during the 1980s I wrote to Kalmbach Publishing Co (now Kalmbach Media)—publishers of *Trains* magazine, one of the world's most widely-read rail industry interest journals—suggesting they consider a special exposé on the subject. I received

---

3   An upgraded solid-state version of Radiation's early 105, and before that, 103 and 102 versions.

a polite response in which the editor thanked me for my interest and suggested the magazine might consider such an article at some time in the future. A feature that included some excellent graphics did eventually appear in the September 2010 issue, by which time I was well-advanced with the skeleton of this book. It should, though, be noted that *Trains* has featured the subject of distributed power in various ways in articles in October 1964 (Morgan), June 1971 (Morgan), February 1983 (Cavanaugh), and in August 2001 (Wheelihan and Murray).

Don Selby is known as the 'father' of LOCOTROL and acknowledges the role of two managers of the Southern Railway[4] in the United States—the late Henry Taylor, Chief of Motive Power, and M Worth Hewitt, Communications Engineer. Selby's involvement, though, was in danger of being lost to history. He has stated to the author:

I did not believe it was possible to file a patent on a concept that was already in use [a reference to North Electric's SCADA[5] systems already being used for electric utility and pipeline network control]. *Turned out it didn't matter… LOCOTROL became the only game in town.*

In 2018, GE Transportation—celebrating the 50th anniversary of LOCOTROL—released this:

> LOCOTROL is deployed on roughly 20 000 locomotives across 54 customers in 16 countries. It has transformed the industry as a truly locomotive-agnostic system, interfacing with nearly every type of braking and locomotive control system on diesel-electric and electric locomotives. LOCOTROL has the unique distinction of running the world's longest and heaviest train [see page 203] and the solution continues to evolve with new functionality to meet important industry needs.
>
> LOCOTROL is currently in its sixth generation, called LOCOTROL XA for its Expanded Architecture. LOCOTROL XA is architected for the future of digital rail communications and provides benefits such as enhanced diagnostics, increased processing power and improved radio communications to reduce associated train delays. With over 700 units deployed and an additional 1000 units on order, LOCOTROL XA is backward-compatible and interoperable with previous generations of LOCOTROL solutions.
>
> As the industry standard, LOCOTROL has become synonymous with distributed power, but the solution enables much more than that. We continue to focus on other capabilities supported by the LOCOTROL platform, including Tower Control that increases efficiency at mines and ports, and Remote-Control Locomotives or RCL, that improve productivity for yard operations by enabling remote control of locomotives in railyards.

This preface is not intended as a LOCOTROL advertisement for GE Transportation, who are uniquely capable of marketing their own product. But the above information is necessary, I think, to provide context for what this technology was and for what it would become. It is my view that between the mass introduction of the diesel-electric locomotive and entry into regular operation of ECP braking, no single technology has so-advanced freight railroading as has unmanned distributed power, most particularly, LOCOTROL. It is for this reason that I embarked upon this project and I hope readers will agree with me and enjoy the journey as much as I have.

Fergus Moffat
West Melbourne, Victoria
Australia
May 2021
femoffat@gmail.com

---

4 Now a part of the Norfolk Southern Railroad.
5 See Glossary.

1

EARLY HISTORY

Steam locomotives had often been used in multiple, requiring a crew on each locomotive. With the advent of diesel-electric motive power, railroads could connect individual locomotive units for multiple operation from one driver's cab. This facilitated an increase in freight train sizes, as locomotive groups [called CON-sists] under the control of one locomotive engineer could be used as Helpers and Pushers, located at the head-end of a train, distributed within it, and at its hind-end. Inevitably, it became obvious that if these helpers and pushers could be automated—in other words, remotely-controlled—even greater operational economies could be realised.

The Southern Railway System under President D W Brosnan was one US railroad ruthlessly intent on cutting costs.

TMU

It was into this environment, in the very early 1960s—thanks to the development of computer technology and advances in Space Radio telemetry—that the General Railway Signal Company of Rochester, NY (nowadays part of the Alstom brand), dipped its toe with its Train Multiple Uniter [UNIT-er] system. TMU involved the installation of a computer onto a modified locomotive. This computer was connected to a pair of strain-gauge transducers fitted to the rear coupler of the TMU locomotive—which, of course, needed to be the trailing unit of the head-end consist. Whenever these transducers sensed an increase in draft force—that is to say, if the 'drag' force of the trailing load should begin to increase—the microprocessor control system would transmit a 'throttle-up' request to the Remote locomotive, and vice-versa if there was a *decrease* in draft force. The 'Remote locomotive' could be a multiple-unit consist and be located within the train as required. The brake pipe remained continuous, and any brake pipe pressure reduction caused the Remote locomotive to throttle down to Idle.

This head-end TMU system could only sense a draft or 'pull' force and was not designed to sense or respond to any 'push' (buff) force being transmitted through the couplers from the body of the train. Neither did the system have the capacity to assist with air brake applications or releases. Trials on the Pennsylvania and Louisville & Nashville railroads in 1964—as well as a few others—were made, resulting in numerous modifications. These included adding some operational flexibility by relocating the strain-gauge equipment to a modified freight car instead of requiring the locomotive to be specially-equipped.

On 28 March 1966, Southern Railway boxcar 5999, with the GRS strain-gauge equipment fitted, was attached to a Remote consist cut into train 155 from Potomac Yard to Spencer Yard near Salisbury, NC. Wabco air repeater equipment[6] was also installed in the boxcar to supply the rear portion of the brake pipe: this equipment being adjusted to repeat the brake pipe pressure of the car immediately ahead of it.

---

6   Air repeater units were used primarily by northern railroads in winter to facilitate operation of longer trains by maintaining brake pipe pressure towards the rear of a long train.

Despite the rapid addition of 'power delay' programming and modified logic, TMU simply could not effectively meet the requirements for stable remote control of pushers and mid-train helpers. The system's major deficiency was its inexorable promotion of a synchronous pattern of dynamic slack action that rapidly degraded to the inevitable point of a train separation. The concept proved unviable in operation and was shunned by the railroad industry.

Selby relates that a Southern train, operating with TMU, ran out of control and was chased through several towns with the assistance of the Highway Patrol, and with no fatalities arising. The author has been unable to confirm this incident, although the Southern experienced a diesel-hauled runaway and the wreck of a conventional train on Saluda Grade in 1964. Selby wrote to the author, *'All telephone relay hardware… hardly the thing to put on a vibrating monster.'* The Southern was not impressed with TMU, and it was never further considered.

Despite this, it will never be known to what degree the TMU experiences of the General Railway Signal Company and its engineers might have contributed to the thinking of others on the concept of the remote control of distributed power. TMU was a bold first step into this realm and deserves its place in any written history of the technology.

RMU

Meanwhile, during this period, the Westinghouse Air Brake Company and its affiliate, Union Switch & Signal[7], had conceived and introduced a new distributed power system. Their Remote Multiple Uniter [again, *UNIT-*er, not *UNITE-*er] was the first system to successfully employ radio telemetry to pass control commands and feedback between two locomotives not physically connected.

Dick Fisher was an engineer with Wabco when, early in 1962, the Southern Railway approached both US&S and the Electro-Motive Division of General Motors to develop a system utilising radio communications as the multiple-unit link between the lead ('Master') locomotive and a remotely-located ('Slave') locomotive. Fisher provided much of the following detail courtesy of his personal involvement and private notes as well as archival research conducted at the Southern Railway Historical Association[8]. Along with others mentioned in this narrative, he acknowledges the early involvement on the RMU project, of Southern Railway Air Brake Instructor – Lines West, John D Corriher.

Fisher would later be hired by L Stanley Crane (at the time, Assistant Chief Mechanical Officer) of the Southern Railway, who wanted personnel who would report directly to him. Fisher states that the prototype system utilised control of the throttle and dynamic brake and the locomotive Independent air brake. The equipment was road-tested on Southern Railway coal trains between Bulls Gap, Tennessee, and Asheville, North Carolina – the results showing promise that led to refinements.

Dick recalls that he first became involved with the project while waiting at the Spartanburg, NC station on 5 December 1962 to catch the *Southern Crescent* for Washington, DC (and home). He received a pager message from his boss in Washington and was instructed to proceed forthwith to Chattanooga to meet with SR officers, J G Moore (Asst VP Mechanical) and Howard G Heinz (Air Brake Instructor – Lines West) on the

---

7   In *History of Railways* (New England Library), a monthly partwork from the early 1970s, there is an unconfirmed reference—in connection with these early DP systems—to an 'Arnold Electric Company' of Pennsylvania. The author wonders if this might be Arnold Electric Supply of Croydon, PA, established in 1945 and perhaps involved as a contractor.

8   Some of this detail has previously been published in the November/December 2009 issue of the Southern Railway Historical Association journal, TIES. The author thanks the Association for their permission to reproduce it.

subject of the next iteration of air brake equipment to be incorporated into the new remote-control system. The major improvement was the addition of Automatic air brake functionality (Service and Emergency). The Master locomotive was LOP&G[9] GP9 № 302 (later renumbered 298). The locomotive's 24-RL brake schedule was to be replaced with the new 26-L and Heinz and Fisher were required to ascertain the details of what was required to effect the conversion.

Wabco had previously furnished brake equipment for automatic train operation[10] at the Columbia-Geneva Steel Division of US Steel in Utah and this was adapted to the SR application. In short order, the air brake equipment was ordered from Wabco, shipped, and installed on the 302 at the Chattanooga Diesel Shop late in December. The GP9 and F7B № 4415 were then moved to the roundhouse in the John Sevier [Suh-*VERE*] railyard in Knoxville, TN, for application of the radios and radio coding hardware—both supplied by US&S—and relay interface equipment supplied by the Southern. US&S had earlier purchased the mobile radio product line from the Bendix Corporation (now part of Honeywell International Inc.).

Not everything, though, could be achieved quickly. Fisher recalls:

Because of the long lead time quoted by Wabco on the cable assemblies, Southern mechanical personnel requested that we ship the cables and connectors separately and they would assemble them. I can still see [SR] Diesel Supervisor Tom Gilbert trying to solder wires to a 20-pin Cannon connector with an old pistol-type soldering gun. I think the job ended up in the radio shop.

One major problem arose when we first tested the locomotive and Slave in Sevier Yard. The coding equipment was digital whereas a brake pipe reduction required an analogue signal to be transmitted. This was a poor time to discover the error. When he was told it would take 3 days to rectify the situation, Richard E Franklin, VP Mechanical, asked if that was 8-hr days. When told 'yes', he replied, 'Then you have 24 hours.'

The air brake equipment consisted of a Function Selector Unit which contained the brake relay logic; a Brake Control Centre consisting of solenoid valves on a common manifold block, which performed the pneumatic brake functions on both the 'Master' and 'Slave' unit; and an Air Brake Control Console, which provided the locomotive engineer's Automatic and Independent push-button air brake functionality. On a later version of the Brake Control Centre, Super Spool Valves from Wabco's Industrial Products Division were utilised.

Manipulation of the 26-C Automatic brake valve handle by the locomotive engineer was not required or desired and so the system software was programmed to ensure that its use when the system was active would cause a penalty brake application. For this reason, its handle was secured by a latch in RELEASE position (see Figure 3, page 12). Reset of a penalty brake application required the ABV handle to be unlatched and moved to SUPPRESSION position as per normal.

The inaugural test trip from Bulls Gap, TN to Asheville, NC on a unit coal train of 70-ton hoppers was made on 15 January 1963 – the Slave consist being situated two-thirds back in the train. Additional units were MU'd to the F7B Slave unit, and caboose X2913 was coupled to them to accommodate observers. One difficulty that occurred during this trip was a break-in-two towards the rear of the train, that—due to the resultant high Emergency brake force developed at the Slave consist—caused another separation immediately ahead of it. To alleviate this situation in the future, the system was modified to provide the locomotive engineer with quick-release (bail-off) of the Automatic brake should an Emergency application occur, and a reduced-pressure Independent application.

---

9   Live Oak, Perry & Gulf Railroad—a Southern Railway (Florida) subsidiary between 1954 and 1994.

10   The author thinks this to be a reference to an EMD SW8 (Geneva № 42, ex-Southern Pacific № 1128) having been converted to remote control in 1954.

Figure 1 & 2: An early Southern Railway 'radio' test train negotiates forested country between Catawba and Terrell NC, en-route to a Duke Power (now Duke Energy) plant with 60 loaded coal hoppers.
Robert E. Abernethy, Southern Railway Historical Assn.

This test trip resulted in the application of the first of numerous refinements. One was the 'fail-safe Emergency feature', whereby a loss of electrical power would de-energise a magnet valve and cause an Emergency brake application. Fisher recounts that a high-ranking mechanical department officer managed to momentarily cut the system power on a moving train, and then—as the train screeched to a halt—questioned why anyone would design such a stupid system. The logic for this function was reversed soon after.

Further test trips—not always without incident—satisfied SR in serving to indicate the potential of the system. On 6 February 1963, RMU went into service between Birmingham and Greenville, South Carolina. On 14 February, Fisher attended a meeting at the Swissvale, PA plant of US&S. Attending were Southern's L Stanley Crane (Asst CMO), D Ruff (Asst VP Communications & Signal), and Tom Gilbert. The purpose of the meeting was to review operating problems and to stress the high priority for developing solid-state hardware. In addition, Ruff had concerns about the radio frequency being used. He wanted to secure Federal Communications Commission approval for 27 megacycle control channels for the Slave locomotives rather than using the 160 Mc road frequency[11]. This latter was used for voice communication and Ruff was concerned that the FCC would deny its use for railroad remote control requirements. US&S advised that there were no 72 volt/27 Mc radios available, so SR decided not to pursue the issue and to hope the FCC would not take any action against their use of the railroad band.

Tom Gilbert—at the time of writing, in his 80s and living in retirement in Fayetteville, GA—recounts:

I began my career with the Southern Railway System in 1954. I was hired out of Georgia Tech into the Mechanical Department of the Southern. Under the leadership of President Brosnan, the concept of

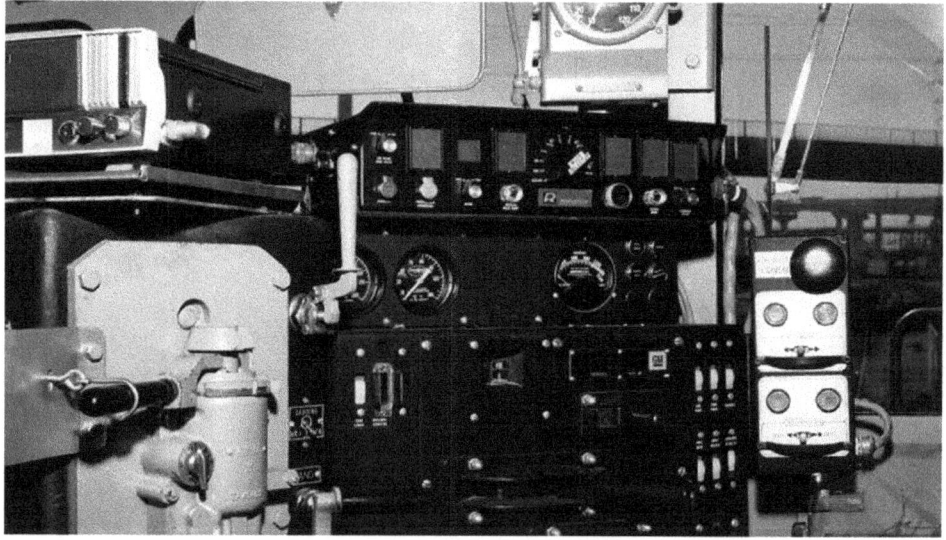

Figure 3: Early Locotrol I engineer's Control and Air Brake Consoles installed on a UP EMD locomotive. Union Pacific Railroad, courtesy Kalmbach Media

---

11 See Glossary.

operating remote locomotives in the middle of trains was being discussed in the 1960s. R E Franklin (Asst VP of the Mechanical Department) had charged Mr Crane (Asst CMO and later VP of Research & Development) to pursue the concept. At that time, I was working out of the Chattanooga Diesel Shop. I was dispatched to the EMD facility at Jacksonville, Florida, where US&S radio-control equipment was being installed on an F7B and a GP9. After a few months, testing of the system began on a coal train out of Bulls Gap, TN to Asheville, NC.

When testing moved to operating from Asheville to Old Fort, NC, down Blue Ridge Mountain, the

Figure 4: Fairbanks-Morse H16-44 locomotive. Railfan.net: The Tom Daspin collection

US&S equipment proved to have some design flaws. At that point in time, the Communications Department of the SR was charged to develop radio-control system design specifications. The North Electric Company [NEC] in Galion [GAL-yin], Ohio, were contracted to build two sets of the redesigned equipment. This company later relocated to Melbourne, FL. By the late 1970s, the Southern had about 50 sets of the equipment in operation on many coal trains and on a few manifest trains.

On 26 March 1963, an order was placed with US&S to equip seven Master and four Slave units plus spares, and two sets of tunnel relay systems. Air brake equipment was ordered from Wabco. To provide more flexibility in use of the Slave units, it was decided to install the equipment in a separate, non-powered vehicle. The first two such units were built from converted Fairbanks-Morse road switcher shells, ballasted with concrete, and numbered R-1 and R-2. The air brake operating valve installed on these units was a standard AB freight control valve. The only additional locomotive equipped at this time was F7 № 4254. On 31 July, the system comprising the 4254 and R-1 was tested between Asheville and Spencer, NC, revealing numerous problems with the US&S coding equipment and radio communication.

Southern stopped further delivery of the US&S order and kept one set in service to acquire train operating experience with the radio equipment. US&S continued to modify the RMU system but to no avail until the Southern finally ended their association with the project in December 1963. US&S did not follow through with a solid-state system until 1966 and SR—who had, since March, been considering another solution following the approach made by Crane to NEC—placed an order with them in November for solid-state control equipment, with hardware specified to incorporate the same logic as that used in the US&S system. An order was also placed with Motorola for the VHF radios and repeaters for the Swannanoa Tunnel[12] – delivery promised in 3 months. Delivery

Figure 5: 'Radio car' R-1 is at Knoxville's Sevier Yard roundhouse in July 1963. Southern Railway Historical Assn.

---

12 The Swannanoa Tunnel near Black Mountain (west of Marion) on the Western North Carolina RR was holed-through on 11 March 1879. At 1800 feet (550 m), it was the longest of the seven tunnels on the railroad between Old Fort and Asheville, NC.

of the Wabco air brake equipment was permitted to proceed since it could be used with a solid-state system.

North Electric coding equipment was installed on GP30 № 2600 and Slave car R-3, an FTB shell. North Electric project engineers Don Selby and Gary Southard were directly involved in this installation and its testing, and Dick Fisher was requested to observe the inaugural run from Birmingham to Atlanta, GA on 11 March 1964.

The North Electric remote-control system performed as intended; however, this was the first exposure of NEC's equipment components to the rugged railroad operations environment and there were numerous failures of soldered connections. After correcting the soldering problem, the equipment was assigned to the Southern's premier freight train, № 153, between Potomac Yard in Alexandria, VA, and New Orleans, LA. It was about this time that the North Electric Co was acquired by Radiation Incorporated of Melbourne, Florida.

At this time, the Southern—perhaps not unexpectedly—faced the threat of union action over the use of unmanned remote distributed power. A compromise was reached with the brotherhood(s) when President Brosnan agreed to the 'locking' of the engineer's Control Console such that the Independent Motoring[13] operating mode would be precluded. This apparently placated the craft unions on the vexed subject of two trains being operated by the one locomotive engineer, although the functionality WAS permitted for the operation of unit coal trains descending Saluda Grade.[14]

Having the use of the Independent Motoring mode on Saluda provided the road foremen of engines—the engine service supervisors assigned the responsibility of handling these 'radio trains' down the 'hill'—with enhanced control as the train crested the grade to commence the descent. This was critical to preventing the train from parting due to overstressed drawgear as it was 'draped' over the crest of the grade. The use of Multiple Unit mode was permitted by the unions for all other operations.

After TMU, the railroad industry's experience with RMU (in this case, specifically that of the Southern Railway)[15] could perhaps be described as another foundational step in the early process of developing the technology of remote control of distributed motive power. The product being developed by NEC, following that parallel approach to them by the Southern Railway, eventually came to be called LOCOTROL, and to become, as Don Selby has stated, '... *the only game in town.*'

⊂ЗВО

Don Selby

Selby was a young electrical engineer with North Electric at the genesis of distributed power and is one of only a few people still alive associated with the formative development of this technology. Don justly describes himself as a 'Father of LOCOTROL' – sharing this epithet with two Southern Railway managers of the day: Chief of Motive Power, the late Henry Taylor and Communications Engineer M Worth Hewett.

---

13 *Independent Motoring* is the use of the engineer's Control Console to operate the Remote units in Power while the Lead units are in Dynamic Brake. *Independent Control* is the use of the Control Console to separately control the Remote units when both Lead and Remote groups are in either in Power or Dynamic Braking.

14 Opened in 1878, Saluda Grade is the 3-mile length of Norfolk Southern (originally Asheville & Spartenburg RR, then Southern Railway) main line between Saluda and Melrose, NC. The steepest portion of the grade is said to exceed 5%, making Saluda—in part—the steepest mainline railroad grade in the USA. NS suspended operations down the grade in December 2001.

15 Don Selby states that the Louisville & Nashville Railroad also tested RMU with similar results. He recounts, 'All telephone relay hardware. Hardly the thing to put on a vibrating monster.'

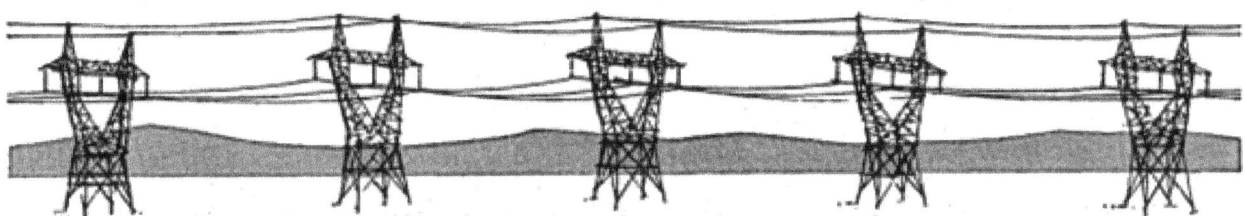

if you must MOVE it...

from one place to another...

North Electric can CONTROL it

Our story is as simple as this... North Electric will engineer a Paricode® Supervisory System to control whatever you move... whether it goes merely inches or thousands of miles.

Paricode is already controlling movement of natural gas, water, crude oil, sewage, electric power, railways, and products on assembly lines. And is doing it with an economy and reliability backed up by 40 years of supervisory control system experience.

Major industries, pipeline companies, utilities, and the government have long recognized that only a company with unmatched capabilities could solve their control problems... and have chosen North Electric to do the job. That's because North Electric has the necessary years of experience, and also assumes complete responsibility to recommend, engineer, produce and install the system best suited to the particular application. These are good reasons why you should get in touch with North Electric when you next consider the need for remote control or automation ... for your transmission and/or transportation requirements.

OTHER DIVISIONS:
DEFENSE SYSTEMS — Electronics • Command and Control Systems
POWER EQUIPMENT — Custom Power Systems • Battery Chargers
TELECOMMUNICATIONS — Public Telephone Systems & Equipment

Figure 6: Early North Electric advertisement for their Paricode supervisory system, from an industry journal. Datamation June 1964. Author's collection

Figure 7: Early North Electric advertisement for Locotrol, from an industry journal. The lack of railroad familiarity by the copy- or tech-writer is evident from the language used. Datamation November 1964. Author's collection

The first railroad distributed power system drafted to the scheme that would eventually become known as LOCOTROL was designed and built in Galion, OH, by the North Electric Company. As previously mentioned, the development of this equipment resulted directly from a Southern Railway request of NEC (presumably at about the same time that Wabco and US&S had been approached for their solution) to see if their emerging SCADA technology (Paricode) might be a vehicle on which to base a radio remote-control system for locomotive operation.

Initially (and perhaps unsurprisingly), the nascent remote-control technology was referred to by NEC as their Radio Control System (RCS), but once it became apparent to the company—sometime later—that they had a marketable creation on their hands, a defining product name was sought. Multiple Consist Control (MCC) was used for a brief period but retired Radiation electronics engineer, Ralph Leffingwell recalls that an office competition was held at NEC, and a young lady in the typing pool created railroad history by contributing the now-iconic 'LOCOTROL' label. This comment may exist as the only public recognition of her achievement.

Selby states:

The original patents for RCS carry [electrical control systems engineers] *Gary Southard and Ron Gottbehuet's names, as they remained at North Electric for some time, and I did not believe a patent was fileable on a concept that was already being used for the electric utilities and pipeline SCADA systems. I refused to write or sign the patent for LOCOTROL, as I considered it merely a re-application of technology with which I was already working in pipeline control systems. Harvey Ellis came to work with me from North Electric's Pipeline Group which eventually adapted many of the engineering and PC board designs we did for LOCOTROL. Turned out it didn't matter...*

Multiple Consist Control

As previously mentioned, this new remote-control system wasn't always called LOCOTROL. At some early stage in its marketing, either by North Electric or Radiation, the product was offered as Multiple Consist Control. When—during the late 1960s—it became clear to regulators that North Electric (with LOCOTROL), and Wabco (with their competing product, RMU), had pioneered a new technology to industry, the Federal Railroad Administration became interested and requested a briefing from the two companies. The author asked Dale Delaruelle about it. He replied, '*We presented the common features as well as the significant differences for the FRA. While I don't specifically recall it, 'MCC' was probably our term for distributed power in those days.*'

Let's have a look at an early proposal written for a potential client. The author has this original handwritten Harris document but has 'typed it out' here in the interests of clarity and readability. This specification serves to describe 'how LOCOTROL worked'.

The purpose of this proposal is to provide your company with a system that will be referred to as a Multiple Consist Control, which will be directly plug-compatible with the present radio control trains [RCS] as operated on your network. This system will be designed from the latest technology in integrated circuits and will consist principally of CMOS[16] *logic, using the technique generally referred to as large-scale integration (LSI). The equipment will be housed in a package physical size of that referred to as Logic Cabinet Two (three-card shelf high) cabinet. System compatibility will be maintained such that the MCC cabinet can be substituted for both logic cabinets at the Lead or Remote (there will still be a Lead-and-Remote type of system) allowing the new logic cabinet to be placed in location 2 and the plugs attached from the present system utilising the present console and the present power supply.*

---

16 See Glossary.

This system as proposed will, however, have a console and a power supply provided with it. The power supply will not be physically able to provide power to the older-type system: however, the console may be used for either the old radio-controlled train operation or for the new Multiple Consist Control system. The interface cards in all probability will remain discrete-type components; the logic referred to as power sensor will be changed to a new technology of one of three designs we will submit to your engineering staff as alternatives. Those will be an optic-isolated design, a magnetic core design, or a magnetic coupled Hall-effect[17] design. The general logic design of the system will be abstracted directly from the logic prints as exist now for the radio-control train system, and the new design will be placed on your format vellums[18] and signed off when final by your engineer.

In this proposal we are requesting and will provide an office for a resident engineer from your company to advise of progress of the new Multiple Consist Control system design and fabrication. This engineer as requested may be present at our facilities at a time designated by your company's supervisor. We wish to utilise the frequency-shift-keying (FSK)[19] system as presently used in your radio-control train system. We will use this design as presently used by your company with the exception that we will place it upon print circuit boards of physical size compatible with the remaining PC boards as housed in the new MCC system design.

There will also be a second set of FSK equipment in each system designed to operate at 700 Hz for 'space' and 1700 Hz for 'mark' signals to aid in expanding the address capability of your present operation from 126 addresses to 252 addresses. The MCC system will have the ability to communicate directly with the old radio-control train system design when the correct address is selected, and a general outline is contained in the section 'Purpose of equipment' as we understand the system presently operates. 'Vibration and temperature' testing is to be performed at your facility before installation of the system on one of your trains. It is suggested that when the system is installed in operation it will be placed in a high-utilisation train (i.e. a coal-train operation). For protection of the operation of your system (that is, the railway system) we are suggesting that one of the old systems be carried and maintained on board until your company is satisfied that MCC meets and exceeds the criteria of the previous radio-control system. Our design engineer is to be present during the environmental testing at your facility and a field service engineer will be provided for two consecutive weeks of on-board testing of the MCC system. Field engineering support for further testing can be provided at cost as designated in the financial portion of this proposal. Total documentation will be provided upon acceptance of the system and will consist of an operations manual, all schematic and logic diagrams, physical PC board layout, wiring documentation, and mechanical structural drawings. All components, electrical and mechanical, used in the system will be purchased from suppliers such as Motorola, RCA, etc. that we feel will remain in the industry and not allow obsolescence to defeat the new MCC system LSI configuration.

Purpose of equipment

The Multiple Consist Control system is designed to provide synchronous automatic control of the consist of locomotives located at some Remote position in a train from a manned control locomotive located at the front of the same train. The MCC system provides control of the Remote consist through command signals initiated at the manned Lead unit and transmitted over a radio link to the Remote consist. Design of the MCC system is such that trains equipped with MCC can operate within radio range of one-another with no danger of crosstalk between trains.

---

17 See Glossary.
18 See Glossary.
19 See Glossary.

Physical description

An MCC system contains a Lead station and one Remote station. The Lead station is composed of three major components:
- Logic cabinet
- Power supply
- Control Console

The Remote station is composed of two major components:
- Logic cabinet
- Power supply

Power supply

Dimensions and mounting information are provided under 'System Specifications'. Major components of the control system are briefly described as follows:

Control console: Contains Remote status indicators and alarms. Also contains the switches necessary to isolate and independently operate the Remote consist.

Logic units: Completely solid-state modular constructed units, which contain the trainline sensing circuits and control logic. Composed of printed circuit boards.

Power requirements

The MCC system is designed to operate from the standard locomotive battery voltage source of 72 volts DC (plus 8 V, minus 15 V) ungrounded. Maximum current drain for each power supply is 5 amperes. The MCC power supply provides voltage to the logic cabinets of +14 V DC.

Communication link

Communication between Lead and Remote units is via VHF radio. The use of this type of equipment requires operating licenses issued by the Federal Communications Commission to the user. Data transmission by the radio is in the form of an FSK signal. Transmission frequency is in the 152-to-174 Mc industrial band. The type of transmission is FM Half-Duplex carrier with a 3 kHz bandwidth. The mark and space modulation frequencies are 1500 Hz for space and 2500 Hz for mark, or 700 Hz for space and 1700 Hz for mark.

Interface

Both input and output circuits have interfacing, which provides isolation between train and MCC electrical systems with transient protection devices incorporated to prevent spurious operation or damage to the equipment.

The input interfacing is composed of sensing circuits connected to the trainline and air brake push-buttons. The presence or absence of the 72 V DC locomotive battery power source is sensed on the trainlines as control signals. The sensing circuits are set up to read a no-voltage signal as 'control not operated' and the presence of battery voltage as 'control operated'.

The output interfacing consists of relay drivers in the logic cabinets and associated 72 V relays in the relay interface cabinet. These control the 72 V trainlines and air brake functions while providing DC isolation between the MCC system logic circuitry and the 72 V locomotive battery.

Principles of system operation

In conventional train operation (non-MCC), the locomotive is controlled from the locomotive engineer's position in the cab by manipulation of levers and controls. These controls are connected through the trainline and control *the applied voltages to the respective locomotive control functions. When two or more locomotives are coupled together for multiple-unit operation, all can be simultaneously controlled from a single set of controls located in the cab of the Lead unit. This is accomplished through jumper cables connected between the units when they are coupled together. In this operation the locomotives are electrically coupled together.*

With an MCC installation, the separation of locomotive power (that is, the placement of locomotives at the front and at some other point in the train consist) makes available full potential of the locomotive power. However, the train control line between the separated units no longer exists. It is here that the principles of digital control employed by the MCC system utilising radio equipment as a telemetry link provide the precise timing necessary for coordinated operation of the Lead and Remote units. This permits division of the locomotive power into separate units operating in unison, resulting in increased power efficiency and faster, smoother, safer stops and starts. It also increases the efficiency of the initial air brake line charging and consequent air brake reductions and releases.

The digital control used in the MCC logic provides all the capabilities required for this type of Remote control. In the binary operation of the digital logic used, only two states are recognised—'true' or 'not true'. These are further defined as '1' or '0' (bits).

Each function of the locomotive, insofar as we are concerned here, is controlled by the presence or absence of a train voltage, 72 V. If a switch is actuated, 72 V is applied to operate the function (dynamic braking is an exception, as discrete sensible circuits in the MCC logic detect—in steps of percentage—the applied portion of full-scale 100% application). When the application of 72 V for a specific function is detected by the respective power sensor circuit, an output is generated which indicates that function has been initiated. This output is a binary '1' as opposed to a binary '0' (which is the output when no function is generated). These binary 1s and 0s are further defined as 'bits'. Thus, it can be seen that if a specific function is initiated, a '1' bit is generated in the logic, otherwise, the output is a '0' bit.

All control functions—actuated or not ('1' or '0')—are conveyed on each transmission. This means that the bits of control information are transmitted serially (that is, one after the other) on each transmission. Each transmission is called a message. Each function is designated a specific bit position in the message whether it is a '1' or '0'. For example, the Engine Run function might be designated to appear in the 10th bit position. Thus, in each message the Engine Run bit is always transmitted as the 10th bit. In this manner, the Remote unit knows the 10th bit of every message it receives is the Engine Run information. Therefore, with each function bit occupying a specific position in the transmitted message, the Remote unit can determine which function is actuated and act accordingly.

The individual '1' bits are handled separately within the MCC logic. They control the logic circuits to operate the associated relays to control the Remote locomotive operation by applying 72 V to the trainline as in conventional operation. In actuality, a message is transmitted periodically so long as no control change or system status change is initiated. If a control or the status is changed, the system senses the change and immediately sends a complete message, with the change included in the message.

Messages are also transmitted from the Remote unit to the Lead unit. Each message received in the Remote unit, with one exception, generates a transmission from Remote-to-Lead which contains the actual commands performed. The exception to this is an air brake function from Lead-to-Remote which, due to the method of air brake transmission, will not cause a reply from the Remote unit. The Lead compares the Remote status with the present status of the Lead. If the two messages are not identical, or the Remote unit fails to reply, the Lead unit transmits again.

Safety and backup circuits are included in the logic. For example, if communication is lost between the Lead and Remote units for more than a preset period of time, a 'No-Continuity' response is generated which is indicated on the locomotive engineer's Control Console. Radio changeover logic is provided which is automatically activated under a 'No-Continuity' condition.

The Remote unit will transmit the message when an alarm condition exists in the Remote unit. This message illuminates an indicator and initiates an alarm on the Lead unit. Thus, the status of the Remote unit is continuously monitored and displayed to the locomotive engineer.

Functional description

A simplified block diagram of the system follows. The system can be operated in either of two modes: 'synchronous automatic' operation or 'independent' operation.

The Lead locomotive operating commands, controlled by the locomotive engineer, are sensed by the MCC logic sensing circuits and digitally encoded. They are then transmitted as an FSK message via FM radio utilising the 3 kc bandwidth. The transmitted coded commands are received at the Remote unit, processed by the logic and applied to the trainline and air brake control circuits by means of interface relays. In this manner, the Remote unit operates synchronously with the Lead unit.

Each message received by the Remote unit initiates a status report back to the Lead unit. This is stored in the Lead logic for comparison with the Lead trainline power sensor circuits. The power sensors are continually sampled and compared with the last received message from the Remote unit. If the data does not compare, the latest control information is transmitted to the Remote unit.

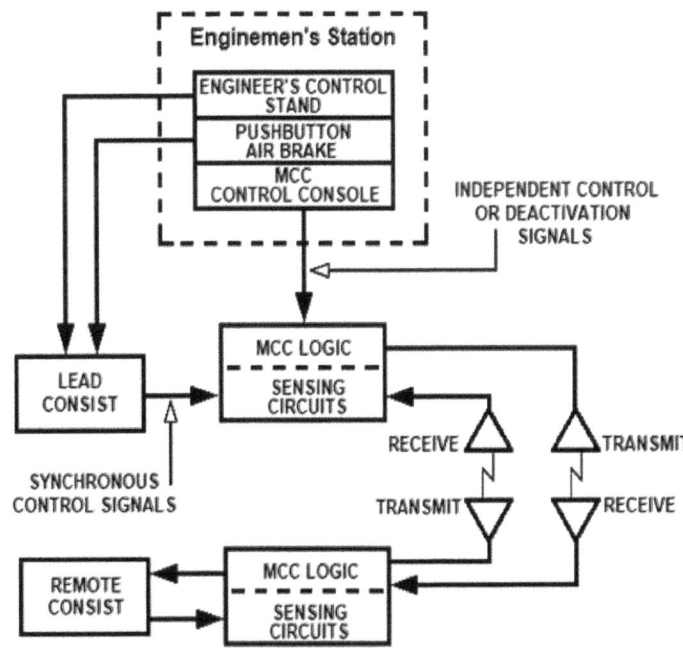

Figure 8: MCC – Simplified block diagram. Harris Controls & Composition

A 6-bit error code is transmitted with each message to insure against the possibility of picking up errors through interference or a faulty component. If an error exists, the message is rejected.

The use of an address or identity code ensures the integrity of the system and permits the use of the same radio frequency band by other systems in the same area. All trains equipped for Remote control operation by MCC can operate in the same area with no danger of interference to one-another.

When non-synchronous operation of the Lead unit is desirable, an 'Independent Control' capability is available to the locomotive engineer. This allows independent degrees of throttle or dynamic braking in the Remote unit to be set by the engineer from the Control Console located on the Lead unit. When required, the Remote equipment can be isolated from the engineer's Control Console, making a 'dummy' of the Remote unit. Status indicators and alarms are included in the Lead equipment for monitoring of the Remote status.

A state of 'Continuity' exists if transmissions are received by both stations at a regular rate (i.e. if the communication link is good). In this situation the console CONTINUITY lamp is lit.

If the Lead station fails to receive a message for a period of 45 seconds, a 'No-Continuity' condition is established, and the NO-CONTINUITY console indicator is illuminated. If the Remote unit fails to receive a message for a period of 45 seconds, it changes to a 'No-Continuity' condition and a relay is de-energised to prepare the Remote unit for 'throttle-down'. The air brake feed valve relay is de-energised and the throttle/dynamic braking is stepped down to Idle at a 1.5-second rate.

The loss of continuity sensed at the Remote unit causes a 'No-Continuity' bit to be transmitted to the Lead unit. If the Lead unit receives this bit, it lights the console NO-CONTINUITY lamp and causes radio changeover. A loss of continuity sensed in the Lead unit causes an 'Inhibit Reset Radio Timer' bit to be transmitted to the Remote unit. This causes the Remote unit to lose continuity and changeover radios but does not cause a feed valve cut-out or throttle/dynamic braking stepdown.

The radio changeover function from Operating to Standby occurs as follows:
- The Remote radio changes from Operating to Standby if no transmission has been received from the Lead unit for a period of 27 seconds
- The Lead radio changes from Operating to Standby if no transmission has been received from the Remote unit for a period of 33 seconds
- The Remote radio changes from Operating to Standby if no transmission has been received from the Lead unit for a period of 39 seconds.

The above sequence and rate (i.e. 12 seconds) of radio changeover continue until continuity is re-established. When continuity is re-established, the combination of radios in operation at that time remains in operation. Any subsequent loss of continuity starts timing as previously described, and radio changeover starts from the present status. A switch is provided to manually force a radio changeover. When continuity is restored, the Lead unit resumes control and the throttle relays on the Remote unit step up until the Remote unit agrees with the Lead unit. This set-up is at the base clock speed (10 kc).

Timing for the changeover function and for loss-of-continuity is inhibited during an air brake application to prevent undesirable radio changeover in throttle/dynamic braking stepdown.

Equipment specifications
Environmental

The equipment is mechanically designed to withstand, with a comfortable safety margin, the shock and vibration encountered in locomotive operation. It has been subjected to environmental tests similar to those given to train locomotives. The logic cabinets and power supplies are furnished with angle-iron frames, which are bolted or welded into position in the locomotive. The MCC cabinets are then inserted into frames and secured in place by steel plates and wing nuts. The electrical connection is made through Bendix pygmy connectors (or equivalent). The Control Console is designed for mounting to the locomotive control stand. The system is sealed against dust, locomotive gases, and oily fumes. It is not weatherproofed and must be located inside a locomotive or radio-control car.

Electrical

Power Requirement:	72 V DC ungrounded, 5amp
Operating Voltage:	+14 V DC
Type of Transmission:	FM Half-Duplex carrier with 3 kc bandwidth, frequency shift-keyed
Transmission Frequency:	152–175 Mc industrial band
Temperature Range:	-30°C to +65°C
Display:	Front panel-mounted indicators

Mechanical

Dimensions:	Lead Unit – 20.32 in high × 23.2 in wide × 10.3 in deep
	Remote Unit – 20.3 in high × 23.2 in wide × 10.3 in deep

	Control Console – 5.5 in high × 22.0 in wide × 9.0 in deep
	Power Supply Frame – 9 in high × 25 in wide × 12 in deep
Mounting:	Logic Cabinet – angle-iron frames
	Control Console – quick-disconnect
	Power Supply Converter – angle-iron frames

Theory of operation
Scope
A 'theory-of-operation' of the system is presented in a block diagram for the reader interested in a general overall picture.

Block diagram analysis
The block diagram analysis is discussed in terms of train operation and logical operation. A block diagram of the train operation [follows].

Train operation
Referring to the block diagram, the Lead consist is controlled from the engineer's throttle stand, over the trainline. The air brake system of the Lead consist is controlled by the push-button Air Brake Control Console. When synchronous operation of both Lead and Remote units is desired, the system is switched to the Multiple-Unit mode and the Remote unit is controlled automatically from the Lead unit. The command signals provided by the throttle stand controls to the trainlines and air brake systems are sensed by the MCC sensing circuits and presented to the MCC logic circuits. The logic circuits encode the command signals into a suitable format for radio transmission and the command message is transmitted to the Remote unit. The command message is received at the Remote consist, decoded in the MCC logic and drives the locomotive control relays which control the operation of the Remote locomotive.

When operating conditions require a degree of individual control of the throttle or of dynamic braking, these functions at the Remote consist can be operated from the locomotive engineer's Control Console in the Lead locomotive independently of the Remote unit. The necessary sensing circuits from the Lead trainline are disabled. The control of the associated operating signals to the

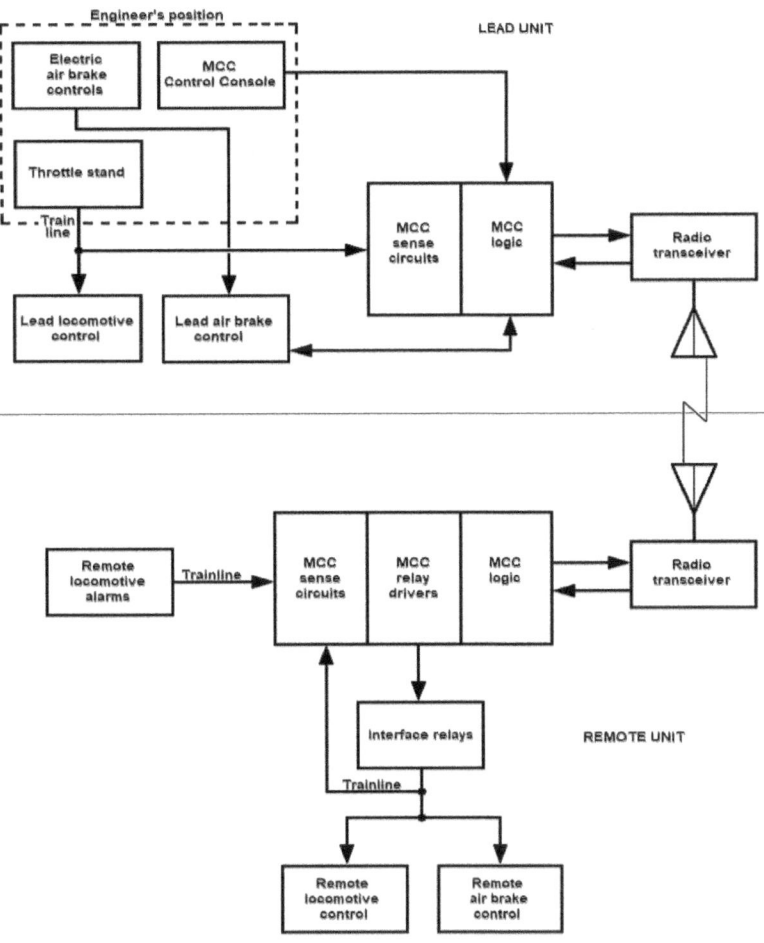

Figure 9: MCC – Train operation block diagram. Harris Controls & Composition

Remote unit is transferred to the Control Console switches.

Isolation of the Remote unit is accomplished by operation of the Control Switch on the console. The air brake feed valve can be cut out by operating a separate switch on the console.

The status of the Remote trainline and all alarm conditions are displayed on the locomotive engineer's Control Console on the Lead unit. Status displays include such information as throttle position or amount of dynamic braking. Alarm displays are the result of Remote consist malfunctions such as Hot Engine, Low Oil, and No Power. Data in the MCC logic drive the warning display lamps on this console.

North Electric[20]

In 1884, George W Drumheller and Charles N North, both with several years' experience in telephony, established Drumheller & North, a small electrical business for the manufacture and repair of telephone equipment in Cleveland, OH. In 1889 Charles North succeeded to outright ownership of the company and the firm's name was changed to the North Electric Works. In 1899, in association with George C Steele, he incorporated the business as the North Electric Company. Under the management of Messrs North and Steele the company forged rapidly to the front, and from the small repair shop on a side street it became one of the prominent manufacturing concerns of Cleveland, employing over 300 workers.

In 1912, the Telephone Improvement Company was acquired and in 1918 the North Electric Company was reorganised under the laws of Ohio, as the North Electric Manufacturing Co. The new company acquired all the property of the old North company, including the factory at Galion, and a large group of important patent rights covering machine switching telephone systems. Charles North remained with the new company as president; however, due to declining health, he retired in 1921. He remained associated with the company until his death in 1926.

North Electric Company became the largest North American supplier of crossbar switching equipment to the independent telephone market and had large orders from the US government for military telephone equipment and systems. The company was incorporated into the Ericsson group in 1951 and eventually became the oldest continuous manufacturer of telephone equipment in the USA. In 1960, a factory was bought to manufacture power supply equipment for computer and telephone exchanges. In 1966 the company was purchased by United Utilities (a holding company for a group of telephone operation companies that had become the third largest in size in the USA after AT&T and General Telephones) and in the 1980s was sold to Alcatel who then discontinued the 'North Electric' name.

Radiation Incorporated

Radiation Incorporated had been founded in Melbourne, Florida, in 1950 by electronics engineers Homer Denius and George Shaw when the space race was beginning to accelerate. Located conveniently close to what is now NASA's Kennedy Space Centre at Cape Canaveral, Radiation produced miniaturised electronics, tracking systems, and pulse-code modulation (PCM)[21] technologies—all crucial to aerospace programs. This involvement included equipment for communication and weather satellites as well as for the Atlas, Polaris and Minuteman missiles, not to mention the first Apollo mission to the moon. Significantly, in 1963, Radiation Inc. had developed their first working semiconductor[22] for use in their communications equipment.

In 1964, Radiation acquired the rights to some North Electric Company products. These included

---

20 The author acknowledges the contribution of K Bushell (USA).
21 See Glossary.
22 See Glossary.

some emergent NEC products that spurred the creation, in 1965, of Radiation's Control Systems Division, producing automatic control systems for pipelines, utilities, industrial facilities, process industries, and (eventually) railways. Along with other similar companies of the time, North Electric, had been developing SCADA technology (although, at the time, that generic technology moniker had yet to be coined). NEC had established—and trademarked—several proven digital command and telemetry systems. These included a supervisory control system (*Paricode*[23]; eventually becoming their Continuous Scan and TF Quiescent control systems) for pipeline, utility, and industrial applications and a Direct Digital Control System for industrial processes (such as subsequently adapted for LOCOTROL).

Emerging then, as a market leader in both digital and space communication, data management, and computer-based control systems (including satellite tracking), the company was, by 1967, a well-established government contractor for both military and civilian projects, and one of Florida's largest employers. Radiation Inc. was soon seeking to expand into the commercial sector and had decided to find a merger partner. Serendipitously, at about this time, Harris-Intertype was also actively seeking to expand – into the field of electronics.

Harris Controls

In the early 1890s, brothers Alfred and Charles Harris were running a jeweller's store in Niles, OH, and indulging their creative desires by experimenting with an automatic sheet feeder to eliminate the laborious job of hand-feeding printing presses. By the late 1890s they had formed the Harris Automatic Press Company in Cleveland, OH, to market a revolutionary printing press they had designed and manufactured. Responsible for many printing innovations during the early 1900s—including the first commercially successful offset lithographic press and the first two-colour offset press—the company acquired two other printing businesses, the Seybold Machine Company of Dayton, OH, and Premier & Potter Printing Press Co Inc. of New York, NY. As the Harris-Seybold-Potter Co, they went on, in June 1957, to merge with Intertype Corporation of Brooklyn in New York City (and in the UK), a manufacturer of hot metal typesetting machines; morphing this time into the Harris-Intertype Corporation.

Intent on becoming an equipment manufacturer in the broad field of electronic as well as printed communications, Harris-Intertype soon acquired the Gates Radio Company of Quincy, IL (later to be the Harris Broadcast Division) and then, in 1959, PRD Electronics of Brooklyn, NY—a manufacturer of microwave test equipment. Harris-Intertype printing machines were still mechanical, though, and the company knew that electronics was the way of the future. In 1967, the company demonstrated its commitment to the growing electronics industry (not to mention its need for electronics expertise) by acquiring Radiation Inc. as previously mentioned, and thus became the owners of the nascent LOCOTROL brand.

Cementing its reputation for innovation, Harris-Intertype now embarked upon the critical management strategy of 'technology transfer'—evolving commercial applications for technology originally developed for government projects. One early example was electronic newsroom technology, a direct result of a previous study made for Harris by Radiation of how to update Harris's mechanical printing presses. Acquiring RF

---

23 Providing precise and reliable supervisory control of a number of devices and operations at any of various remote locations. The respondent status—being all phases of control, indication, telemetering, data-logging, and status reporting—of conditions of these remote systems was conveyed via reliable transmission systems (i.e. microwave, power line carrier, line wire, telegraph circuit, and tone equipment). Paricode was marketed for oil-and-gas pipeline, electric utility, hydro-electric generation, liquid and gaseous matter distribution, and production processes, conveyor systems, and data acquisition & processing. An engine-sequencing function also provided fail-safe operation of auto startup & shutdown, load control, and overload protection.

Communications of Rochester, NY (a manufacturer of point-to-point radio equipment) in 1969 and Farinon Corporation (who produced microwave radios and Digital Telephone Systems switchboards) and then telecommunications equipment manufacturer Dracon Industries in 1980, Harris-Intertype plunged into worldwide communications and moved to secure its market position.

Meanwhile, the vital aerospace contract work continued, with the company being responsible for the production and development of the data-handling systems for the pre-flight check of the Apollo spacecraft and for the digital command-and-control computer of the Gemini spacecraft. In 1972, General Electric's TV broadcasting equipment product line was purchased and shortly thereafter the University Computing Company's Dallas, TX subsidiary, Communications Systems Inc. (producing computer terminals and communications subsystems for the data processing industry) was also acquired.

In 1974, Harris-Intertype acquired Datacraft Corporation, a producer of superminicomputers, and changed its name to Harris Corporation. In 1978, acknowledging that its earlier Radiation Inc. purchase had created a new technical focal point, the company moved its headquarters from Cleveland to Melbourne, FL. This move signalled a new phase in the growth of Harris Corporation and it was at this time that the company began to develop into the ruthless but highly commercially successful corporate entity that would so dominate the global electronics and communications industry. Harris Corporation's Controls & Composition Division absorbed LOCOTROL and some other railroad technology developments and created their Railroad Product Line. They would soon recruit a Canadian railroader to manage it.

Harris had become the largest producer of printing presses in the USA but in 1983 it sold this business to concentrate exclusively on electronics. In 1988, the company acquired General Electric's semiconductor operations, significantly expanding their participation in this sphere, and creating an association that would eventually culminate, in 1991, in a partnership with GE Transportation in (and ultimately GE's outright purchase of) the Harris Corporation's Railroad Product Line, including LOCOTROL.

GE Transportation is now owned by Wabtec, and LOCOTROL continues to be manufactured—nowadays by GET's Intelligent Control Systems Division—in Melbourne, FL.

○❀○

The development of LOCOTROL

Figure 10: Dennis William (Bill) Brosnan, 1903-1985.

'The Southern Gives a Green Light to Innovations.' The Southern Railway was one transport company that—during the 1950s and 60s—had been keeping abreast of emerging technology. Driven by a prescient vision and a consuming economic imperative, its irascible president, Bill Brosnan, was intent upon reducing the one major operating cost over which he had a measure of direct control: the wages bill. This new remote-control technology—if it could be made to work reliably—would be an important component of his strategy to rejuvenate the Southern's financial state. The SR approached NEC (see Tom Gilbert's previous comment), who accepted the challenge and its engineers went to work on the project. Don Selby relates:

Sometime in 1962, Stanley Crane, SR Vice President Research & Development [and later to be President] *at the Southern Railway came to North Electric with a Request-For-Quote for a system that would provide for the radio remote*

Figure 11: John Sevier Yard, Knoxville, Tennessee. Built in 1925 by Southern Railway, and became America's first fully automated hump classification yard in 1951, handling 35 trains per day. Illustration courtesy David Patten, Environmental Operations, Norfolk Southern

control of diesel locomotives. The SR specification was very basic: what they were asking for was to apply North Electric SCADA technology to rail transportation. At North, we reached into our crystal ball and gave the railroad a dollar figure, which they accepted!

The first RCS system was developed and delivered in March 1963 and duly installed, at Knoxville's John Sevier Roundhouse, in a new GP9 as Master unit whilst the Slave control car was a stripped-down F4B unit. The F4B was chosen because it was being phased out and had all the air brake piping (including sanding hose connections), MU wiring and jumper sockets, as well as a park brake. We had to add battery connectors to supply control power from the attached locomotives to the equipment in the Slave car. With the prime mover and main generator, etc. of this F4B removed, a large concrete block was installed to bring the centre-sill back to level and the coupler back to the correct height above rail.

We got our noses bloodied on the prototype; components shook loose on the printed circuit boards. The equipment was returned to NEC for remedial work, which included hand-soldering of components and conformal coating of the PCBs[24].

In 1965 SR placed an order for 16 sets of solid-state coding equipment, which were built in Melbourne, FL by Radiation, plus 16 air brake sets from Wabco. Although Radiation had established a commercial relationship with

Figure 12: A 'radio car' under construction at Marion NC on 23 December 1965. Ken Marsh

---

24 See Glossary.

1. Brake Valve
2. MU2A Valve
3. A-1 Charging Valve
4 Brake Control Center
5. H-5 Relay Air Valve
6. NS-1 Reducing Valve
7. IBOS Pressure Switch
8. RSS Pressure Switch
9. PC Switch
10. Flow Meters
11. Minimum Reduction Reservoir
12. VB-28 Control Valve (Behind Brake Control Center)
13. ORBS Pressure Switch (On Tunnel override cars only)

Figure 13: Early Locotrol air brake rack (Wabco) in Southern 'radio' car. The Brake Control Centre at the bottom of the rack converts the electrical signals to pneumatic through the solenoid valves. Markings show it has received a COT&S in January 1980. The 26-C engineer's brake valve has the handle removed. It appears that with the equipment in this Southern configuration, the Remote unit could not be cut in if 'dry-charging' of the train (i.e. Lead unit not coupled to train) was required in the yard. In this case, someone had to enter the Remote Control car and cut the brake valve IN; hence its presence on the brake rack. Mac Connery, courtesy Southern Railway Historical Assn.

New York Air Brake, Dick Fisher relates that the Southern chose to remain with Wabco. Fisher believes that if Wabco and US&S had not been corporately related, Radiation would have chosen Wabco as their supplier for LOCOTROL brake equipment.

The Remote[25] control cars were built by Marion Machine in Marion [MARY-in], NC. Installation of the equipment onto the Lead locomotives (GP-35s 2650–2653 and SD-35s 3000–3012) and remote-control cars 5950–5965 commenced in May 1965 at the Industrial Electric Company of Charlotte, NC. Dick Fisher recalls that Henry Taylor collected money from those present to purchase padlock keys for the remote-control cars. These keys were to be distributed to the SR master mechanics based between Alexandria and New Orleans, and since Taylor did not expect Asst VP Mechanical, J G Moore to authorise the expenditure, he (Taylor) did not desire to bear the total cost himself. As well as the above-listed locomotives, GP-30 № 2600 and GP-9 №s

---

25 The industry eventually accepted that the terms 'master' and 'slave' were politically problematic, and 'Lead' and 'Remote' were adopted instead. They became formal technical terms.

298 and 299 were in RCS service, the latter units being geared for mountain operation.

Since the SD-35s were equipped with the Wabco 26-L Unitised air brake arrangement, there arose a problem. In applying the Brake Control Centre portion of the RCS equipment to a Lead locomotive, it needed to be connected to existing air brake piping. Since these locomotives had a unitised air brake manifold rather than piping, a special filler block was fabricated to provide for external pipe connections to the manifold. Made aware of this dilemma at the last minute, Wabco managed to courier the first filler block to Charlotte on a Sunday afternoon, and it was duly installed.

Fisher says that although the operation of this equipment was mostly trouble-free, several matters required improvement:

- The 12 V Motorola transistor radios required a 72 V-to-12 V inverter… which proved to be failure-prone. To solve this problem, Motorola developed a 72 V all-transistor radio, thus eliminating this troublesome requirement.
- Heat-sensitive transistors in the logic circuits of the RCS equipment failed intermittently. When the temperature returned to normal, the transistors operated as intended. Since the faulty components were all from one manufacturer, Radiation arranged a new supplier and a retro-fit on all units.
- The Wabco Super Spool valves used on the Brake Control Centre had rubber seals and were solenoid-actuated. It was found, during extended brake applications ('maintaining' or grade-braking), that if equalising reservoir pressure leaked past the seals, the brake pipe reduction set by the locomotive engineer would continue to increase [in other words, BP pressure would continue to decrease]. The Norfolk & Western—also an early user of the 26-L schedule—experienced a similar problem on their eastbound grade from Bluefield, West Virginia, this leakage inevitably occurring in the extra piping installed by EMD for the dual control stands. The equalising reservoir in a 26-L system is of much lesser volume than in the previous 24-RL schedule, and what was tolerable leakage in the latter was greatly magnified with the 26-L. For the Southern, this leakage problem manifested on the Blue Ridge grade[26]. The rubber seals were changed out several times before a satisfactory compound was developed; the work eventually being performed by Wabco engineers at their Birmingham Motel following union complaints about it being carried out in the Chattanooga Shops. Technical staff made daily trips to the airport to pick up new seals, and to Norris Yard to install the modified valves on the units.
- The mechanical strength of the Function Selector Unit (relay box) was in doubt regarding the vibrations encountered in service. This fear was confirmed at a vibration test conducted at an Alexandria, VA testing laboratory. The box was strengthened, and Lord[27] shock-mounts added.
- Moisture caused by condensation in the brake equipment aboard the remote-control car would freeze. This was alleviated by installing drain cocks to the brake rack piping and by encasing the rack in an electrically-heated steel housing that immortalised Diesel Superintendent H C Taylor by attracting the affectionate moniker of 'Henry's dog-house'.

For ease of identification, SR LOCOTROL-equipped locomotives carried number boards with black numerals on a white background. Depending upon intended direction of travel, the radio address codes were set for either 161.475 or 161.565 MHz.

---

26 The Carolina Division (Asheville to Spencer) of the Southern Railway in the Blue Ridge Mountains of western North Carolina, between Old Fort and Ridgecrest, having a maximum ruling grade of 2.2% (by which a change in elevation of 1000 feet in 13 miles of track is achieved). It includes the 1832-foot-long Swannanoa Tunnel.

27 LORD Corporation (global HQ in Cary, NC and now a Parker Hannifin company) is a diversified technology and manufacturing company providing adhesives, coatings, motion management devices, and sensing technologies for aerospace, automotive, oil & gas, and industrial applications. The company's products include anti-vibration mounts and flexible couplings.

Figure 14: Leonard Stanley (Stan) Crane 1915-2003.

A common problem (especially prior to the development of more advanced control valves) in long train operations was undesired emergency air brake applications (UDEs)[28], and the SR 'radio trains' were not exempt. SR technical staff had informed VP Crane that they suspected the problem was originating within the LOCOTROL brake equipment. Fisher believed the most likely initiator to be the telephone-type relays in the Wabco Function Selector Unit, and states that Crane confronted Wabco, who agreed to investigate the problem. It was arranged that Wabco personnel would observe the operation of one set of LOCOTROL equipment on train Nºs 153 and 154 between Potomac Yard and Birmingham over a 30-day period. Special brake pipe coupling hoses were installed behind the Lead consist and at each end of the remote-control car. These hosebags incorporated a transparent portion containing a fixed indicator 'wick' which would bend under the influence of the exhausting air flow resulting from an Emergency application, thereby pointing in the direction from which such an air brake application had been initiated.

The tests commenced in February 1967, with Wabco observers aboard both the Lead and Remote consists. These observers rode the train from Alexandria, to Charlotte, where they were relieved. A fresh crew of observers continued aboard the train to Birmingham, where they alighted and waited for the train to continue to New Orleans and return. Fisher recalls that the 'north end' was manned by Wabco Washington office personnel, and Charlotte was chosen as the relief point rather than Spencer Yard for logistical convenience when air travel was required. No UDEs occurred during the 30-day observation period and Crane was satisfied that Wabco had fulfilled their obligation.

Fisher recalls that Wabco did go on to produce a solid-state FSU which was installed on a locomotive at Atlanta and placed in service between Chattanooga and Cincinnati on 13 July 1967. This FSU required a separate power supply and—although it performed satisfactorily—there was no apparent advantage and the SR lost interest in it.

Selby continues the narrative regarding the Southern's first big order to NEC:

These 16 sets were the first production systems and the Remote portions were all installed in gutted F4B units. Standard operating practice at SR dictated that a hostler delivering locomotives to the fuelling point was to place the MU Control Switch to OFF and set the park brake, which prevented a runaway. Further, the Slave car was normally left uncoupled from the locomotives. If the Slave car was brought in from the yard, it was usually parked near the communications shack for review and set-up by the technicians for the next run.

Prior to moving the SCADA product line to Radiation Inc., a revised specification was issued by the Southern Railway. They wanted us to reduce the weight of the equipment cabinets (there were two cabinets of PCBs plus a Power Supply).

The original versions of these cabinets required two people to lift one cabinet or Power Supply onto the locomotive walkway, and the most significant part of the Spec: '... give us 2047 individual System Addresses.'

---

28 See Glossary.

Eastbound trains were to have even-numbered addresses below 511; westbound trains were to have odd-numbered addresses below 511; northbound trains were to have even-numbered addresses above 512; southbound trains were to have odd-numbered addresses above 512.

What a selling point! This allowed for a possible 255 total LOCOTROL systems on the SR and their joint operations with other railroads. SR intended installing these new LOCOTROL systems on new SD35s and GP30s. As I remember, these were the last General Purpose-series locomotives purchased by SR until after 1970.

Figure 15: Don Selby testing early Locotrol equipment. Harris Corporation pamphlet, author's collection, courtesy Dale Delaruelle

A new-concept PCB cage was designed and built from punched 0.020-inch aluminium: it was a disaster. We put the LOCOTROL system on Harris' NASA shake-table, programmed-in the EMD locomotive's profile, high-lighted the front with a strobe light, and started the table. About halfway through the profile we were rapidly catching PCB pieces as they literally flew out of the cabinets. We called in a Harris Aerospace mechanical engineer and it took him all of a day to solve our problem. The solution was to make lightweight 'cake tins' that were pop-riveted to the PCB card cage and filled with air-cure foam. It worked… made them strong enough you could stand on them without bending them. We now had two cabinets and a Power Supply that could be handled by one person. Remember, these vintage LOCOTROL systems were discrete components on PCBs. The equipment today is microprocessor-based and only the size of a large shoe box: even the air brake equipment is microprocessor-controlled, and the MU connections

Figure 16: L-R: Harvey Ellis (dec), Dale Delaruelle (partly obscured), Ed Dimmerling (rear), Don Selby (crouching). Harris Corporation pamphlet, author's collection, courtesy Dale Delaruelle.

Figure 17: Early Locotrol equipment suite for Southern railway Lead and Remote units, including the engineer's Control Console. The radios are not shown. Harris Corporation pamphlet, author's collection, courtesy Dale Delaruelle.

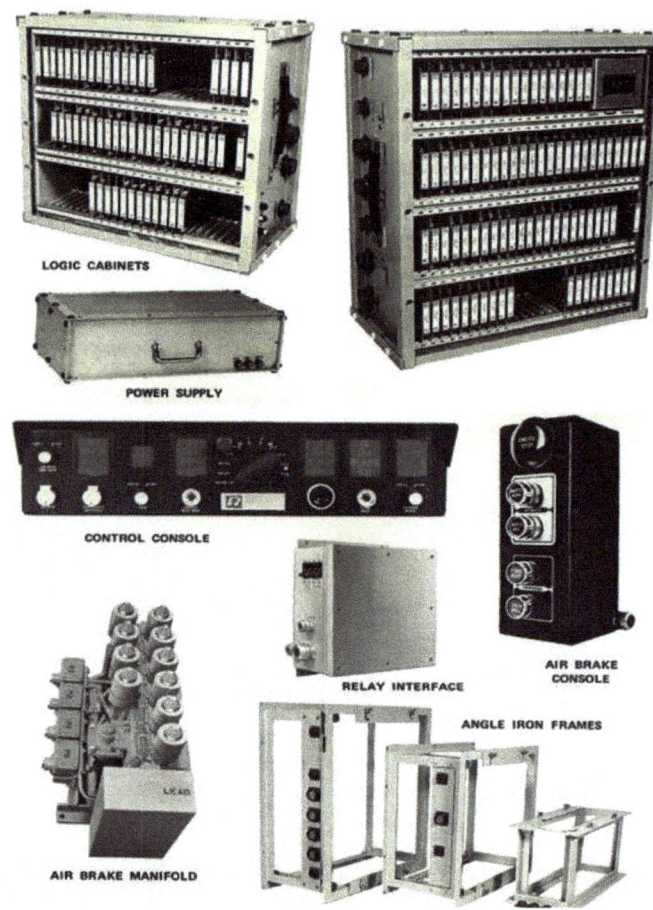

Figure 18: Early Locotrol 105 equipment elements for Southern railway Lead and Remote units, including the engineer's Control Console. Radios not shown. Not to scale Harris Corporation pamphlet, author's collection, courtesy Dale Delaruelle.

are probably SSRs[29]. This vibration retest was completed with flying colours.

The original tests were run under the direct supervision of SR President Brosnan, although I was Radiation's Director of Locotrol Engineering & Test Trains. I changed his running orders from four Master units + two Slaves to 3 + 3 and then to 2 + 4, with which configuration we finally had the most successful runs. In the Diesel Shop after making this historic run, Mr Taylor received a telephone call from Brosnan asking who authorised the motive power configuration changes. Mr Taylor said, 'This young man standing right here...' and handed me the phone, whereupon DW let me know HE was running the Southern and my ass was FIRED (this is a mild statement, toned down because I would need to look up the spelling of all the swear words DW used). I told him I didn't work for the SR.

From 1963 through 1965, Ron Gottbehuett and I ran many test trains. In (I think) early 1966, Southern ran a 300-car test train from Chattanooga to Cincinnati. In 1966, Dale Delaruelle was a new hire at Radiation—fresh out of school, I believe. He and Ron were involved.

Later in the project, a second Remote was added for the same run, with Dale and Ron very much involved. We had a Master with two Slaves before anyone had ever heard of multiple Remote consists[30]. The two Remotes could not be individually controlled, but one could be placed off-line if required.

Dick Fisher recalls that in 1966 the Southern placed an order for 50 additional LOCOTROL systems. The remote-control cars (converted boxcars) were built by Berwick Forge & Fabrication at Berwick, PA. The equipment was installed by the Cleveland Electric Company at Southern's Constitution Yard, Atlanta, GA, to 26 SD-35s (Nºs 3050–30750), 16 GP-30s (2562–2577), 8 GP-7s (2180–2187, with 12-tooth pinions for mountain operation), and 50 remote-control cars (5900–5949). VP Operations W H Moore apparently suggested equipping GP-30s and-35s instead of the GP7s but Brosnan insisted on using the GP7s.

Don Selby continues the narrative:

We started deliveries and assisted with the installations. Ray Britt came on board and started tech writing for me, and I have one of his original service manuals. The buyout of Radiation by Harris Controls was completed. Now we were in the big league and I was installed as Marketing Manager.

---

29 Solid-State Relays. See Glossary.

30 It is not widely known that this early LOCOTROL model had the capability to control Remote motive power at two separate locations within the train (albeit both Remote groups could only be operated in unison). This configuration would later be available from the LOCOTROL III model onwards, with all distributed consists able to be independently-controlled.

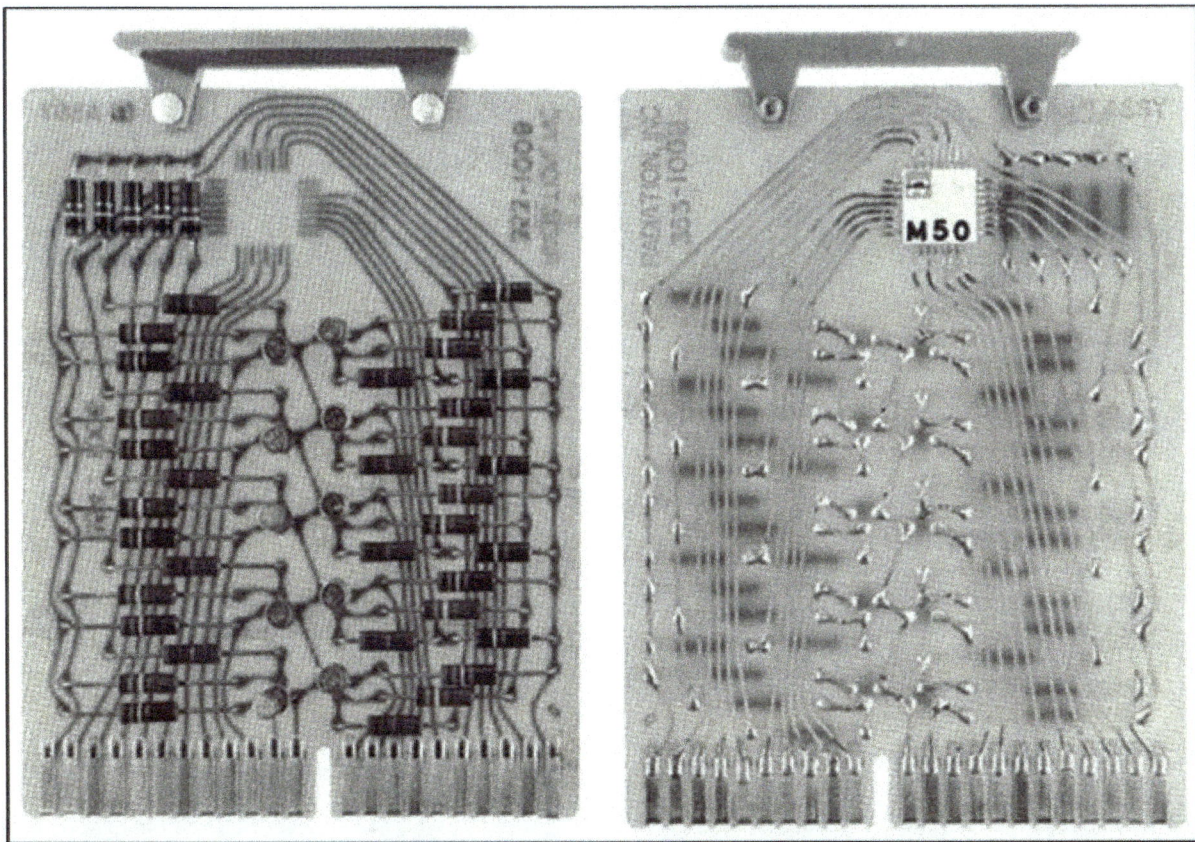

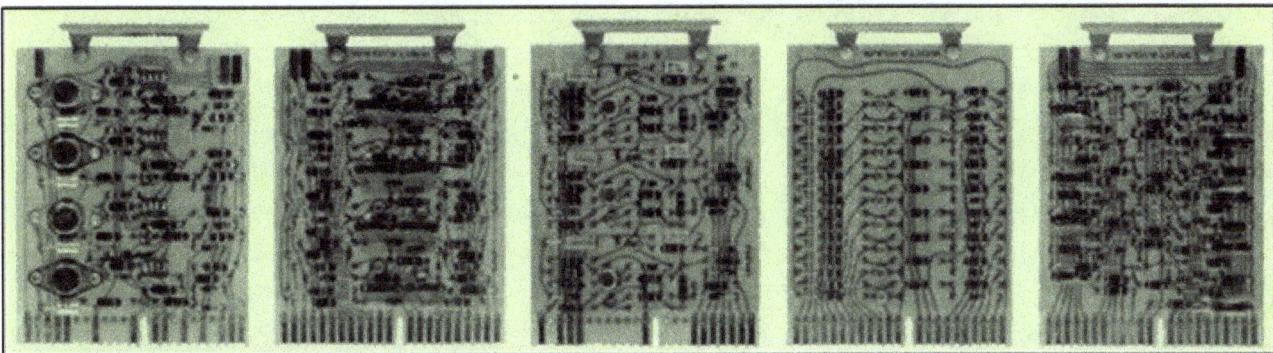

Figure 19: Locotrol PCBs. State-of-the-art Integrated Circuit technology of the 1960s. Harris Corporation pamphlet, author's collection, courtesy Dale Delaruelle.

Figure 20: A Locotrol system undergoing vibration and shock testing at Radiation Inc. Harris Corporation pamphlet, author's collection, courtesy Dale Delaruelle.

We expanded our offering with NYAB brake equipment and our own Relay Interface Cabinet. At last we now had a complete LOCOTROL system to offer other railroads. Of course, the next railroad user was the Kansas City Southern with four systems on GP40 Master units and SD40 Slaves—all painted white and referred to as the 'White Power'. With all of this came presentations to the FRA and NTB on safety issues, while attempting to market to other railroads.

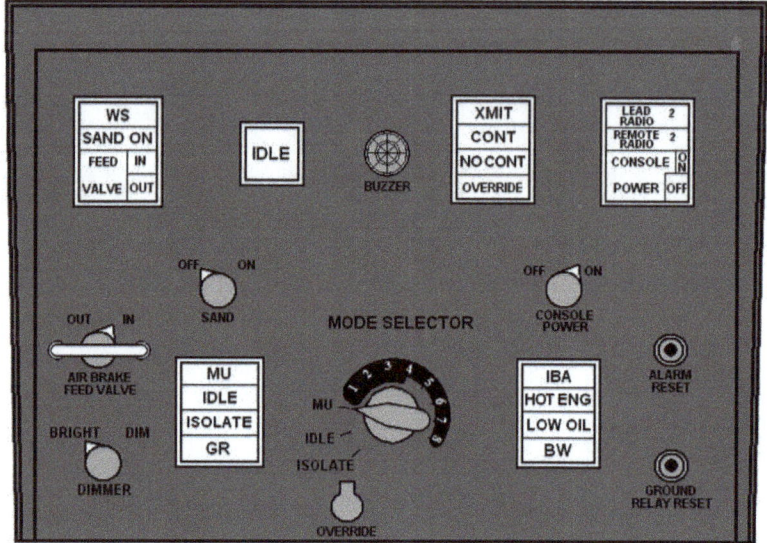

Figure 21: Early Southern Railway RCS/Locotrol Engineer's Control Console type-D. Note the lockout block fitted around the Mode Selector switch to prevent the use of Independent Motoring. This modification was to appease the craft unions. Artwork, the author

As I've previously mentioned, H C Taylor [Henry Clay Taylor, Diesel Supt.] taught me about diesel locomotives, and Worth Hewett worked out the communication losses. Chief Road Foreman of Engines, Glenn Goldstein, taught me how to railroad and there was another gentleman, whose name I cannot remember, who taught me air brakes.

The original electric-to-pneumatics air brake package on the prototype and first production systems was built by Wabco, who were competing at the time with their Remote Multiple Uniter system. So, when I moved from Engineering Product Manager to Marketing Manager, I approached NYAB to build an E-P braking system. Instrumental on this at NYAB was Chief Engineer Charlie Hart and a young engineer, Dean Scott Mitchell.

We slipped delivery on the KCS order and their VP called me to Kansas City for an ass-chewing on his interest cost for the locomotive financing. Those unit installations were supervised by Harvey Ellis. Dale Delaruelle or Ron Gottbehuett were directly involved with me in the test trains.

The main use for LOCOTROL on the Southern were the unit coal trains: the first, states Richard Kimball, being the Parrish-to-Yellowleaf service (the Alabama Power Co plant at Wilsonville, AL) on the Alabama division. Following merger with the Norfolk & Western, the use of LOCOTROL continued until the late 90s, when tonnage ratings were updated and rules changed to allow additional head-end power. The Norfolk Southern Communications & Signal department at Chattanooga applied several updates to the LOCOTROL equipment. This eventually rendered the original Southern equipment incompatible with that—such as

Figure 22: KCS 'White Power' GP-40 at LaQuincy LA in December 1985. Railpix, Gary Morris

LOCOTROL II—being introduced to market by Harris Controls, and it was phased out of operation. The NS's interchange of Powder River coal trains with the Union Pacific and Burlington Northern Santa Fe railroads soon required the introduction of modern equipment, which has come to include ECP technology. At this point, the focus of LOCOTROL installation moved to Canada and this will be discussed in the next chapter, except that no historical account such as that intended by this book would be complete if it neglected to record the 500-car test train (plus caboose) operated by the N&W on 15 November 1967 (see Chapter 6 - Demonstration trains).

ॐ

Other US LOCOTROL customers[31]

[Author's note. This account is not comprehensive. At the time of publication, not all information sought had been obtained.]

The Atchison Topeka & Santa Fe railroad installed Radio Control Systems (LOCOTROL) in 1967 on York Canyon, NM, coal trains destined for the Kaiser Steel mill at Fontana, CA. The Union Pacific railroad installed it in 1968 in two SD45s (3622 and 3623) and a DD35A/DD35 set (82 and 82B) as part of a test of radio control operations on the system, ultimately equipping 19 even-numbered SD45 Leads and 18 odd-numbered SD45 Remotes. This placed both ATSF and UP at the forefront of the development of RCS operations.

UP 82 and 82B were the only non-SD45s to receive RCS equipment. They received their installation to compare their characteristics with those of the two SD45s. The DD35 units were equipped, apparently, for use on an ore train between Salt Lake City, UT and Butte, MT.

Rob Leachman has written:

These [DD35s] were the only LOCOTROL-equipped double diesels. The intent was to power the seasonal Silver Bow ore train with this set. I have never seen any pictures of that... must have been an incredible show. They removed the RCS equipment from the 82 and 82B within 1–3 years and put them in SD45s, but the 'table-top' antenna platform remained many years afterwards, maybe until the 82 was scrapped.

Leachman also recounts that when he rode UP 82 in 1971, the RCS equipment was operable but when he next rode the unit, in August 1975, the equipment had been removed. Company roster listings of RCS units during 1975 confirm the 1975 removal date. UP 82 and 82B were retired in December 1979 and January 1980, and both were scrapped by Precision National in late 1980.

Installation of the RCS radio equipment was in the short hood of the SD45s. This necessitated the relocation of the Coded Cab Signal equipment box to the conductor's-side walkway, behind the cab. The initial antenna configuration consisted of a pair of 'firecracker' antennas mounted directly on the locomotive cab roof. Reception difficulties resulted in a redesign of the antenna system to include an antenna platform, or groundplane, mounted on risers above the cab roof, with a pair of 'can-style'[32] antennae mounted on the platform. The normal voice communication 'firecracker' antenna was relocated to the back of the long hood, between the rear-most radiator fan and the sand filler hatch. Several SD40-2 orders were purchased with 116-inch short hoods, (called 'snoots' by railfans), supposedly to house RCS radio equipment, but none are known to have been so-equipped.

---

31 The author retrieved this material from www.utahrails.net, and acknowledges Don Strack as its source. The terms 'RCS' and 'LOCOTROL' appear to be used interchangeably.

32 *Firecracker* and *can-style* radio antennae—so-named for their physical shape—were types originally associated with the Wabco RMU distributed power system.

Like the SD45s, the RCS radio equipment on the UP 82 was installed in the unit's short hood. For the 82B (being cabless), it was installed at the centre of the unit, on the interior wall of the centre crossover walkway. The initial antenna configuration consisted of a pair of 'firecracker' antennae mounted directly on the locomotive cab roof on the 82, and on the roof over the centre walkway on the 82B. When photographed by Rob Leachman in DP operation near Bitter Creek, Wyoming, in August 1969, the unit was equipped with the later ground plane 'table-top' antenna.

UPs radio control operations were initially tested in Utah's Weber Canyon, using units 3622 and 3623. By September 1968, units 3600 and 3601 were also equipped for RCS operation. The number of UP RCS-equipped units continued to increase, with 3600-3607 being equipped by May 1969. The two DD35s also remained in RCS operations, reportedly for use on an ore train between Salt Lake City and Butte, MT. Throughout 1968 and 1969, RCS operations were in their test phase, using the ten SD45s (3600–3607, 3622, 3623) and the two DD35s (82, 82B).

By early 1971, six more units, 3608–3613, were equipped for LOCOTROL operations. By August 1971, another six units, 3614–3619, were in this fleet. In April 1972, the fleet had grown still more, and included SD45s 3600–3631, along with the 82 and 82B. The continued success of LOCOTROL operations, along with the delivery of new DDA40X Centennial locomotives for the road's fast intermodal trains, eventually brought the SD45 LOCOTROL fleet up to its final size of 37 units—3600–3636—in October 1972. The RCS equipment was removed from the two DD35 units during 1975.

UP C40-8W units 9400–9403 (GE Dash 8-40CW) were equipped in May 1994 at Salt Lake City, UT, for testing with LOCOTROL III (thus becoming GE Distributed Power Units – DPUs) and were assigned to trains in the Blue Mountains. During August 1994, the set was tested in stack-train service between Chicago and Los Angeles[33]. All 22 SD90AC (EMD SD90MAC-H) units, supplied to UP between 1996 and 1999 were equipped with LOCOTROL III, thus also becoming DPU's.[34]

With this much larger fleet to support distributed power operations, the radio-control units were spread to several locations across Union Pacific. Distributed power operations were found to be difficult in Weber Canyon—on UP's famous Wasatch Grade—due to the deep cavity, several short tunnels, and numerous curves. LOCOTROL was soon moving trains across Wyoming and in the Blue Mountains in north-eastern Oregon, between Huntington and Hinkle, where it was regularly used on wood chip unit trains, known as 'Chip Rack Specials'.

Leachman adds the following remarks about how the units were used in the Blue Mountains:

In 1974, there were about half-a-dozen SD45 LOCOTROL sets (even-

Figure 23: Union Pacific chip train CRSW at Nordeen OR, Oct 1974. Rob Leachman

---

33 Milt Deno recalled making these trips—see Summary timeline at the end of Chapter 2.
34 This was Deno's first contract with the Union Pacific Railroad—see Summary timeline at the end of Chapter 2. Also, see Mike Iden's comments, p38

numbered Lead, odd-numbered Remote) assigned to Hinkle-Nampa service to get big trains over the Blue Mountains without helpers. With the onset of the 1975 recession, not so many units were needed, and then sometime around 1977 or 1978 they started using helpers in the Blues again.

During late 1980 and early 1981, LOCOTROL was being used in service on coal and grain shipments between Salt Lake City, and Yermo, CA.

Chicago & North Western used LOCOTROL equipment twice, the first being when they acquired the Chicago Great Western. CGW had two of its nine SD40s equipped with early LOCOTROL 102 systems. One was the Lead unit and the other was set up as the Remote. The equipment was not used by C&NW until about 1968 when the LOCOTROL Lead unit (927, ex-CGW 407) was reassigned to Escanaba, Michigan, for iron ore service. The equipment, however, was removed from the other SD40 and installed inside a former CGW F7B unit—number 104B—converting it into 'RCU' (Remote Control Unit) car X262401. This RCU car had been in the process of being converted into a slug for GP35s. The RCU permitted the SD40 to remotely control any other locomotive (typically a Fairbanks Morse H16-66), which was conventionally MU'ed to the RCU car. The equipment was usually operated on 130-car iron ore trains. The LOCOTROL experiment lasted only about one year and was terminated because the EMD SD40 was too slippery in ore service.[35]

C&NW re-examined LOCOTROL beginning in 1982, while planning the western coal operation. By 1992, interest was sufficient to borrow two BC Rail C40-8Ms (4607 and 4617) for testing.[36] Three C&NW SD40-2s were given to BC Rail in exchange for the Dash 8s. After receiving the BCR units from the Duluth, Winnipeg & Pacific Railroad at Duluth, MN, C&NW moved them to Proviso, UT and then west to WRPI[37].

The first test train was a 150-car behemoth handled from Shawnee Junction, WY to South Morrill, NE. This test train consisted of 35 cars from a unit coal train (which had 115 cars) coupled to a second complete 115-car consist, making a 150-car, 19 800-ton train. Locomotives for the train consisted of a BC Rail C40-8M and C&NW C42-8 number 8577 (the 'Safety and Reliability' slogan unit) on the head-end, and the second BCR C40-8M operating as a Remote unit MU'ed to C&NW C40-8 number 8542 (the 'Wyoming Centennial' unit), mid-train. The most difficult part of the test was stopping and restarting the entire train on the 1% grade on Lost Springs Hill. Several days later, the BCR units had been relocated to Fremont, NE and a similar test train was operated as far east as Boone, Iowa, handling the train over the stiff Arlington Hill east of Fremont.

In early 1993, C&NW purchased three sets of LOCOTROL II equipment, and installed it on C42-8s 8550, 8551 and 8552 at Marshalltown, IA. In late September, the three Dash 8s were reassigned to Escanaba, on Michigan's Upper Peninsula. Historically, C&NW had always operated 108-car trains of short ore 'jennies' between the Empire Mine (one of the two surviving northern Michigan iron ore mines) and the railroad's Lake Michigan dock at Escanaba. The trains consisted of one C42-8 on the head-end, followed by 100 loaded cars, a second C42-8, and the last 50 cars. The third C42-8 was always used to shuttle loaded cars out of the mine up to the main line: an arriving empty train always exchanged one C42-8 LOCOTROL unit for the mine job's C42-8. The operation was generally successful but was hampered by the inability of the mine to consistently load enough cars to allow uninterrupted operation of one 150-car train every 8 to 10 hours.[38]

In December 1994, the three C42-8 LOCOTROL units were pulled from Escanaba ore service, equipped

---

35 See the comment in Chapter 2 referring to Henry Hoagie and the eventual introduction by EMD of its IDAC wheelslip control system.

36 Deno recalled his involvement in these trials.

37 Western Railroad Properties Inc (WRPI), was formed in 1982 as a jointly-owned company of the C&NW and UP to acquire—in 1983—a half interest in the Burlington Northern Powder River Basin (Wyoming) coal line from Shawnee Junction north to Coal Creek Junction. Following its acquisition of C&NW, UP merged with WPRI in August 1995.

38 Deno recalled his and Gene Smith's involvement.

with dual C&NW/UP cab signals, and reassigned to handling 128-car coal trains between WRPI in Wyoming and Midwest Power System's power plant at Sergeants Bluff, IA (near Sioux City). Thus began the first regular operation of larger-than-normal coal trains with distributed power in the Powder River Basin. When the merger occurred, Union Pacific suspended LOCOTROL operation in the Basin, while making plans for restarting the operation of 128-car trains later in 1995.

In early 1995, C&NW purchased five sets of LOCOTROL III equipment for application to C44-9Ws 8726–8730—again for Powder River service—handling 140-car coal trains to Commonwealth Edison near Chicago. Because of the merger, this LOCOTROL equipment was diverted to Elko, NV where GE applied it to standard-cab Union Pacific C40-8s as part of the expanding UPRR distributed power program. The 50 C&NW AC44000Ws ordered by the railroad before the merger (for delivery to UP) have been equipped by GE with LOCOTROL III and are believed to be in Powder River coal service.

Mike Iden, then General Director of Car & Locomotive Engineering for the Union Pacific Railroad, wrote to the author early in 2010:

I first met Milt Deno in the spring of 1992 when, as Director Motive Power Engineering at the Chicago & North Western, I borrowed two GE C40-8M LOCOTROL-equipped units from BC Rail for several weeks of experimentation. Milt accompanied us on both test runs: they being a 150-car loaded coal train with 132-gross-ton cars, three Dash 8s on the head-end and two on the rear-end – one being isolated as a 'safety unit'. We ran first in Wyoming and Nebraska—about a 110-mile trip—but including the longest and highest 1.0% grade on the C&NW, and then reassembled a similar train the next week for a 120-mile run from just-north of Omaha to mid-Iowa. The BCR units served as LOCOTROL Lead and Remote units. I required that the entire train be stopped on the ruling grade in Wyoming to verify that four Dash 8s could restart it. [see comment on previous page re Lost Springs Hill]

Our road foreman was having some difficulty starting the train, and Milt reached over, put the Remote BCR in Notch 8 and wiped the Lead throttle to 8. The loadmeter needle walked over and pegged at 1800 amps (which we had seen and used many times on non-LOCOTROL coal trains). After about 30 seconds, we finally got movement. On a sunny day with dry rail, we did restart but it was a real grunt lift.

A year later, after equipping three C&NW Dash 8s with LOCOTROL II, Milt conducted a training class in Escanaba, Michigan, for all of our RFE-types and myself. We started using the units in taconite pellet service, running 1 × 1 (100 cars + 50 cars) instead of two SD60s with 110 cars. A year later, the trio went into coal service from Wyoming to Iowa, running 125-car trains where they'd previously been 115 cars.

An associate and I at C&NW wrote a 60-page report in 1992 on the LOCOTROL tests with the BC Rail units and I sent a copy to the UP in Omaha. About a month later, I was talking to Fred Routledge (Harris' LOCOTROL salesman) who informed me that Union Pacific people were flying into Melbourne, FL to sign up for 100 sets.

I joined UP through the merger in July 1995, and our DP fleet has grown since then to 2800 units out of a total fleet of 8350. In 2009, 65% of our train starts were DP operations.

A key combination of technologies was AC traction and DP, which from 1996 onwards allowed us to convert the entire UP Wyoming coal operation from 115/117-car head-end trains to all-DP today: 128–150 cars (almost all 143-ton cars). Major track additions were also required in unison (one stretch of the joint BNSF-UP main line in the coal basin is now four-track). Our record so far has been, I believe, 40 loaded trains in one 24-hour day. The coal trains join our transcontinental main line near North Platte, NE. On the triple track east of North Platte we have moved upwards of 130+ trains (all types) in one day.

I had worked with Richard Kimball when I was a management trainee at Southern Railway. (One reason I went to Southern versus [opportunities] at two other US railroads was… LOCOTROL). I was at Southern for 2 years, then EMD for [more than] three, then C&NW until 1995 and UP since. Basically, I took my fascination with the potential for Remote consists along throughout my career. I persisted but made little progress with it at C&NW for 14 years but by then western coal was flowing fast and I finally had a VPO who also saw the potential (hence buying the first three sets) in 1992. I now see so many DP trains it seems incredible that back in the 1990s I would think about what the railroad would look like if we had distributed power. Now it is difficult to find trains without Remote consists!

2

MILT DENO AND THE GLOBAL APPLICATION OF LOCOTROL

The railroad career of Milt Deno and the story of LOCOTROL are inextricably linked. As a railroader, Deno was surely among the last of a generation at the end of a unique era of railroading – in North America, as well as many other countries of the world. This was an era when operating managers were selected by their superiors from assessment of technical competence as much as for their seniority, and who succeeded or failed on their continuing ability to demonstrate specialist expertise as well as their capacity to organise themselves and their work. It was no doubt this connection—the culture of the Canadian Pacific Railway and his learning and experiences within it—that prepared Deno for his role and for the challenges of bringing LOCOTROL to CPR and, eventually, to many other railways throughout the world.

Without doubt, Deno's awareness of the winds of change beginning to sweep through both the huge transportation organisation and the industry of which he had been a part for more than half a working lifetime figured large in his eventual decision to accept an offer from Harris Corporation to move to Florida and specialise in the development and application of this equipment. For more than a decade, Deno—the only career railroader at this huge electronics company—would be the vital link between it and the railroad industry who were its Railroad Product Line customers.

During his career with Harris Corporation, Deno installed and tested LOCOTROL I, II, III, and a few Tower Control systems on many railroads throughout the world. In Canada, he installed, and tested LOCOTROL on every major railroad except one. In the United States, he installed and tested it on every Class I railroad except one, and on a few Class II railroads as well. Throughout the world—other than Canada and the United States—Deno installed, and tested various models of LOCOTROL in Iran, Algeria, Mexico, Morocco, Australia, (Mount Newman Mining/BHP Iron Ore, Queensland Rail, Robe River Iron Associates), India (at two different locations: one air-braked Broad-Gauge, the other vacuum-braked Metre-Gauge), Brazil, and Germany. One of these—his first overseas project—will be covered in this chapter.

☙❧

I first met Milt Deno in 1987 at Port Hedland in Western Australia where I was a Locomotive Crew Foreman (North American translation: Road Foreman of Engines) with iron ore producer Mount Newman Mining Co Pty Ltd, and Deno and colleague Gene Smith were on-site to install some new LOCOTROL II systems. I was assigned by our railroad manager to work closely with this team for several months in crew training and in the production of a customised engineman's instruction manual—something that Harris Corporation had hitherto not provided to their various customers. Subsequently, Deno offered me the opportunity to join a LOCOTROL installation project in India, functioning as a crew trainer—both in the classroom and over-the-road—for a period of 4 months at two sites.[39] Following completion of this project, he asked me to accompany him to Iran

---

39 See Chapter 4.

for a 12-month project: however, before our contract negotiations for this mission could be completed the Harris business was purchased by General Electric—who, in light of the political climate of the time vis-à-vis the Islamic Republic of Iran, had a different commercial imperative—and the project was cancelled (leaving me 'high and dry' since I had quit my job with the iron ore miner in anticipation of this project). I moved on to other things, eventually returning to railroading, and Milt and I remained firm friends. The cancelled project was to be the second Iran Railways installation, the first having been completed back in 1978.

Many people, of course, have conducted expatriate work for an employer, sometimes for considerable project periods: however, in the field of transport equipment sales and installation, there can be few who travelled quite as extensively and worked as consistently in as many countries for one employer as Milt Deno. Others in Florida were intimately involved in designing, building, marketing[40], and planning for LOCOTROL, but it took the dedicated and specialist effort—over several decades—by Deno and usually one or two technical assistants (one of whom was, more often than not, Gene Smith) to constantly travel to foreign destinations and to work there—often with improvised facilities in remote locations—to install the equipment and instruct local railroad personnel in its operation and maintenance.

The author was privileged to be able to join Milt Deno and a small project team on one of their international odysseys to bring another foreign railway system into the fold. What I learned on that project was that Deno had only a few simple rules for his team: to do the work that Harris had contracted to do (and to accept professional responsibility for that work), to treat the customer with deference and respect, to act with integrity, to establish a professional and courteous relationship with the railroad and its personnel, and to have fun! Deno was a railroader and this was railroading. Oh, and the litre of Johnny Walker Black Label he carried in his suitcase would *not* be opened until the project had been carried to successful completion.

One further dimension to the man's character might be mentioned. M C Deno was an old-school railroader who had risen through the technical ranks of his company, and he was a CPR 'company man'. From his superiors he had learned—and he greatly valued—qualities of loyalty, friendship, integrity, commitment, professionalism, and respectfulness. His railroad career had afforded him the valuable opportunity to develop and practise advanced leadership skills. These were essential to the success of the many projects he worked on and led, both as a CPR master mechanic and with Harris Controls. He could be short-fused and brusque, but he always remained open to challenge (although you had better be right… or at the very least be prepared to prosecute a robust argument based on fact or logic). His ability to detect hubris, mendacity, and deceit was impeccable. Those who passed these tests of character won Deno's admiration and lifelong respect, those who ran afoul of them, his studied contempt.

These personal traits coalesced to the immense benefit—not only of the CPR—but also to that of Harris Corporation and the LOCOTROL product. It is for these reasons that this man's personal journey is critical to the greater 'LOCOTROL story'. This account is the product of personal observations and many hours of discussions over our years of friendship.

---

40 During this period, Fred Routledge was Harris Corporation's sales director. He and Deno frequently travelled together, but Routledge's high-level role preceded the commencement of an installation project and inevitably saw him in-and-out to the client's head office and staying in 5-star city accommodation. Not to diminish his contribution—which was a significant part of the LOCOTROL story—but he did not have to spend weeks or months in remote localities, working 'down-and-dirty' with the locals. Fred eventually retired and operated a furniture store in Melbourne, FL.

The family line

Milton Carlyle Deno was born in Wilkie, Saskatchewan, in central Canada on 9 January 1933, the son of Carlysle [note different spelling] Frank Deno (known to many people as 'Jimmy') and Ethel Nora Clarke. Ethel had emigrated from England with her family in the early 1900s and hailed from Moose Jaw, SK, where her father was a minister in the Anglican church.

Frank Deno—who was born in 1921 and whose family had originally moved up from Saginaw, Michigan in the US in the mid-1920s—attended the University of Wisconsin at Madison, graduating into the Great Depression with a mechanical engineering degree. Securing work as a fitter/machinist with the Moose Jaw Electric Railway[41], Frank Deno later moved to the Canadian Pacific Railway in the same capacity at the Moose Jaw locomotive roundhouse in 1924. In 1931 he was elevated to Night Foreman at Sutherland, SK. A promotion to Acting Locomotive Foreman at Wilkie soon followed, leading to a transfer back to Moose Jaw in the same capacity in 1934. Frank was promoted to General Foreman at Moose Jaw in 1936, then eventually to Divisional Master Mechanic, and later moved to Regina, also in that capacity. His next promotion saw him move his family to Winnipeg as District Master Mechanic (and in 1942 to become Assistant Superintendent of Motive Power – Western Lines). It was during a field trip to Moose Jaw in March 1946 that Frank Deno died from a heart attack, aged 44. He would have undoubtedly achieved even greater eminence within CPR had he lived.

Figure 24: Frank Deno. The author's collection, courtesy M.C. Deno

Within the North American railroad industry, the position of Master Mechanic was, for two generations, utterly pivotal and had acquired a long and highly respectable history. Since the earliest days of American railroading, the Master Mechanic was a road's foremost mechanical manager, the ultimate overseer, and in many respects the final arbiter of things to do with motive power, its acquisition, and its repair and maintenance. Everyone who had anything to do with locomotive operation worked, ultimately, for the Master Mechanic.

In later years, a railroad mechanic started his career with a technical apprenticeship, and a special few rose through the ranks of supervision to run a repair shop or roundhouse, then several roundhouses and backshops across an operating division, then perhaps all roundhouses on a railroad. But a master mechanic was a lot more than this. He managed the division's motive power requirements, he organised and project-managed train wreck recovery, snow clearance operations, new technology testing and trial, and he managed all the people involved with these functions. Many were considered by their railroads in the very early years

---

41 The Moose Jaw Electric Railway streetcar operation was conceived in 1909 and ran first in August 1911. Operations ceased in October 1932. With the eventual creation of the Moose Jaw Transportation Company, public transport in Moose Jaw became a motor bus operation.

Figure 25: In 1939, as a divisional master mechanic, Frank Deno was assigned to the westbound Royal Train conveying Their Royal Highnesses King George VI and Queen Elizabeth. He rode engine 2850 between Moose Jaw SK and Kamloops BC. After this tour, these semi-streamlined, oil-fired 4-6-4 Hudson-type locomotives were bestowed with the 'Royal Hudson' class name.
E.Towler/Bud Laws Collection. http://www.trainweb.org/oldtimetrains/photos/cpr_steam/Royal_train_1939.htm

to be pupil engineers[42] by virtue of their knowledge and experience, and a reliable master mechanic was considered by the professional engineers who rose to executive positions to be a vital link between them and the operational 'real world' of the railroad and its people. Some master mechanics—such as Milt Deno, and his father (who *was* a professional engineer) before him—rose to become Superintendent of Motive Power or similar, only one hierarchical step below that of Chief Mechanical Officer. Nowadays, the title 'Master' has fallen somewhat from grace and the railroads have invented a slew of pretentious but more fashionable 'new age' job titles that cover the functionaries once known as Master Mechanic, Road Master, Track Master, Train Master, and Yard Master.

Mentoring was a crucial and well-established process at the Canadian Pacific Railway, and once a tradesman's dependability attracted the attention of the 'higher-ups', that person could expect—should their performance and reliability continue to impress—to be listed in someone's succession plan. For this, a mutually respectful relationship with that important 'someone' was essential, and if the lot of a tradesman with such a rapport might be a fortuitous one, it was also likely to be challenging. Master mechanics were not selected to be accorded an easy life on the railroad.

In Winnipeg, Milt attended the Sir Isaac Brook [public] School and after his father's death, the boarding school at St John's Ravenscourt, achieving good grades and excelling at football and hockey (making the senior football team at 13 years of age, playing 'half-back'). In Grade 10 in 1948, aged 15 and with a summer job (providing what he felt was a lucrative income) working for a construction company in Moose Jaw, Deno—much against his school's advice—decided against Grade 11 and liberated himself. Ethel Deno had meanwhile moved to Vancouver and eventually convinced her son to come out west and return to school. Enrolled at Kitsilano Secondary in Vancouver, Deno immersed himself in hockey and football and, with a waning interest in academic pursuit, distinguished himself there by his mediocre grades. It was also at Kitsilano that Deno met Geraldine (Gerri) Muscroft, who would become his first wife.

With a maternal anxiety born of her son's evident lack of career design, Ethel finally contacted Bill Stewart,

---

42 The earliest of those people we now call 'engineers' (from the Latin *'ingeniare'* – to create, generate, contrive, devise – and *'ingenium'* – cleverness) were pupil-trained novices who learned the practical skills of their nascent crafts as apprentices and associates of close relatives or mentors. The formal tertiary education process that nowadays produces an engineer did not exist when the early railway pioneers began to conceive, calculate, and construct. The early master mechanics were pupil-trained engineers and were often the railroad's locomotive drivers. The North American term 'locomotive engineer' derives from this.

her late husband's successor as Assistant Superintendent of Motive Power – Western Lines, and an old and firm friend. Thus, in March 1950 Deno found himself reporting—along with one Walter Mills, another potential apprentice and someone who would come to work for Deno in later years—to the Chief Clerk in the Divisional Master Mechanic's office in Moose Jaw. The Chief Clerk stated, *'I have two apprentice positions open in Regina—one for a mechanic and one for an electrician. Mr Deno, what is your choice?'* (Deno would always wonder if the reason for him receiving the privilege of first choice was his family connection to the CPR… it probably was). Nonplussed, Deno hesitated… *'I don't know.'* The Chief Clerk turned to Mills. Same answer. Frustrated, the Chief Clerk took the initiative: *'Alright, I'm going to flip a coin—Mr Deno* (that concession again), *you call it!'* Deno called and won the toss and the Chief Clerk said, *'Okay, Mr Deno, you are an electrical apprentice.'* And so, Milt Deno embarked on the career that would take him through the technical ranks of the Canadian Pacific Railway to an eventual senior position, and that would take LOCOTROL to the world.

Undoubtedly, the respected Deno name opened doors for Milt, but these were doors that would have quickly slammed shut in his face had he not been able to maintain the trust and respect of his sponsors. It would seem obvious that as he grew as a man and as a railroader, Milt Deno came to be seen within CPR as possessing the same qualities that had seen his father progress to a senior mechanical position.

☙❧

In the beginning

If not through the Canadian Pacific contract—Harris Controls' foremost LOCOTROL sale to that point—one wonders how the adoption of distributed power globally might otherwise have happened. But we do know how it *did* happen. CPR secured a massive export coal contract, decided they needed distributed locomotive power to move the tonnages required, made a bulk purchase of the equipment, and assigned a diligent young up-and-comer to oversee its establishment onto the railroad. This young man would eventually oversee the introduction of LOCOTROL onto many other railroads.

Milton Deno commenced his working career as an electrical apprentice with CPR at the Regina roundhouse in March 1950. In June 1951, he was transferred to the Drake St Shops in Vancouver and in September 1952 to the newly-built Alyth Diesel Shop in Calgary, AB. In July 1953, he was transferred to Nelson, BC where—in April 1954 upon completion of his 10 600-hr apprenticeship—he became a tradesman electrician. In July 1955 he was posted to the roundhouse at Field, BC as Afternoon Shift Foreman Mechanical and by June 1961—aged 28—Milt was the Canadian Pacific Railway's Diesel Inspector at Medicine Hat, AB.

One day in the fall of 1966, Milt's mentor, Art Reynolds—Assistant Chief Mechanical Officer – Pacific Region—called Bill Flett, Superintendent of CPR's Medicine Hat Division, and told him that Deno was required in Revelstoke 'ASAP' because of a problem they were having with their brand new GMD[43] SD40 locomotives. Little did Deno know that in order to correct this problem he'd be away from Medicine Hat for the better part of a year! Deno recounts:

The initial complaint was that they could not keep sand in the locomotives while they were negotiating the Mountain Subdivision. Sand was carried on these locomotives in large quantity—56 cubic feet per locomotive— and was used both manually and automatically to assist with adhesion. In about 30 to 35 miles going up and down the mountain they would run out of sand. I had worked on wheelslip a little bit, but what I was now

---

43 See Glossary.

hearing, I thought, was ridiculous, so in my inimitable fashion I set out to find out what was going on. I certainly didn't anticipate what would be involved. My initial findings were that these locomotives had the lousiest wheelslip system of all the locomotives Canadian Pacific owned. Speaking to General Motors personnel about the system and its inadequacy in operation was fruitless.

Not long before, I had read an article in a railroad magazine about a guy in Baltimore who was electronics-oriented and who had designed a locomotive rate-of-change wheelslip detection system. He was not a railroader and had come up with the idea while he was sitting in a barber's chair across from the Baltimore railroad yard and had heard the screech of wheels slipping on locomotives in the yard. He must have had friends at the Baltimore & Ohio Railroad, because he'd looked at some locomotive wiring diagrams and had come up with his design after doing some research. Art Reynolds was in Revelstoke at this time so I told him about this article, and because General Motors wasn't paying us much attention, I recommended that we get a-hold of the guy and ask him to come up and have a look at our situation. Reynolds agreed.

The man's name was Henry Hoagie [or Hogie... the author is uncertain of the spelling]. I contacted him and he arrived in Revelstoke a few days later, ill-prepared for our winter weather. I took him under my wing and got him equipped with all the winter clothing he needed plus safety boots. He was very excited about going to work with our railroad. He had never been on any locomotive before (which surprised and kind of concerned me). I needn't have worried; he was an electronics genius.

So that we could study the effects of wheelslip, I had equipped one of our SD40 locomotives with small flexible pipes that were secured and trained on the area where the leading wheels contacted the rail: the wheel/rail interface. These pipes were connected through electric valves to a reservoir in the nose of the locomotive. With a push-button I could activate the solenoids and feed oil under the front wheels of the locomotive. With the locomotive under full load and hauling maximum tonnage, the resulting wheelslip was extreme: on occasion to the point that all six axles were spinning synchronously. This, of course, was never supposed to happen and General Motors said it couldn't happen. I told them I would make them believers... and I did.

I was riding a lot of trains between Albert Canyon and Glacier, and Beavermouth and Glacier. Both runs were on 2.2% grades, some of the steepest on the CPR. I was getting tired with this regime, so I asked Mr Reynolds if I could have some help. He told me to get whomever I wanted, so I chose Bill Ruzek from Alyth Diesel Shop whom I knew to be a smart person with electrical systems. Mr Reynolds also put one of the railroad's private cars, with a cook at Glacier, at our disposal so we could stay on the job and not be faced with driving back and forth to a motel in Revelstoke.

Henry Hoagie was with me for about a week. At that time, he thought he had enough information to go home, tweak his design and then return with a prototype that we could install on a locomotive and measure as to its effectiveness in controlling wheelslip. In the meantime, Bill and I continued to test and observe what the heck was going on with these locomotives. Fairly quickly, we found that the excessive wheelslipping was causing the locomotive wheels to heat up to such a degree that the surface metal on the tread was peeling off, a condition called 'spalling'. This was a serious consequence of what we felt was the deficient wheelslip system on the SD40. It was extremely costly to have new wheel-sets fitted, and it was unsafe because of the possibility of wheels cracking and causing a derailment.

Mr Reynolds also informed me that the diesel shop in Calgary had been changing out an inordinate number of turbochargers on these new locomotives. At that time, bad wheels and turbochargers were not costing CPR any money; they were on warranty. Reynolds' concern was that when warranty was over, CPR would be faced with a huge expense for these items. He asked Bill and I to try and find out why the turbos were failing so often; and we found out real quick!

When wheelslip takes place on a locomotive, electrical power to the traction motors and subsequently tractive force to the wheels is reduced until the slipping stops, whereupon power is restored. When the locomotive is at full power, as they are on mountain grades, the turbocharger is being driven by the engine exhaust; however, if the wheels lose adhesion, power is reduced, the fuel supply to the engine is decreased, and the turbocharger loses speed. Eventually, the clutch mechanism will engage with the locomotive gearing and drive the turbo until the exhaust gas flow is again sufficient to take the turbo drive off the gearing and permit it to freewheel.[44]

We found that wheelslipping on the SD40 was so persistent it was causing the turbocharger to constantly alternate between being connected to the clutch and freewheeling, subsequently causing the clutch and other mechanical parts to fail. When I reported this evidence to Mr Reynolds he immediately transmitted it to Montréal. The reaction was pretty darn swift.

Bill and I were told to continue testing and to apply Henry Hoagie's prototype as soon as he returned. The dynamometer car was to be transported from Angus Shops in Montréal to Alyth Diesel Shop in Calgary, and be prepared for testing on the Langdon, Mountain, and Shuswap subdivisions. I (but not my family) was temporarily transferred to Revelstoke until further advised. This turned out to be about 9 or 10 months.

Henry returned from Baltimore, and when I described what was happening, he was anxious to get started. We applied his prototype wheelslip system to an SD40 locomotive and all were eager to get out and test it. Charlie Duncan, General Motors' Calgary representative, wanted to come on this test run. Now, even though I liked Charlie, I told Mr Reynolds I didn't think it was a good idea. They'd not helped us, so why let them in on what we were doing right now? He agreed.

With full tonnage for a single unit, we took the train to Albert Canyon. Mr Reynolds came with us and had arranged with the dispatcher for an uninterrupted run to Glacier, this being mostly on 2.2% grades. Henry's wheelslip system functioned exceptionally well. Even when I applied oil to the rails, the axles did not go into an uncontrolled spin, power was reduced and reapplied incrementally, and the turbocharger didn't oscillate back and forth on-and-off the clutch. We were elated, but I don't think as much as Henry was. He was like an excited kid! Mr Reynolds sent Henry home to Baltimore, and we built three more of his devices so that we could eventually have four locomotives equipped.

Bill and I returned to work; there were still a couple of things that needed modifying on the SD40s. Traction motor shunting was a problem, so we went to work on fixing that. There were some other smaller modifications that we made, then we were sent to Calgary where we supervised and helped install all our modifications on another three locomotives. The dynamometer car was coupled and connected mechanically and electrically to the four modified locomotives.

The dynamometer car is an information-gathering vehicle. It is capable of measuring and dynamically recording locomotive horsepower, tractive effort, wheelslip, wheel-to-rail lateral forces, and numerous other factors. In summary, it paints a graphic picture of what is going on with many parameters on the locomotives while they're pulling a train. It has sleeping and rudimentary bathing facilities, a kitchen, and a dining room.

---

44 Horsepower produced by naturally-aspirated engines (including Roots-blown two-stroke engines) is usually derated 2.5% per 1000 feet (300 m) above mean sea level; an enormous penalty at elevations of the order of 10 000 feet (3000 m) over which some Western US and Canadian railroads operate (and amounting sometimes to a 25% power loss). Turbocharging effectively eliminates this derating. EMD 645 engines—as were installed in SD40 locomotives—utilised either a Roots blower or a turbocharger for their cylinder air supply and scavenging. For turbocharged engines, the turbocharger—at low engine speeds (when exhaust gas flow and temperature alone are insufficient to drive the turbine)—is gear-driven as a blower via a centrifugal clutch and at higher speeds de-clutches to be purely exhaust-driven. This design improves high-altitude performance (potentially up to a 50% increase in maximum rated horsepower over a Roots-blown engine for the same engine displacement) and may also provide fuel consumption benefits.

About a quarter of the forward end of the vehicle is reserved for test and recording equipment. Bosses get to ride in the dynamometer car, as did Henry. Bill and I got to ride on the corn-binders, periodically being invited to have a bite to eat.

We were now going to go into qualified test mode. Charlie Parker [Assistant Superintendent of Motive Power – System] *from head office in Montréal, accompanied by technicians to operate the dynamometer car, Mr Reynolds, Harry Piper—Master Mechanic Revelstoke—and last but not least, myself and Bill Ruzek, were officially assigned as test crew on the SD40 project. Obviously, Henry Hoagie was a vital component of the crew! Our objective was to evaluate the performance of the modified SD40 locomotives against the performance of the standard GMD model. It was now only a couple weeks before Christmas 1966, so it was decided by Mr Parker and Mr Reynolds that for the sake of continuity, we would not begin testing until the beginning of January.*

On 1 January 1967, Bill Flett informed me that I was to be District Diesel Inspector of BC, located at Revelstoke, starting 1 March, and reporting directly to Mr Reynolds. I was not replacing the incumbent Diesel Inspector but would be assigned solely to special duties. Around the middle of January 1967, we all regrouped in Calgary, including Henry Hoagie. The four modified locomotives and dynamometer car were coupled to a tonnage train of wheat and moved to a yard just west of Calgary so that we could do some preliminary static testing before heading into the mountains and commencing dynamic testing in earnest. We initially tested on the Langdon Sub between Lake Louise and Stephen, which has a 2.2% grade with a couple of nasty curves that were tight, on-grade, and equipped with flange lubricators. These lubricators squirted grease supposedly on the inside face of the rail head where the wheel flange makes contact: however, these greasers inevitably manage to get some on top of the rail as well.

On the Mountain Subdivision, we tested between Beavermouth, and Revelstoke in both directions. This was the most rugged area of all and had numerous curves with lubricators and long stretches of heavy grade. It also included the 5-mile Connaught Tunnel wherein the rails were prone to be slippery due to the effects of locomotive exhaust and water dripping from the roof. On the Shuswap Subdivision, we tested mostly at Notch Hill.

The data collected from the dynamometer car was carefully analysed and compared to data taken from standard SD40 locomotives, the conclusion being that the modified SD40s outperformed the standard locomotives in all respects. Replacing GM's proprietary wheelslip system with Hoagie's system solved the main problems. The locomotives no longer ran short of sand, wheelslip was well controlled (turning a tiger into a pussycat), and wheel and turbo drive-train damage was arrested. These were major improvements, but we tweaked a few other characteristics of the locomotives that also contributed to improved operation.

When we returned to Calgary, I deemed it now appropriate to invite some General Motors sceptics to the Alyth Shop to give them a demonstration of how deficient their wheelslip system was. I lined them up to one side of the 5557 that still had my oil-application device and asked them to observe all six wheels on that side. Off to one side I told Bill Ruzek, to watch the wheels, and tell me when I had all of them in a synchronous spin (which GM said couldn't happen). With the locomotive Independent brakes applied, I advanced the throttle to Run 4, and—manipulating the brake—I let her creep slowly forward under load, squirting oil until I had slicked all of the rail beneath her. I then released the brake and opened out to Run 6, and the locomotive achieved simultaneous wheelslip of all axles. For dramatic effect, I let her slip for a few seconds until Bill started waving his arms, whereupon I throttled back. Unable to resist, I looked down at the GM experts and asked, 'How do you like that?' I didn't even bother to go down and talk to them.

Our Chief Mechanical Officer, Harold Hayward, summoned Mr Reynolds and I to a meeting with Mr Snow, the President of GM Canada, and his staff in Montréal to discuss our test results. The meeting was somewhat rancorous. GM's Chief Electrical Engineer told Harold that I was not qualified (they must have done some research on me and my schooling) to undertake modifications on the SD40s, and if Canadian Pacific continued making these modifications on other units, the warranty agreement would be rescinded. That took the lid off her. I had never been in a high-level meeting before and didn't know much about 'professional protocol'. I informed the CEE that it seemed that GM Canada weren't aware of the inherent deficiencies of their locomotive.

Harold told everyone to take a break except me. He then set about enlightening me on my behaviour, saying that he thought I was right. He advised me that when the meeting resumed, he would handle discussions around any shortcomings on the part of GM, including said Chief Electrical Engineer. GM Canada said they could not acquiesce to our request to have them duplicate our modifications across our SD40 fleet. Mr Hayward responded, in polite terms that would have been beyond my abilities, 'If you won't do our bidding then you can take your warranty and place it where the sun don't shine.' The fallout of this meeting took place a couple of years later when GM lost a large order for CPR locomotives to Montréal Locomotive Works.

Meanwhile, GM's Electro-Motive Division in the States schmoozed Henry Hoagie, convincing him they would buy his invention. That never happened and he got screwed. He was so guileless that he'd shared with them how his rate-of-change system worked. They took that, made a few changes, and produced a wheelslip system they called IDAC. When their uprated model, the SD40-2, was subsequently introduced, it included all the modifications—in one form or another—we'd made earlier. I personally never had any faith in GMD ever again.

During the time that we spent testing wheelslip systems, Henry and I spent a great deal of time together and became good friends. I'll always remember an evening when we were having a drink together in Vancouver. He pulled some coloured paper from his suit jacket, and began to meticulously fold it, the result being a small bird. Offering it to the waitress, he pulled its tail and its little wings flapped; she was enthralled. He said, 'It's called Origami... I taught myself. On a few Christmases past, Macy's department store has hired me to make large origami figures for their window displays.' Henry later became ill and passed away before his time.

Author's note: It may be that Henry Hoagie had already introduced his nascent wheelslip control system to the Baltimore & Ohio RR prior to his involvement with CPR. Some records state 'B&O had developed a wheelslip system that EMD later adopted and designated as their IDAC[45] Wheel Slip Control System.' Around the time that EMD introduced its new 645-powered line of locomotives, there had indeed been research into the subject of diesel locomotive wheelslip control conducted by people not connected with the locomotive manufacturers. One recorded concept was the use of direct-measuring coils mounted on the traction motor leads that could detect small but instantaneous changes in traction motor current. This system was developed and tested (it is said, by outside companies and various railroads, one of which may have been B&O), but EMD apparently adopted it about a year after its new locomotive model was introduced.

IDAC (later packaged as 'WS10') worked by modulating the traction motor field excitation. The system detected wheelslip by the change in resistance due to increased speed of a traction motor (the electrical resistance of a series-wound traction motor is inversely proportional to its speed—resistance decreases as speed increases). The system then automatically dropped sand and momentarily cut power to all axles. Wheelslip was corrected by the instantaneous reduction of locomotive power together with automatic sanding. After adhesion was regained, a timed application of sand continued while power was smoothly restored. The system functioned entirely automatically, and no action was required of the locomotive operator.

Numerous North American railroads retro-fitted 40-series EMD/GMD locomotives with IDAC in the early 1970s prior to introduction of the Dash 2 model. EMD claimed that a railroad could count on an SD40 developing 18% adhesion 99% of the time or 24% adhesion 95% of the time regardless of weather conditions: and an SD40-2 could be expected to develop 21% adhesion 99% of the time or 27% adhesion 95% of the time,

---

45 **Instantaneous Detection And Correction.**

regardless of weather conditions. The subsequent introduction by EMD of their Super-Series wheelslip control technology utilising Doppler radar overtook IDAC.

Was Henry Hoagie indeed the first person with this concept? Had he revealed it to B&O (and EMD) even prior to his involvement with Milt Deno and the CPR? Had the American railroad industry realised that Hoagie was onto something, quietly filched his idea, and closed ranks to elbow him out? Whatever the answer, IDAC went down in history as a new technology developed by GM's Electro-Motive Division.

ぐ3இつ

A coal contract—Deno's career advances

On 1 January 1967, Bill Flett had informed Milt that Art Reynolds had called from Vancouver to advise—as Milt has explained in his preceding account—that he was to be appointed Diesel Inspector at Revelstoke, starting 1 March 1967, and reporting directly to Reynolds. Milt would not be replacing the incumbent Diesel Inspector but would be assigned solely to special duties.

Contacting Milt from Vancouver, Reynolds told him it looked as if 1968 was going to be a busy and interesting year. CPR had just signed a lucrative contract with the Japanese for moving coal from south-east British Columbia to the Roberts Bank export coal terminal[46] just south of Vancouver city. It was reckoned that in order to meet the railroad's commitments, the CPR would need a new type of railroad car, new locomotives, and—in order to alleviate congestion on the Mountain Subdivision—to somehow make trains longer and heavier.

Milt was informed that he would be travelling extensively over the Pacific Region, fulfilling assignments directly related to the movement of coal to Japan. He was now a member of a team formed to devise a plan, and eventually a working model, of the systems and processes that would fulfil the requirements of this contract, the team having, at best, 16 months to accomplish

Figure 26: Roberts Bank Coal Port, 1983. The Roberts Bank export coal terminal is a man-made island in the Strait of Georgia, 25 kilometres, (15.5 miles) south of Vancouver. It is linked to the mainland by a 6.4 km (4-mile) causeway. The first CP Rail unit train brought coal from southeastern British Columbia to the port for export on 20 Apr 1970. Author's collection

---

46 Now also a busy container terminal.

this. Charlie Parker in Montréal would head the team, with Reynolds as second-in-command, and other members co-opted as necessary.

Reynolds stated that the necessary personnel on the Pacific Region had been advised and instructed to provide Milt with whatever assistance he might request. Were Deno to encounter any resistance, he was to call Reynolds immediately. Milt's first assigned task was to evaluate the MLW[47] 3000 horsepower Century-Series 630 locomotive as the prime power source for these new coal trains. Secondly, he was to learn all he could about the remote-control equipment currently being used in Virginia by the Southern Railroad and manufactured in Melbourne, Florida by Radiation Incorporated.

Although surprised by this turn of events, Deno was determined, as ever, to give it his all. Revelstoke would now be his home base and the focal point of the equipment testing to follow.

The head of the Rollingstock Department at head office in Montréal was Mr Teoli. He was the engineer who designed the brand new 'bathtub' coal-car. Its tare weight was 32 tons and it had a capacity for 105 tons, larger than any other railroad in North America. These cars were built for us by Hawker Siddeley Canada in Trenton, Nova Scotia. Also, I didn't know too much about these new MLW 630s, so I went down to Montréal and visited their plant. I became satisfied that this locomotive would do just fine pulling the new coal trains and communicated this to Mr Reynolds, thus fulfilling that first request. I then commenced to investigate what the heck this new remote-control equipment was all about.

☙❧

Figure 27: *Road Foreman of Engines testing Locotrol 105 at Alyth Shops, Calgary AB.*
Author's collection, courtesy Dale Delaruelle

Charlie Parker had already held preliminary conversations with both Radiation and the Southern Railway, with the result that CPR would lease one Lead and one Remote set of electronic equipment—including air brake manifolds—to be installed on a locomotive and a suitably-ballasted remote-control car at the Angus backshop[48] in Montréal. The consist would then be dispatched to Calgary, where the dynamometer car would be added.

Deno was requested to come to Montréal to meet with some of the Radiation people and to be involved with the installation and initial testing. He recounted that his initial enthusiasm at being chosen for this project diminished rapidly when the Radiation people showed him the cabinets full of solid-state electronics and mysterious-looking air

---

47 Montréal Locomotive Works.
48 A 'backshop' is a specialised workshop found at locomotive depots.

Figure 28: Back-to-back testing at Alyth Diesel Shop, Calgary. Author's collection, courtesy Dale Delaruelle

Figure 29: Original test set at Vancouver Drake St shops. Author's collection, courtesy Dale Delaruelle

Figure 30: Johnny Wallin, Diesel Inspector Revelstoke, at Alyth Diesel Shop. Author's collection, courtesy Dale Delaruelle

brake equipment. His discomfort only worsened when they pulled out the logic diagrams of the equipment.

I knew how to read wiring diagrams for locomotives, but the symbols and names on these logic diagrams were like a foreign language. The guys from Radiation were great, though, and in the short time we had together they showed me as much as they could.

With the installation completed at Angus, the locomotive and Remote car were attached to a fast freight bound for Calgary. Diesel inspectors at every point along the way were instructed to ride with this equipment and ensure that absolutely nobody got near it. Reynolds told Deno to go to Calgary to meet the train.

When it arrived, we had a crew for the dynamometer car: Mr Parker, Harvey Peterson from the Westinghouse Air Brake Company, the Superintendents of the Langdon Subdivision and Mountain/Shuswap Division, Mr Reynolds, and me. It was broadly outlined by Mr Parker what the purpose of the testing would be and the responsibilities of the team members. My job was to look after all the electronic equipment (about which I knew little) and the air brake equipment.

With the commencement of test running, I was to be stationed in the Remote car and everybody else would be located in the dyno car (complete with cooking facilities, food, a cook, beds, showers, and the amenities of life). In contrast, the Remote car had been ballasted with cement and supplied with a mattress and an oil stove for heating.

Figure 31: CPR Locotrol test personnel at Field, B.C. 1969. Top (L-R), W.R. Haden – Asst. Supvr. Electrical Equipment, Vancouver; F.V. Hooley – Chief Road Foreman of Engines, Vancouver; C.W. Parker – Asst. Chief of Motive Power & Rollingstock, Vancouver; W.R Thompson – Mechanical Asst, Montreal; M.C. Deno – Asst. District Diesel Inspector, Revelstoke. Bottom (L-R), G.A. New – Mechanical Asst, Montreal; J. Hewitson – Asst. Supvr. Standards & Methods, Montreal; C.S. Major – Engineering Asst. Telecommunications, Montreal; A.T. Reynolds – Asst. Supvr. Motive Power & Rollingstock, Vancouver; L. Greco – Gen. Yardmaster, Field. Author's collection, courtesy M.C. Deno

Communication with the Lead locomotive and the dynamometer car would be by walkie-talkie. Since there was no way to keep food on the Remote car for any length of time, I would be advised from the head-end as required, to be alert at the next yard switch or signal post where food would be left, so that by hanging from the lower step on one of the Remote locomotives—and if they were going slow enough—I could retrieve it. Unbelievably, most of the time it worked, and when it didn't, I went hungry. I'm sure the bears or some other critter enjoyed my lunch or dinner occasionally.

The team pursued two main goals in order to successfully undertake the movement of coal in the quantities required by the Japanese contract:

1. Individual train tonnage had to be substantially increased; and

2. Individual train length had to be increased to overcome congestion on the 2.2% grades and through the single-tracked Connaught Tunnel.

Deno continues the story:

We achieved both objectives in late 1968: however, it took us many months of trial-and-error and blood, sweat, and tears (mostly mine) to get there. I must mention that Charlie Parker was the key person who toiled over many hours to come up with formulae on how to successfully set tonnage ratings, and the placement of Remote locomotives within the train. Those formulae continue to be used to this day. The advice we had received from the Southern and from Radiation as to where to place the Remote units in-train was only applicable on territory that had no hills, dips, or mountains (i.e. on flat track).

On the railroad at this time, there was a lot of wheat to move out to Vancouver, so naturally we coupled up to a grain train of approximately 100 cars, that being about 20 cars more than normal. The brains trust was all in the dynamometer car behind the Lead consist. Me, the gopher, was back in the Remote car looking at a diagnostic tool called a lamp monitor—which I did not yet understand—and waiting attentively for any instructions over the radio.

Vick Hooley, the Chief Road Foreman of Engines from Vancouver, had been brought out to supervise and when necessary operate the train. I'd been told that when time might permit, I should train him on the operation of LOCOTROL. I reckoned somebody better train me first.

Vic radioed and asked if I was ready to go. I said 'Yes.' He then asked me to tell him when I started to move. This became routine throughout all our testing. The Remote car and coupled locomotives were two-thirds of the way back in the train, that having been the ideal location for them according to the Southern Railway and Radiation people. They were both entirely wrong. On this very first test run, I believe we broke the train in two on six occasions. When a break-in-two happened, that's when—most of the time—I was put to work.

An example: Walkie-talkie comes to life…
'Milt, can you see where the break is?'
'Yes sir, I can. It appears to be around 15 to 20 cars ahead of the Remote locomotives.'
'Good, grab a knuckle, and get up there and fix it. Let us know when you're ready for us to back up and re-couple.'

Before we were finished testing, there is no question in my mind that on all of the Canadian Pacific Railway, no-one could hold a candle to me on the time it took to change a knuckle and get a train back together. Before our very next test run, Mr Reynolds made sure the Remote car had a large quantity of new knuckles placed in it. He didn't want me to run out!

Over the course of our testing we didn't just break knuckles, we tore drawbars out of cars, and on two occasions we tore an end out of the car. But our most famous destruction took place when we tore a car completely in half on Stoney Creek bridge. That bridge—a couple of miles to the east of Connaught Tunnel—is formidable: curved, and quite long, and at least a couple hundred feet above the creek bed.

Most of our train break-in-twos took place while on the severe grades of the Mountain Subdivision: however, there was a place called Choate on the main line about 99 miles from Vancouver, that exacted its toll on our test train. The track through this area passed over a long series of wave-like undulations so significant that when you looked back along the train you could only see the roofs of cars in the dips while other cars were in plain sight. We tried many different operational procedures, with varying throttle settings on the Lead and Remote and both sets with the same throttle settings. We eventually found the correct combination, but before that took place—on about four occasions—we didn't just break the train in two, but into three and even four pieces.

We had another problem that didn't involve the train breaking in two. It happened right off the bat as soon as we started testing the equipment, and it drove me nuts. The scene was like this: eventually we would—one way or the other—successfully ascend the 2.2% grade on the east side of the Selkirk Mountains and reach a siding called Stoney Creek where we would stop to cut off the pusher engines. The grade from Stoney Creek to approximately three-quarters of the way through the Connaught Tunnel was just under 1% ascending, abruptly changing at the western portal to 2.2% descending.

As a result, because of the reduced grade departing Stoney Creek, we would accelerate to approximately 25 mph before entering the eastern end of the tunnel. After the whole train was in the tunnel (about 1½ to 2 miles), the Remote consist would suddenly throttle back to Idle and the train would stall. That was undesirable, since we had to divide the train to remove it from the tunnel; not easy to do in the dense blackness. Also, it delayed other trains until we had the mess cleaned up.[49]

Questions were immediately forthcoming, like, 'Milt, did you do something to cause this?' I reiterated to those who would listen that I didn't know enough about this equipment to cause anything. I suggested that we get someone from Radiation to come out and ride the test train. That eventually took place but before it did, we stalled in the tunnel on two more occasions.

Two engineers, Dale Delaruelle from Radiation, and Wayne Barber from New York Air Brake were dispatched to investigate our problems. This was great; they travelled on the Remote car with me, so I had the opportunity to interrogate both about the electronics and the air brake systems. They were superb guys and taught me a lot. We worked together a great deal over the years and have remained good friends to this day.

The whole thing about shutting down in the tunnel was complex to fix, but these two guys quickly found out what was causing our problem. In short form, it went like this: when we entered the tunnel, the train acted like a plunger and created a partial vacuum within those confines. This resulted in an increase in train brake pipe leakage and caused a charging airflow in the brake pipe to maintain its pressure. This was sensed by the LOCOTROL flow meter and resulted in the Remote consist throttling back to Idle, as designed in software for safety reasons.

I explained this to Mr Reynolds and told him we had to lay-over someplace so that the logic boxes could be removed and modified, and a maintaining reservoir added to the air brake equipment. When we got to Vancouver, Dale and I took the electronic boxes from the Remote car and Lead unit to our motel where he performed the software modifications. Wayne and I found enough equipment at the Drake St shops to do the modifications that were required on the Remote air brake manifold.

We stayed in Vancouver for two whole days. What a wonderful holiday. Great food, excellent rum, and lots of interesting learning sessions from those two guys. They stayed with us for another two complete roundtrips, making sure that everything was copacetic, so I learned a lot more about the system.

On one of our trips, the Lead locomotive, 5557, began giving ground relay alarms. We were headed west and were just about at Kamloops. I told Mr Reynolds we would have to stop there and put the 5557 over a pit in the shop so I could get underneath and look at the traction motors. When we got over to the shop, the foreman came out to see what was going on and asked if there was anything he could do to help. His name was Tom Johnson. I told him once we got over the pit, could he help me take off all the traction motor covers so we could check them out. He was eager to help and when he found out that this was the LOCOTROL Test Train, he was all over me, wanting information about this new system. I told him once we found out what was causing the ground relays, I'd be pleased to tell him what little I knew. It turned out that we had a 'bird's-nest' (a whole bunch of burned-off wires) in one of the traction motors and the locomotive wasn't going anywhere—and neither was the train—until we could at least replace it with a new motor or put in a new truck.

Neither of those things eventually happened. We just disabled that armature from turning by cutting off the pinion end, switched the Remote car out of the train, coupled it to the 5557 along with the dynamometer car, and attached everything to a fast eastbound freight headed for Calgary. This took some time to get arranged, so Tom and I had time to talk. I was impressed with him; he was an 'eager beaver'.

From the beginning of the testing I had been given the task of looking after engine 5557 and the Remote car

---

49 See Chapter 1 for a description of the outcomes from the early trials of LOCOTROL operation.

'Robot 1'. I was given padlocks to put on all the doors and nobody but me had the keys. Before each run was made it was my job to get up earlier than anyone else on the test team, go to the shop track, unlock the equipment, help couple them to whichever other locomotives were going to be used, then perform a complete function test before I could give the okay for them to be coupled to the train. When we reached our destination, the Lead locomotives and dynamometer car would be uncoupled from the train and taken quickly to the shop track. Once there, the higher-ups on the dynamometer car would alight and head to the hotel to clean up. Me, I had to wait on the Remote car until a yard engine was dispatched to collect me and the Remote car and consist, then take me to the shop track. I then had to secure all the equipment and get myself over to the hostelry. By that time, I was so tired I usually didn't eat; I just crashed.

We were headed down to Lethbridge. Apparently, there was a backup of wheat cars they wanted moved to Vancouver. It would be the longest train we'd run so far—180 cars. We arrived with the equipment in the late afternoon. Mr Hill, the superintendent, had invited us to join him for drinks and dinner on his private car. He was a

Figure 32: 160-car CPR Locotrol Test Train in the Thompson River canyon, near Spence's bridge, 1967. Author's collection, courtesy M.C. Deno

Figure 33: CPR Locotrol test train with SD-40 locomotives and dynamometer car, Thompson River canyon, 1967. Author's collection, CPR Employee Newsletter

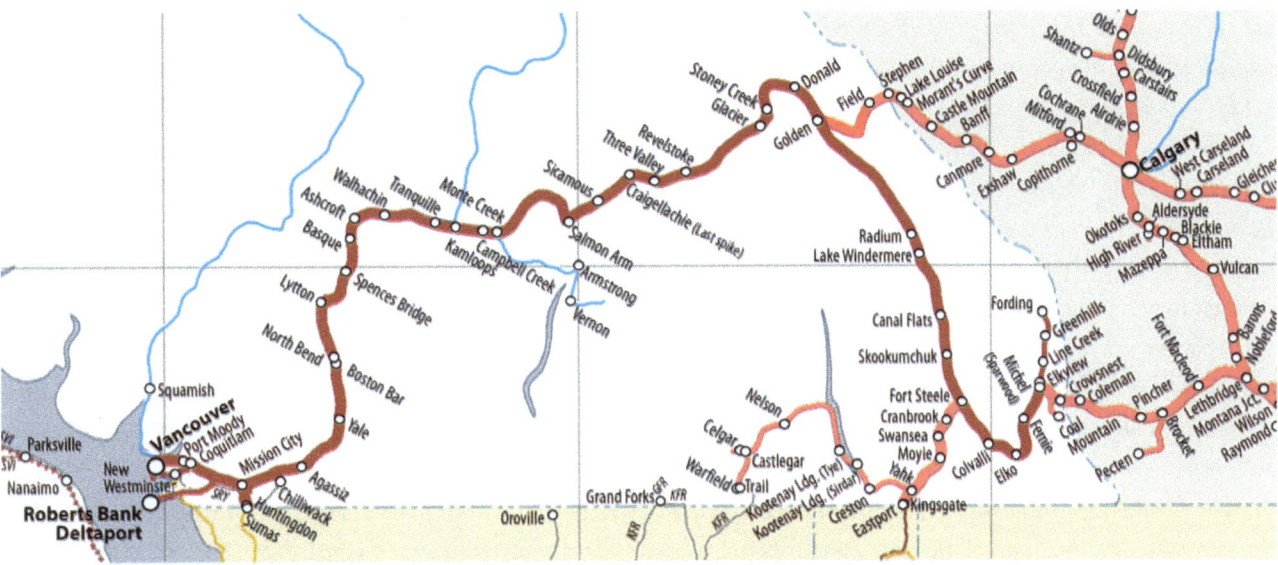

Figure 34: CPR map of coal train route between Fording and Roberts Bank in British Columbia. Canadian Pacific Railway

great friend of Art Reynolds. Dinner was set for 6.00 pm but about an hour ahead of that, a messenger from the dispatch office informed Mr Reynolds, Mr Parker and Mr Hill that they were to leave immediately for Calgary to meet with Jack Fraine, the President of Operations – Pacific Region.

Mr Hill said for me to go over to the private car and he would leave a message for the cook to feed me. I went. The first thing I encountered when I stepped into the car was the fantastic aroma of roast turkey. The cook seated me at a table with sparkling white linen, crystal glassware, and as much silverware as you'd see at any royal table. The meal was fantastic: turkey, cranberry sauce, dressing, gravy, mashed potatoes, a medley of vegetables, homemade bread, and for dessert, ice cream, fresh fruit, or lemon meringue pie.

When I was finished, an interesting thought flashed through my mind pertaining to the lack of good food on the Remote car. So, I said to the cook, 'come on and sit down with me and we'll have some coffee and a cigarette together.' We chatted a bit about this and that, and then I said that it was fortuitous that the bosses got called to Calgary, and he asked why. I told him, Mr Reynolds—Mr Hill's good friend—was terribly allergic to turkey and would've had to leave the dinner party because he couldn't even bear to be near it. The cook was thankful that I had told him this, because he had planned to make the rest of the turkey into sandwiches and put them on the dynamometer car before we left for Vancouver. Well, he would just throw it out.

I told him that would not be necessary if I could find a suitable container to put it in, which I didn't have on the Remote car. He said, 'That's no problem; we'll go over to the dining car store. Ice is no problem—I have lots of it on the private car.' Well, we did that. The container was like one of those Coleman Coolers that you keep ice and beer in. He packed the turkey in the cooler along with cranberry sauce, stuffing, butter, a loaf of homemade bread, and what was left of the lemon meringue pie. He put ice in plastic bags, sealed them, packed them around the food, and told me that should keep everything good for several days. He also gave me one of the old sharp knives to carve the turkey and bread with, and also a knife and fork.

I had bent the truth a little about Mr Reynolds' condition, but who was going to find out? And for once I was going to eat as good as those guys on the dyno car.

The bosses got back from Calgary, and without delay we set off from the Lethbridge yard with the 180 wheat cars for Vancouver. We experienced a couple of broken knuckles on the Mountain Sub, but other than that the train behaved just fine. With the positive scrutiny of everyone's input, we were getting this new type of train operation to be very manageable, so long as certain new processes were followed precisely. All of us agreed that if we continued to make progress like we had over the past 7 months or so, probably within another 2 months we would achieve the goals and a working plan to fulfil our Japanese contract. We achieved our goals just two-and-a-half months later, around the middle of October 1968.

We moved the Lead unit, Remote car, and dynamometer car back to Calgary and prepared them for their subsequent trip to Montréal, so that the LOCOTROL equipment, which had been leased, could be returned to Radiation in Florida and the dynamometer car returned to its home at Angus shops. The whole team had a final meeting in Calgary to summarise what we'd achieved. At its cessation, Mr Reynolds said he wanted to take us all downstairs to the Palliser Hotel restaurant and buy us a sumptuous last supper.

Well, we all got cleaned up, regrouped downstairs, had a couple of drinks and went into the dining room. I've mentioned it before, but I will again: these Canadian Pacific hotels were certainly 5-star. We all sat down, and the waiter came to the table to take our orders. When he got to me, I ordered a New York strip steak—the largest they had—which I think was around a pound-and-a-half. He then asked me how I would like it done, and before I could answer Mr Reynolds piped up, and said, 'This boy has been working very hard, and not eating as well as the rest of us. So just cut off its head, wipe its ass, make sure it's rare and put it right there…' and he pointed to my plate.

That was the first time I had ever heard my mentor use an off-colour word. We finished our meal and said our goodbyes and just before I left Mr Reynolds said to hang back a bit. Everybody was gone when he said to me, 'You did a heckuva good job Milton: however, for your personal edification, I am not allergic to turkey.' With that he shook my hand and for the time being we parted company.

Charlie Parker and company returned to Montréal, reported their findings, and a decision was made based on those reports to purchase five complete sets of LOCOTROL *equipment from Radiation, to be installed on new MLW 630 locomotives, and on new Remote cars.*

Figure 35: *CPR sales team at the Harris plant in Melbourne FL. L-R: Unknown (CPR), unknown (CPR), Fred Routledge (Harris Controls), Dale Delaruelle (Harris Controls), unknown (Harris Controls software engineer).* Author's collection, courtesy Dale Delaruelle

I would be remiss in not mentioning that Tom Johnson had been testing the Westinghouse Company version of Remote control. Poor Tom got a lemon. I don't think he ever had a single trip that something on the Westinghouse equipment didn't fail, and as a result, the CPR did not consider purchasing it. Mr Reynolds returned to Vancouver, as did Vic Hooley. Tom Johnson went back to Kamloops, and I went back to Revelstoke, to do more 'diesel inspector' things.

The decision by Canadian Pacific—based on our testing—to embrace a remotely-controlled in-train consist of locomotives was very cutting-edge. As far as train operations are concerned, it has been compared in complexity to the change from steam locomotives to diesel. A few of us (the test team) understood what a huge step this was, and the amount of training that LOCOTROL *would require of enginemen, firemen, conductors, yardmasters, carmen and yard engine crews, plus snakes [switchmen]. Myriad logistical problems still had to be resolved to make these long, remotely-controlled unit trains function like a well-oiled machine. Me… well I didn't have to worry about these things anymore, because I was back to being a diesel inspector.*

☙❧

It was getting close to Christmas 1968. We had been invited to the Appleyard's New Year's Eve party in Medicine Hat. I had vacation time coming because I never got any while we were testing, therefore I took a couple of weeks. Al Appleyard and his wife were great people: he was Car Foreman at Medicine Hat. We had Christmas in Revelstoke, and a few days later, we headed out for the 'Hat.

I can't remember where we stayed in Medicine Hat, but I do remember the New Year's Eve party. The Appleyards had invited some neighbours and several railroaders. The 'rails' wanted to know about my experiences while testing with the new equipment in the mountains. Everybody had chipped in, so there was lots of food and a more-than-ample supply of booze. They had a spacious rumpus room down in the basement, so there was plenty of room for the shindig. I was having a lot of fun kicking up my heels on the dance floor, suitably lubricated with rum and coke. Midnight came, we tooted our horns, blew our whistles and sang 'Auld Lang Syne', kissed all the ladies, and toasted the New Year… 1969.

Al Appleyard sidled up to me, and over the raucous behaviour of the others said, 'There's a phone call for you, the phone is upstairs in the kitchen.' I said, 'Who is it?' He said, 'I don't know, but whoever it is said it's important that they talk to Milton.' I thought the only people it could be were either Johnny who was babysitting for us, or the railroad.

I should add this… when you were in a position like I was on the railroad, you were on call 24/7, so wherever you went you had to advise the local dispatchers office; that was mandatory. Seeing as how nobody else knew where I was in Medicine Hat it had to either be Johnny or the railroad. Even though I was on vacation, the higher-ups appreciated knowing where you were at, so I complied.

I said, 'Hello?'… and the person on the other end said, 'Happy New Year.' I knew that voice. I said, 'Happy New Year to you too Mr Reynolds.' He asked me if I was having a good time at the party and I told him I certainly was, all the time wondering 'Why is he calling me on New Year's Eve?' I sure as hell didn't think it was only to wish me a Happy New Year. He said, 'Milton I am sorry that I have to cut your good time short, but we have a serious dilemma on our hands. Frank Oleigis [General Locomotive Foreman at Alyth Diesel Shop] has become incapacitated, and will be off work for an indeterminate period. You have been promoted to that position as of now. You need to be at the diesel shop by 7 o'clock tomorrow morning. Give me a call as soon as you get there, and we'll talk about this further.'

In my entire life I don't think I sobered up as fast as I did after that phone call. I went downstairs and told everybody what had taken place, said my goodbyes, and went back to where we were staying to get my shaving gear, and clothes. It was close to 2.00 am when I finally left for Calgary. There were warnings out about black ice on the highway, it was minus 40°F, and I had 183 miles to negotiate before I got there. It was somewhere around 5.30 am when I reached the outskirts of Calgary. I was early, but that was okay; I needed to get some tomato juice, coffee… lots of coffee… bacon, eggs, and home fries into me before presenting myself to the troops at the Diesel Shop. Fortunately for me there was a good truckstop near the Shop.

I walked into my new office around 7.15 and was confronted by a guy (Bob) who asked me what I was doing there. I told him I had assumed the job of General Locomotive Foreman, and he retorted that couldn't be right because he had been called early that morning at home and had been informed by master mechanic Davies that he was being appointed to that job. Needless to say, the next couple of hours were interesting. In short form, it went like this:

1. I called Mr Reynolds and related the situation. He was upset.
2. He called the master mechanic.
3. The master mechanic called Bob, and told him he had made a mistake and that Mr Deno is the new general locomotive foreman. Mr Davies left it to Bob to tell me. He (Davies) never spoke to me, and that was another big mistake he made.

Art Reynolds called me and asked if Mr Davies had straightened everything out with me. I told him everything seemed to be straightened out, but that Davies had not spoken with me. Reynolds said, 'REALLY? Give me a call later when you get settled.' I did, and we talked about a number of things pertaining to my new job… and then he said, 'Give me a call tomorrow morning and I'll explain the grave situation we are in concerning locomotive availability at Alyth.'

In the morning, I called Mr Reynolds and was informed that to keep the railroad functioning properly Vancouver needed Alyth to have a locomotive availability factor of between 92 and 95% and that it was currently at 62%. Customers were upset and you can bet Canadian Pacific management sure as hell was. The weather was *certainly a factor; it had been at minus 40 for about a week now and it stayed that way for an entire month. That wasn't the only factor causing the poor availability. Personnel issues were also playing a big part. Mr Reynolds said, 'Milton, I want you to look at everything and prepare a plan by close-of-business tomorrow, then get me on a conference call and I'll have Mr Fraine and Mr Booth join us to decide on it.'*

I have to elaborate on what I'd walked into. The union ranks and their leaders were upset because no-one had been listening to their complaints, with the result that morale was low. Some diesel foremen were not being even-handed with their work assignments, playing favourites, a definite no-no. On top of that, out of a fleet of 300 locomotives, we had 114 out-of-service.

My right-hand man was Ken Drummond. I had him arrange to assemble the union leadership in my office at 10 o'clock that morning, and also, to find Cecil Lloyd and bring him to the office right now. I knew most of the shop workforce from my days as an apprentice and more recently as a diesel inspector. Cecil had been a lead machinist in my father's time, and was now a day-shift mechanic. He had befriended and helped me when I was just an apprentice boy, he was long-headed, he knew the workings of the diesel shop backwards and forwards, and was well-respected by the tradesmen. He had been asked on many occasions to take promotion, and had always refused, saying 'I'm happy as a mechanic; just leave me here.'

I outlined to Cecil the problems as I saw them, and told him the way we operated in the shop was going to change as of today. I told him I wanted him in the room when I had the union leaders there, and to prompt me if he thought it was necessary, to keep me on the right track. He agreed to help.

I told the union leaders that I had to present a plan to Messers Fraine, Reynolds, and Booth by the end of the day as to how we were going to rectify our current situation. I brought my secretary into the room to record their comments. I then outlined what I had in mind:

1. *Until we broke the back of this thing, I wanted certain key tradesman to stay at the shop 24 hours a day, 7 days a week. To look after their comfort I would have beds brought in and distributed in available space in the shop, I'd have meals provided 24 hours a day, I would make showers available at the railroad bunk house, and we would send their clothes out to be laundered when necessary. The men would be paid overtime for anything-over 8 hours and double-time for working on a 'day off'.*
2. *All salaried personnel (diesel foremen, including me) would put in 12-hour shifts until this mess was cleaned up.*
3. *Two tracks inside the diesel shop would be assigned for the sole purpose of thawing frozen locomotives, of which there were many.*
4. *Locomotives in the worst condition were to be repaired first.*
5. *Unless there was an absolute emergency, all vacations were cancelled (that didn't ruffle any feathers. Heck, it was the middle of winter, and really cold!).*

All-in-all, there was general agreement that these steps had to be taken to get us out of the manure. I told them I still had to get the 'okay' from the top, and as soon as I had that, I would personally hard-copy the union leaders with that decision. I thanked them all for their forbearance.

I called Mr Reynolds with the plan. He said to hang tough in my office... he would get Mr Fraine and Mr Booth, and call me right back. I still had Cecil Lloyd with me because I wanted him to overhear proceedings so he could pass the word down to the troops and affirm that I wasn't short-changing them. Those three senior bosses— to my mind, three of the best railroaders that I would ever meet —said, 'Hello Milton', listened to my plan, asked some questions, made some suggestions, and said that I had their combined support to execute the plan. I told Ken Drummond and my secretary to call the afternoon- and nightshift foremen to have them in my office at 7:30 the next morning. The dayshift foremen I would speak to today.

I broke them up into teams working 12-hour shifts, emphasising that one of the most important jobs was the thawing of units because they couldn't be worked on in their present condition. Most of the locomotives had engine cooling water systems frozen solid and cylinder liners broken because of it. In those days, hot water heaters were employed to heat the locomotive cabs, and these were likewise frozen. I instructed Ken to find as many hot air blowers as possible and if there weren't enough, I'd purchase more. When we pushed those frozen locomotives into the two tracks I had set aside in the shop, it took one-and-a-half to two days to thaw them out. Once that was achieved, we moved them to where the tradesmen could begin the repair work. A few of the locomotives couldn't be repaired at our running shop. They had to be moved to the backshop, because the freezing had caused cracked and broken engine blocks. While this was happening, I received word that Mr Davies was no longer master mechanic at Calgary. He should've known better than to mess with Art Reynolds!

⊂₰⊃

Early in April 1969, I received a message from Vancouver that Fred Booth, the Superintendent of Motive Power – Pacific Region, would arrive at Alyth on an inspection trip, so 'Get the shop ship-shape, Mr Deno.' He would arrive in a couple of days and intended to stay down at the shop for one whole day. I instructed everyone to get into clean-up mode so that he would be dazzled by the orderliness of our premises. I didn't have to say anymore. This stuff was old-hat, and everybody knew what to do.

Fred was a good railroader. He expected the best from his people, didn't suffer fools, and could pick slackers out faster than you could find flyshit in sugar. He backed his people up 100 percent, even though they might have inadvertently been on the wrong path. He would soon put you on the right path, but never in front of anyone, and he couldn't tolerate brown-nosers. He and I got along well.

We walked around together, which was his custom when on an inspection trip. He had a few minor criticisms, but for the most part he was happy with what he saw. He knew the shop like the back of his hand because when I was still an apprentice, he had been the general locomotive foreman. He knew just about everybody and would stop and converse with them. As a result, the inspection trip took the better part of the day. Back in my office he told me not to make any plans for the next morning around 9 o'clock because he would be back and wanted to converse with me. Also, when I finished my day, to come down to his room at the Palliser Hotel; he was going to hold a 'prayer meeting' with me, and some of the guys.

Fred expected everybody to work hard and, when times afforded it, play hard. Prayer meetings were our way of letting off steam. Much booze was consumed, much straight talk took place, and—unlike a lot of office committee meetings in this day-and-age—we made decisions and accomplished a great deal. Fred only invited a select few to these meetings, people who knew when to speak up and when to keep their mouths shut, who had a great love for railroading, and also had exceptional resilience (for instance, attend a prayer meeting early in the evening, leave as late as 4.30 or 5.00 am, and still be on time for work. I became quite proficient at this. Not as a boast; it's just a fact. It wouldn't fly today, though: the Zero Harm managers would have apoplexy.

The next morning, precisely on time, Fred showed up at my office wanting strong coffee before we began any discourse. He wanted to know who, in my estimation, should be the next general locomotive foreman of Alyth Diesel Shop should I ever be moved out of that position. I didn't have to cogitate too much on that one. My right-hand man, Kenny Drummond, was certainly well-equipped to take over my job immediately if necessary.

He said, 'Milt, you are the only person in Canadian Pacific's Mechanical Department that has any working knowledge of LOCOTROL. *The company is about to purchase five complete systems, Masters and Slaves, which will be dedicated to the movement of coal from Sparwood to Roberts Bank. Treat this information confidentially: however, I want you to think about what it will take in manpower and infrastructure to make this operation a success. Presently we are thinking of a turnaround time of 90 to 91 hours. I also want you to make a trip down* to Radiation's plant and find out from them what they consider is a sufficient quantity of spares for those five systems: what tools and test equipment will be needed. If you are willing to undertake this, I will phone Radiation today, tell them what I have in mind, and tell them we will have you down there in a couple of days.' Three days later I was on my way to Melbourne, Florida.

I flew CP Air from Calgary to Vancouver then Seattle, then Eastern Airlines from Seattle direct to Melbourne. I arrived just before dusk and when I got to the doorway to deplane, I thought 'My goodness Milt, you have arrived in paradise.' When I left Calgary, there was still snow on the ground, and lots in the mountains. At night it was pretty darn chilly. In stark contrast I was standing in the doorway of this airplane looking at a beautiful clear-blue sky tinged with pink, a soft warm breeze caressing my cheek, and statuesque palm trees swaying in the slightly moving air. I hadn't stepped on the ground yet, but I was already in love with this place.

Ralph Leffingwell came out to pick me up. He took me to a brand-new Holiday Inn and said, 'After you unpack and get yourself settled, meet me in the bar.' That sounded like a great idea. After a few drinks we had dinner. Ralph told me he had arranged a meeting for the next morning with the people who would be able to answer my questions. He said, 'I'll meet you for breakfast at 7 o'clock, then we'll go to the plant.'

Well, we arrived at the plant, and that surprised me a little bit until he explained the situation. The name across the front of the building was Winn-Dixie, a popular grocery store in the South. Apparently, the company was getting prepared for expansion and had rented this empty grocery store. He informed me the next time I came, the company name 'Radiation' would be across the front of the building. I was introduced to several people, including Dale Delaruelle, whom I had previously met in Canada.

They gave me a tour of the plant, which finished off the day, then took me to the Melbourne Steakhouse. They had arranged for a secretary to get me a flight home the next day, and would deliver the tickets to me at breakfast, and then take me to the airport. I really didn't want to leave that fast because I didn't know if I'd ever be back again and I sure had a yearning to check the area out.

Back at the Alyth Shop in Calgary, I put together a report on my findings and sent it to Fred. Everybody at work wanted to know what Florida was like. My answer was, 'No snow, blue sky, friendly people, soft breezes, excellent rum and coke, cheap cigarettes, and an abundance of beautiful, attendant, secretaries.'

〜

Around the middle of April 1969, Deno received a directive to be in Vancouver for a meeting with Jack Fraine and Fred Booth. Nobody told him what it was to be about, but that was not unusual. First thing in the morning after his arrival, they met in a conference room at head office on the waterfront of Burrard Inlet. Fraine opened the meeting saying that he'd had the opportunity to read the report Milt had sent to Fred Booth. He told Deno, 'We'll talk about that a little later, but in the meantime, I am going to inform you of some things that must presently remain confidential but are necessary for you to know in order to add more specifics and detail to your report.'

Revelstoke had been chosen as the command centre for this new operation. The Revelstoke Division included the Mountain and Shuswap Subdivisions, and because of this new operation it would now include the Windermere Sub and that portion of track going eastward from Cranbrook to Sparwood. This would add approximately 150 miles of railroad to come under the immediate authority of the staff officers at Revelstoke. The inclusion of longer, heavier, bathtub coal cars, and a LOCOTROL operation was going to require the training of people in many different disciplines and would require a lot of planning. Deno was requested to broaden his initial plan in light of this extra detail, paying particular attention to manpower requirements, training of the various disciplines, minimum spares and test equipment requirements—and anything else that might be pertinent to the successful fulfilment of the Japanese contract.

As of right then, Deno was placed on special duties reporting directly to Booth. He was advised to complete his report before the end of the month because there was a lot of work ahead and a short time in which to accomplish it. He was also informed that they concurred with his choice of Ken Drummond as the next General Locomotive Foreman at Alyth, and that all personnel in the region would be notified to assist in any way necessary to effect completion of the report. In the meantime, someone would be assigned to investigate the infrastructure upgrading that would be required—not least the roadbed on the Windermere Subdivision. Because he was on short time to finish his report, Deno was authorised to use CP Air when possible.

I had a secretary get me a flight back to Calgary; and I went down to the shop and congratulated Kenny on his promotion. He wanted to know what went on at the meeting, and that if he was now the boss at Alyth, what job did Milt Deno have. I told him I couldn't tell him anything about the meeting: however, I could tell him it appeared I was in limbo because I was now on special duties, reporting directly to Mr Booth. I asked him to get his secretary to arrange for a berth on The Canadian *for me to Revelstoke tomorrow. Even though I knew Revelstoke well from my days working there, I needed to be on the ground so that I could try to visualise what the needs of this new operation would be at that terminal.*

When I arrived, I went up to the general office in the station and asked if I could meet with the division superintendent. Everybody in that office knew me, so without fuss I was ushered into Mr L A Hill's office. You may recall that this was the same man who was Superintendent at Lethbridge when I purloined the turkey a few moons previous. He welcomed me warmly and inquired as to why I was there. I wasn't at liberty to tell him exactly—I just said that Mr Booth wanted me to check out several of the mechanical facilities on the region. He asked if there was anything I needed, and I replied that it would be helpful if I could be allocated a vehicle. He complied and I decamped to get myself booked into McGregor's Motor Inn.

It took me a couple of days of looking around at the shop area and the yard before I was satisfied that I had covered all the bases. I also spent a day-and-a-half checking out Golden, and because I was on short time, I had someone take the truck back to Revelstoke. I called Mr Hill, thanked him for his hospitality, and hopped a fast freight to Calgary because I had to interface with the head of communications and appraise him of what needed to be done in the Connaught Tunnel.

I explained that for safety reasons it would be necessary to have reliable communication between Lead and Remote units while in the tunnel, particularly because of the changing grade within from 1% ascending to 2.2% descending westbound, the loaded train direction. He said this could be accomplished by stringing an antenna within the tunnel.

I had gathered a lot of information and now it was time to sit down and put the report together, so I borrowed an office from Ken Drummond at Alyth, along with a secretary. The major issues went something like this:

Revelstoke
1. *A location near the shop track would be required for a building to house the* LOCOTROL *test equipment, cabinets and spares, including NYAB test equipment. A shop track should be dedicated solely for the testing of Lead locomotives and Robot cars. This shop track should be just outside the building housing the test equipment.*
2. *The capacity of the locomotive fuel point on the main line just west of the station would need to be increased so that a consist of four locomotives could be fueled in 10 minutes or less. When it was decided where the Remote consist was to be placed in the train, a new fueling workstation must be built next to the main line at that exact location*

3. A roadway would be required on the north side of the mainline track, to accommodate vehicle traffic for the whole length of a train.
4. At the Remote locomotive location, two crossover tracks would be required to connect the main line with the adjacent track to the south to facilitate the removal of bad-order equipment in the Remote location, and the reinsertion of working equipment.
5. A high-capacity sand truck must be purchased with sufficient capacity to sand eight locomotives, and the capability to blow this into the sandboxes on the locomotives. It might be necessary to purchase two of these trucks—one for each consist of locomotives in the coal train—in order to keep terminal time to an absolute minimum.
6. Enough single- and double-rotary coupler coal cars—some to be held in the loaded condition—should be purchased and kept on hand as bad-order replacements.
7. Inspection of trains by carmen at the Revelstoke marshalling yard has always been a very time-consuming and difficult task in the wintertime, due to the heavy accumulation of snow. Even though the spreader tried to clear away most of the snow, walking up and down trains was still extremely difficult. Because these coal trains were to be on a very tight schedule and must move through the yard expeditiously, I suggested that the company purchase five Ski-Doos, complete with trailers, a spotlight to be attached to each one so that necessary apparatus on the coal cars could be checked out at night. The trailer to be loaded with the necessary tools and spare parts, and two carmen would be onboard—one driving, the other checking the train. The train would be inspected simultaneously on both sides to save time.

Golden
1. Two shop tracks needed to be constructed somewhere near the main yard. One of these should have a pit between the rails sufficiently long to be able to look underneath at least two locomotives. Both tracks should be capable of handling at least four locomotives and a Robot car.
2. A small building containing an office and an area for spare parts should be constructed next to the shop tracks.
3. A new yard should be constructed south of the main line, near the south side of Golden, to handle north and southbound coal trains and other traffic. Suitable roadways should be constructed so that trains could be easily accessed.

Connaught Tunnel
The entire length of the tunnel should be wired in such a way as to maintain continuous data radio communication between the Lead and Remote consists.

Manpower requirements
1. Car department staff numbers at Revelstoke would have to be increased in order to handle the increased traffic.
2. The present complement of two road foremen of engines should be increased by at least four, perhaps more, because these people were going to be involved in the training of all engine crews, and the subsequent riding with, and monitoring of them. Before that happened, the road foremen would have to be trained in the intricacies of distributed power train operation.
3. The company was not purchasing a spare Remote car or equipping a spare locomotive with Lead equipment: therefore, it appeared to me that any failure of the LOCOTROL or NYAB equipment would in most cases have to be repaired, or at least an attempt be made to repair on the fly. Presently there was one very good diesel inspector located at Revelstoke. I recommended that we increase the number at that location to five. Also, one at Golden, one at Kamloops, and one at Roberts Bank, for a total

of seven new diesel inspectors. My choice of people for this new operation was because they were salaried and could work on all electrical, mechanical, and air brake, systems—unlike shop mechanics, electricians, pipefitters, etc., who are governed by union and specific work rules. I thought this would save a lot of money in the long term.
4. *I included a list of spare parts that had been recommended by Radiation and NYAB.*

By the last week of April 1969, I was able to advise Mr Booth that my report was complete and was instructed to catch an early flight to Vancouver. Landing at 10.00 am, I got down to the office about an hour later and they had a meeting set up with Mr Fraine, Mr Booth, the Regional Civil Engineer, the Regional Purchasing Agent, and Mr Hill.

Because this was an entirely new endeavour, there were many questions directed at the contents of my report. For instance: Mr Fraine—with carefully selected words—said, 'Mr Deno, you can't be serious about having Ski-Doos running up and down between my tracks in the Revelstoke railroad yard. Think of the safety considerations.'

I said, 'Mr Fraine, you have stated to me that turnaround cycles for coal trains from Sparwood to Roberts Bank and return cannot exceed 90 hours: therefore, my calculations for the round trip indicate that the loaded train—when it arrives at Revelstoke—cannot be in the terminal for longer than 30 minutes. To meet this stringent timeframe, I am resolved that we will need Ski-Doos, fuel stations with increased capacity and sand trucks. But these are just recommendations. I have tried to highlight the areas that I think require attention. The person or persons designated by head office to implement this operational plan may be informed by some of my recommendations.'

They thanked me for my candour, and Mr Booth gave me the rest of the day off. At 10 o'clock the next morning, we again convened in the conference room. Mr Fraine started off by telling me the group had adopted most of my recommendations: however, he said there were some things we needed to discuss further, such as the use of Ski-Doos. I said that perhaps we should wait and discuss everything when they had designated whoever was to be in charge of this new operation so that he or they might be fully informed. Mr Booth answered that by saying, 'We have chosen someone; that would be you, Milt. As of 1 May 1969, you are promoted to Master Mechanic of the Revelstoke Division and are responsible for getting this new project underway.'

I was floored. May 1 was only a couple of days away, and this was a huge promotion, with many exacting responsibilities. Fred Booth and Al Hill took me out for a drink after work. Fred said he had arranged a flight for me leaving early in the morning to get me back to Calgary. Mr Fraine informed me that afternoon that it was anticipated the first coal train would be run sometime in the spring of 1970. I was happy about that because it gave me several months to get my act together, and that was going to take time.

I reported to Al Hill at Revelstoke on 1 May 1969, my family arriving later in the month. One good thing about this move was that we didn't have to look for a place to live; we would have the company house allotted to the master mechanic. I knew this house because I had stayed there with the Davis family when I was just a kid many years ago and had also visited Harry Piper when he resided there during his tenure as master mechanic.

My first order of business on the new job was to interview people who were presently working for Canadian Pacific, to see if they were interested in becoming part of this new era of railroading. Messers Fraine and Booth had given me the authorisation to go wherever I liked on the Pacific Region to interview the people that were needed to fill seven diesel inspector and four road foreman of engines positions. I decided I would start that process in Revelstoke.

Because of my prior years working as a foreman in Field and as a diesel inspector at Revelstoke, I knew most of the engine crews working the Mountain and Shuswap Subdivisions; it was from this group that I would choose my *road foremen of engines. The diesel inspector presently at Revelstoke, Lawrence Takkinnen, was a prime candidate to be interviewed, along with Tom Johnson[50], who was in Kamloops and had worked on the Westinghouse remote system during the test phase of that equipment. I asked my secretary to pull the files of several enginemen whom I thought would be good candidates for the RFE positions, and I also asked her to get me the files on Lawrence and Tom. I brought both in together to be interviewed.*

I started off by telling them that in order to get these new trains running successfully, I envisioned that—if they accepted the job—they would be working many, many, long days riding this equipment because the company had not purchased any spare Remote cars or Lead units. So, for the most part they would have to conduct troubleshooting and repair on-the-fly. I said, 'Before you say yes to the job, I require that you and your wives and your families really understand what you'll be getting into. Then, and only then, come back and see me with your decisions. Lawrence, this job is going to be a far more demanding role than Diesel Inspector, so it is imperative that you have this talk with your wife.' They both came back within a couple of days with total agreement from their families to take the job, and I was pleased. The program had just gained two very competent people.

My next move was to interview prospective new road foremen of engines and diesel inspectors. These individuals were going to play an important part in the success of the operation; they would have to learn the intricacies of operating a train with distributed power. Once competent at this, they would have to train the many engine crews based between Sparwood and Roberts Bank. They would also be required to make inspection runs with these enginemen to ensure they were performing correctly.

I subsequently interviewed several candidates and chose four. Unlike the diesel inspectors, I did not have to forewarn them about working long periods of time, any hour of the day. Being enginemen, they already knew about that one.

With his interviewing at Revelstoke completed, Deno now turned his focus to Calgary, particularly the Alyth Diesel Shop. It was there that he was certain he'd find the other diesel inspector candidates. He recognised that getting these people to leave the city of Calgary for places like Golden, Kamloops, and Revelstoke, which didn't have anywhere near the population or amenities of the big city, was going to be problematic, especially for the families. He resolved this by pointing out to regional management in Vancouver that these people were going to have to work above-and-beyond the call of duty to make this operation a success, and that he would be expecting the maximum consistent output from each of them. The deal brokered was that if these guys and their wives fulfilled their obligations to the company for a period of 3 years, they would all be promoted. Without exception, the company honoured this proposal, and Deno got the six diesel inspectors he wanted.

Milt returned to Revelstoke to confer with the general car foreman on his views as to what extent his staff had to be increased to provide for the coal-train servicing. He also met with Al Hill about the necessity for staff training. One important issue in Deno's mind, was the correct placement of cars having a solid coupler at one end and a rotary coupler at the other.[51] These would be most cars in the train: however, there were some with rotary couplers at both ends and it was imperative that they be placed correctly in the train.

Deno proposed the creation of a new position to be called the 'Coal-train Coordinator'. That person would be responsible for the make-up of every coal train to ensure all cars were oriented correctly, plus any other requirements. The expected costs for implementing this were considerable, so headquarters in Vancouver

---

50   Johnson would later become CP's Supervisor of Diesel Equipment.
51   Since these cars were to be unloaded by being rotated upside-down in a car dumper without being uncoupled from the cars next to them, the unit train needed to be assembled such that when it arrived at the Roberts Bank terminal, the cars were marshalled so that each group being rotated in the dumper cells had a coupler at its outer end that had a rotating shank. This was to accommodate the adjoining coupled cars that would remain upright outside the dumper cells.

baulked at the suggestion. The reason for Deno's concern was that in being unloaded in the car-dumper at Roberts Bank, the cars—still coupled to those adjacent—were secured by hydraulic arms and rotated upside down. As Milt was wont to say:

If you didn't have the rotary couplers where they needed to be, bad things could happen. Well, headquarters okayed the hiring of a Coal-train Coordinator after the second train we ran had the wrong coal car next to the caboose (having a fixed coupler instead of rotary), so that when that car was rotated, it flipped the caboose on its side. Fortunately, there were no crewmembers aboard at the time. The dumper operator was pretty shaken for a while because he thought he had caused this situation.

I'm ahead of myself by about 9 months but it was appropriate, I thought, to describe this issue at this time. Now that we had a pretty good handle on the staffing requirements for this new operation, I needed to take some time to figure out what was required for training. Diesel inspectors would have to be trained first, because they would have to train the road foremen of engines on the new operational procedures required. In turn, the RFEs would have to train all their engine crews.

Deno called Fred Booth, explaining the training requirements and that he wished to set up training classes for the diesel inspectors and himself, first at New York Air Brake in Watertown, NY and then at Radiation in Florida. He expected this would take around two-and-a-half weeks.

This was a big deal. In the first place, railroad air braking was a specialty subject and was referred to on most railroads as a 'black art'. In even the largest railroad shops there might be two or three mechanics who understood the 'ins and outs' of a locomotive's air brake system. The LOCOTROL system used electronically-controlled air brake valves on the Lead unit, and on the Remote car (which controlled the trailing locomotives coupled to it). The diesel inspectors and I knew diddly-squat about air brake systems, but we were going to have to be knowledgeable in a doggone hurry!

Well, it was a big deal in Fred Booth's mind. He immediately wanted to know why people couldn't be dispatched from NYAB and Radiation to train us in Revelstoke. I told him I didn't think this was feasible, but I would find out and call him back. I did find out and it was simple: we didn't have any electronically-controlled air brake or LOCOTROL electronic equipment on-site that was operational, therefore meaningful classes could not be held at Revelstoke. I asked both companies to send Fred a telegram explaining this.

I could understand his reluctance in sending so many people so far away to school. This was something new. The railroad always was very long-headed and took a somewhat sceptical position of all things new. In my opinion, Fred's biggest problem was that he did not understand the complexity of this new technology. He called me back saying that he had received the telegrams, and he was going to have to discuss this situation with Mr Fraine.

Well, Fraine okayed the request without hesitation. I connected with the two companies and explained that I wanted my group to attend the NYAB school first, followed immediately by the school at Radiation. Our schooling would begin on 4 August 1969 at NYAB, running through until 15 August. Because there had been some hesitancy on the part of Fred Booth and some others in his group about us going away for a month to school in the States, I was determined that all of us were going to put our noses to the grindstone and come away with a good report card from both NYAB and Radiation. So, before we left Canada, I gathered the group together and instructed them that after school each day we would go to my room and for 2 hours study the subject matter from that day. If we had problems that we couldn't resolve among ourselves I would call the instructor and have him come to the motel to help us out. There would be no exceptions to this instruction: anyone failing to abide by this rule would be on the next plane home and be demoted back to wherever they came from. Weekends would be entirely dedicated to having fun or whatever the individual wanted to do.

Wayne Barber met us at the airport, settled us into the hotel, and said he'd see us Monday morning and take us to the plant and that he'd be our instructor. We were a lively crew. All the guys were in their lower-to-mid 20s. I was the old man at 36. We were all excited about getting this chance to demonstrate how much we had learned about the new product we'd be dealing with. No-one except me had ventured far across the Canadian border before, so everybody was eager to experience these new places.

We had an excellent reception at NYAB. Wayne exhibited a wonderful demeanour, even though he was bombarded with questions from we electricians who had absolutely no insight into the esoteric discipline of the air brake. His patience paid off, because between the questions in the classroom, and studying in the evening, we all absorbed a great deal.

We didn't get to see much of Watertown: however, on two separate occasions they took us out for dinner and drinks. We also took a tour of the plant, where we saw for the first time their air brake rack, which is where they did extensive testing of car and locomotive brake systems. For the uninitiated, these air brake racks are quite something. They are a vast assembly of piping, valves, and reservoirs that can simulate up to a 200-car train.

We completed our fortnight on air brakes and departed Watertown on an early flight for Melbourne, FL on 16 August, arriving there in the early afternoon. Ralph Leffingwell met us at the airport and took us to a new motel called the Host of America. To our surprise the Marquee outside the motel read: 'Welcome to Milt Deno and His Group from Canada'. School would begin at 8 o'clock on Monday 18 August. Dale Delaruelle would be our instructor for the next 2 weeks ending on Friday 29 August.

Ralph mentioned there were several people from the plant who wanted to meet us, so he had taken the liberty of booking a small meeting room at the motel. He told us where the room was, and said, 'Once you have freshened up, come on down, I've arranged an open bar and some finger food.'

❧

Well, for some guys from Canada with hayseed in their hair, this was some introductory meeting to Florida! Huge bowls of peeled shrimp with different sauces, oysters on the half-shell, pieces of steak on skewers, and various types of sandwiches; it was a feast. But it didn't quit there. They had a bartender and an open bar with unlimited choices. We were in hog heaven! They supplied us with several brochures and recommended places that we should go, and where to eat, etc. Ralph, just before leaving, put the icing on the cake. He handed me the keys to a '69 Chevrolet convertible, saying it was ours to use while we were in Florida. Throughout our entire stay, Radiation's hospitality was outstanding.

The next day we all decided we should go on a scouting tour of the area. We started off by heading south on US Route 1 down to the Melbourne Causeway, which would take us over to Highway A1A. The causeway traversing the Indian River was a two-lane road with a swing-bridge roughly at its centre that allowed watercraft to proceed up or down the Indian River, an element of the Intracoastal Waterway. Once we arrived at A1A we took a left and headed north, our destination being Cocoa Beach. The drive up the coastline was magnificent. Except for a few businesses, and a few homes, the view to the beaches and the ocean was unobstructed all the way—it was beautiful. We had our swim gear with us, so we stopped at Cocoa Beach for a swim. We were used to summertime water temperature in the lakes of Canada being about 72°F [22°C]. To our surprise and delight, the ocean felt like getting into a warm bathtub.

We headed west across the causeway connecting Cocoa Beach with Highway US 1, and that got our collective attention. We had located paradise: a line of bars, and nightclubs stretched the entire length of the causeway. After swimming in the saltwater, we were parched, so it didn't take much prodding to get me to stop at one of the watering holes. The following weekend between Friday night and Sunday night, I don't think we missed exploring

Figure 36: Map of Brevard County, Florida. The city of Melbourne is much more extensive than depicted here, but this will help to orientate the reader. Harris Corp., from the author's collection

any of them. It was during that same weekend we visited a nightclub in Cocoa Beach called the Mousetrap. People at Radiation had highly recommended it for dinner and a show. We were seated at a round table close to the stage with an excellent view of the performance. At one point a comedian took over. He looked out over the crowd and saw us, a bunch of guys sitting with no ladies, and made a somewhat demeaning comment about our sexual inclinations, continuing by asking us where we were from. An astute young member of my team yelled back at him, 'We're from Moose Jaw, Saskatchewan.' After a suitable pause accompanied by much laughter from the crowd the comedian asked, 'Moose Jaw! Where the hell is that?' With only a nanosecond of hesitation, my young disciple stood up with outstretched arms, surveyed the crowd, turned back to face the comedian, and said loudly and clearly, 'About five feet from the Moose's ass.' Much to the chagrin of the comedian, the house erupted in laughter and applause. From that moment on we couldn't buy a drink. People came over to talk to us, we just had a great time. The comedian's act was over!

The first week of school was tough. Dale was a good teacher, but this digital logic stuff was a brand-new technology to us. System logic diagrams contained symbols for 'And Gates', 'Nor Gates', and 'Flip-flops'. These elements and how they functioned were a challenging mystery to us. By the end of Week One, we were confused but by the end of Week Two we were starting to comprehend. Dale came over several evenings for our study sessions. That helped us out a great deal and we really appreciated it.

After that first week of school our minds were mired, to say the least, so after the study session on Friday night we climbed into our soft-top and headed for Cocoa Beach and the causeway in an effort to unclog. By the time the weekend was over—and just about every club and bar on the strip had been visited—whatever brain cells remained in our head were resting in peace. Food had not been a high priority but the folks from Radiation had highly recommended a place called Ramón's, practically right on the corner of A1A and the causeway road 520. As usual, they were right. It was superb. It was also a major hang-out for the astronauts involved with the Saturn rocket shots.

At the end of Week Two we were smart enough to be a little bit dangerous, but to my mind not smart enough for us to be proficient at the job that was ahead of us. I spoke to the General Manager of Radiation. He informed me that some of the new equipment should be in Revelstoke within a month or so along with the simulator, and at that time he would send up an engineer called Dick Sears to give us a hand and get us started on test procedures. I said that would be a good idea, and it would give us all a chance to determine how much we had learned at school. In my mind I thought that working on the actual equipment would generate many more questions on our part and it was my intention to ask management at Canadian Pacific to send us all back to school a second time. I had no doubts I was going to have one hell of a time convincing them that it was the thing to do.

Our final evening in Florida—29 August 1969—was momentous. Radiation had a special agreement with the Host of America Motel. A very large area with the bandstand, dance floor, bar, and buffet tables, was partitioned off. We were told to be there at 6.00 pm. Along with drinks, Radiation personnel made some speeches and we CPR people each received a Certificate of Proficiency along with a compact multimeter. I thanked them all for their diligence and hospitality.

Then the festivities began. Several secretaries from Radiation showed up, so there was dancing, schmoozing, and lots of drinking, with one interruption. Most of us were sitting at one large table with some of the Radiation guys and secretaries, when a lovely little cocktail waitress presented herself, and over the raucous noise asked, 'Who's the leader of this Canadian group?' I never got a chance to open my mouth, when several fingers pointed at me. She gave me a stern look and said, 'You had better go out near our front door and extricate one of your tribe who is fornicating with some lady out there, which is not good for business and must stop.' I went out and sure enough there was one of my young disciples thinking he was partially hidden behind a palm tree performing as if he was behind closed doors. I put a stop to it and told them both to take their act into seclusion. The next morning, prior to getting on our plane, I confronted this youngster about his lack of decorum, and how this may have affected the views of the people at the Host of America about the barbarity of Canadians in general, and Canadian Pacific Railway personnel in particular. He agreed his behaviour had been despicable and promised it would never happen again. He turned out to be one of the better ones!

Unknown to me, the general manager of Radiation Corporation had penned to Mr Hayward—the CPR chief of motive power located in Montréal, Québec—a glowing letter of our accomplishments at school. He passed it on to Jack Fraine and Fred Booth.

◆

'Vancouver' informed us the new building to house LOCOTROL, and brake test equipment should be in place by the end of November. Also, the new shop tracks and office building in Golden should be complete sometime in December. New Lead locomotives and Remote cars would start to arrive in Revelstoke around the middle of October. It couldn't happen too soon for me because the hands-on training of the diesel inspectors on real equipment needed to take place quickly so their shortcomings could be noted and brought up when we attended the next school at Florida. That's if I could swing another school with management.

The building of the new locomotive fuel point near the main line was going to be delayed for who knew how long. British Columbia government regulations for locomotive fuelling stations were rigorous, so it looked like when we started to run trains in the spring, we'd be fuelling the Remote units from trucks. I had previously talked to the guy who owned the fuel business in town about this, and the possibility of him increasing the delivery capacity on his trucks.

The divisional engineer, Danny Danaluk, had been working at putting in the crossover tracks connecting to the main line in the Revelstoke yard, and the construction of a new yard just to the south of Golden. He told me he was quite concerned about the main line track on the Windermere Subdivision, in particular the increased weight of unit coal trains on 100-lb rail, attached to fairly-old wooden ties, supported in his mind by inadequate ballast, resting on what he called 'Loon Shit'—an earthen content of malleable red clay. He was going to recommend that large sections of this subdivision be undercut and re-ballasted with concrete ties and 140-lb Continuous Welded Rail.

I mentioned before that the advent of unit coal trains was going to vastly increase the size of the territory that I was going to be responsible for. Thank goodness someone in Vancouver recognised this, because I got a call from the head of purchasing asking me if I had any preferences concerning station wagons. In all innocence I responded, 'what are you asking me for?' His answer '… because I've been instructed to purchase a vehicle for the master mechanic at Revelstoke, and that's you.'

This probably doesn't sound like a big deal, but I was to be the only master mechanic on the entire Canadian Pacific Railway system that had his own company car. Somebody had figured out that I just wasn't going to be able to touch all bases travelling by train—mostly by freight train—which ain't too fast in the mountains. I negotiated with him for a Chevy station wagon, with the largest engine that they supplied for that vehicle, and the Royal Canadian Mounted Police-type pursuit car suspension set-up. In the wintertime, I would load the back of the vehicle with enough weight in order to have better traction. I also had studded tyres fitted. She served me well over the 3 years I spent as a master mechanic on the Revelstoke Division.

In mid-November 1969, Deno wrote to Fred Booth requesting that the diesel inspectors and he should attend another school at Radiation in the first part of the New Year. CPR had received some new Master locomotives and Robot cars, and after some testing on the new shop track had formulated a list of queries for the engineering staff at Radiation. He felt they should all return to school in Florida now that they had acquired hands-on time and some familiarity with the intricacies of the new equipment. Another round of schooling would put the group in a position where they would have a good start at being able to maintain the equipment.

Booth's view was that it wasn't all that complex, and that maybe Deno was just looking for another vacation for the group. That agitated Milt. Some of his close cohorts like Angelo Vulcano and Tony Kruk tended to agree with Fred, and told him, 'Jesus, Milt, you've already had one vacation… you can't expect another!' Deno was so dismayed by this that he decided to take a step into where 'angels fear to tread'.

Every corporate employee knows, or should know, you don't go over your boss' head. I told Fred, 'You and several other people including Mr Fraine wanted me to set things up to the best of my ability to ensure the successful implementation of these new coal trains. I'm trying to do that, and in this case, you disagree with me. Therefore, I'm requesting a meeting with you, and Mr Fraine to resolve this matter.' It was Fred's turn to be pissed-off and he told me so in no uncertain terms, but he did set up the meeting.

I flew from Kamloops to Vancouver for the meeting. Fred had Vulcano and Kruk with him. Mr Fraine sat at the head of the conference table, and I sat on the opposite side from Booth and his cohorts. Mr Fraine opened the meeting by addressing me, saying, 'Milton, I understand that you want to take your group of diesel inspectors down to Florida for another school. Fred and the company don't feel it's necessary, so please explain to me why you think you need to.'

I presented the same argument I had presented to Fred, except I mentioned to Mr Fraine that I didn't think Fred, Angie, or Tony had any real concept of the complexity of this equipment. That caused a furore among the boys. Mr Fraine let them vent for a while, then brought the meeting back to order, saying to me, 'Do you have anything further to add Milt?' I said, 'Yes sir, I do. I think the only way that I can get Mr Booth to understand my request for another school is for him to accompany us and attend the school. He will then be in a more knowledgeable position, as to whether our people need to attend these schools or not.' Fred glared at me. Mr Fraine looked like he was studying me. Nobody was saying anything. Finally, Mr Fraine looked at Fred and said, 'I think that's a good idea Fred, then you can report to me first-hand.' I was elated. I knew that when we were alone Fred would have a piece of me, but that was a mere bag-of-shells. We all were going back to school! That was the whole point of the exercise.

Figure 37: Dale Delaruelle instructs the CPR team. Author's collection, courtesy Dale Delaruelle

Figure 38: The CPR trainees and Harris personnel in Melbourne FL. Dale Delaruelle and Gene Smith are 2nd and 3rd from left and Lewis Cox is far right in the back row. All are Harris software engineers. Author's collection, courtesy Dale Delaruelle

Figure 39: The CPR trainees and Harris personnel in Melbourne FL. Gene Smith, 2nd from left; Lewis Cox in front with maroon tie; Dale Delaruelle, 2nd from right. Author's collection, courtesy Dale Delaruelle

When I got back to Revelstoke, I contacted Radiation about setting up an advanced school. I told them we had been working with the equipment, and felt comfortable navigating our way around it, however we had many questions about situations we'd encountered. Radiation got back to me within a couple of days and said they'd set up an advance school starting 2 February 1970, ending on 13 February. I conveyed this to Fred and Mr Fraine. I also gathered my guys together to tell them that we were going back down to Florida. I instructed them that I wanted each to write down all the questions they had about the new equipment for presentation to the instructors. I also informed them there would be no overly-amorous displays as had happened the last time. They were to conduct themselves in a gentlemanly manner, remembering that they were officers representing the Canadian Pacific Railway Company. During this dissertation I particularly locked eyeballs with the fornicator.

☙❧

Things were generally going along as best as they could in preparation for the startup of the coal trains. The new yard, shop tracks, and buildings were nearing completion, a double-width trailer home would be moved onto its new foundation within the month as the resident diesel inspector's home in Golden. Apparently, there was insufficient funding, Danny reported, to rectify the 'loon shit' problem on the Windermere sub.

The wiring of the Connaught Tunnel was proceeding slowly, but that was to be expected because they could only work when there was no traffic, and that left only brief windows of time in which to work. It was mandatory that the wiring of this tunnel be completed before we ran any remote-control trains, since it was vital the Lead and Remote units maintained radio communication in the tunnel.

At Revelstoke, the facility for testing equipment and the crossover tracks connected to the main line had been installed, and new high-capacity fuel pumps for the head-end locomotives had been installed. The Remote locomotives, for the time being, would have to be fuelled by truck since the British Columbia environmental people had not yet given us permission to build a new fuel station. A sand truck was being manufactured somewhere in the States. It would be ready on time, it seemed.

At Sparwood, the coal silos, loading tunnels, and the entrance and exit loop tracks would be completed in another month or so. The car rotation dumper at Roberts Bank was being installed, the entrance and exit tracks on the man-made causeway out on the Pacific Ocean were just about complete, as were the diesel inspector's and carmen's office and locker rooms.

On Saturday 31 January 1970, we left Vancouver to start school at Radiation on 2 February. This was going to be an advanced school and Dale Delaruelle was again the instructor. A good deal of time was going to be spent learning how to properly read logic diagrams and understand how the various components functioned. Also, Dale was going to show us the radio messaging structure between Master and Remote and how at the sending and *receiving stations ones and zeros were converted into radio waves, and then back to ones and zeros at the receiving station. For us this was heavy-duty stuff to absorb, which ended up with us spending about 3 hours a night in my room instead of the usual two, trying to piece it all together.*

Fred Booth spent 2 days—Monday and Tuesday, the second and third day of February 1970—attending school and studying with us. At breakfast prior to the third day of school he informed me that he would be leaving that day to return to Vancouver. His words went something like this: 'These past 2 days have convinced me you were absolutely right in insisting that our people attend this instruction. When I return to Vancouver, I'll be reporting this to Mr Fraine.' Without any further trouble, I sent people down to school in Florida on several more occasions, although I never attended another.

The first remote-controlled unit coal train was scheduled to run on 30 April 1970. Deno drove to the mine site at Sparwood with one of the diesel inspectors, who was to ride the train back to Revelstoke. Deno wanted to be at the mine to see the loading procedure. Things did not start off well.

Figure 40: Harris Controls logo

To load the train, it passes through a tunnel and beneath three silos. A load-out operator in a cabin inside the tunnel opens and shuts gates to allow the coal to flow into the cars. In his excitement, the operator dumped a whole load of coal on the first locomotive. That stopped the operation, and it took Deno, the diesel inspector, and a couple of mine guys about an hour to clean the coal off the top of the locomotive and from the radiator cooling and dynamic brake fans.

It was up to the locomotive engineer, in radio contact with the load-out operator, to control the speed of the train so that the operation was smooth and continuous. As more and more cars were loaded the engineer found it almost impossible to maintain an even speed. Many times, the load-out operator requested that the train be stopped and backed up so that a car or cars could be loaded to the required capacity.

Deno had seen enough. They needed a fully automated speed control system on the coal-train locomotives. The train finally got loaded and made a successful run as far as Revelstoke. During the train inspection, a sharp-eyed carman identified two coal cars that—due to the orientation of their couplers—were incorrectly positioned in the train. Remember that a rotary coupler on one car had to be coupled to a fixed coupler on the adjacent car to permit the cars to be rotated upside down for dumping at the coal terminal. The errant cars had to be switched out, turned on the shop turntable, and then re-marshalled into the train. It was fortunate this fault was noticed, otherwise there'd have been a dreadful mess at the car dumper at Roberts Bank.

☙❧

It can be noted that there was one major component in the plan to achieve the requirements of the Japanese coal contract with which Deno was not required to be involved. This was the construction of the export coal terminal at Roberts Bank, BC.

Vancouver's first shipping ports were coal terminals located in Burrard Inlet, close to the city. From 1961 to 1966, the annual terminal capacity at these facilities was approximately 13 million tonnes[52]. By the mid-1960s, it had become apparent that demand would soon exceed the maximum design capacity of these facilities. In 1968, Kaiser Resources negotiated a 15-yr contract with Mitsubishi & Company for metallurgical coal that

---

52   1 tonne equals 1.1 US tons.

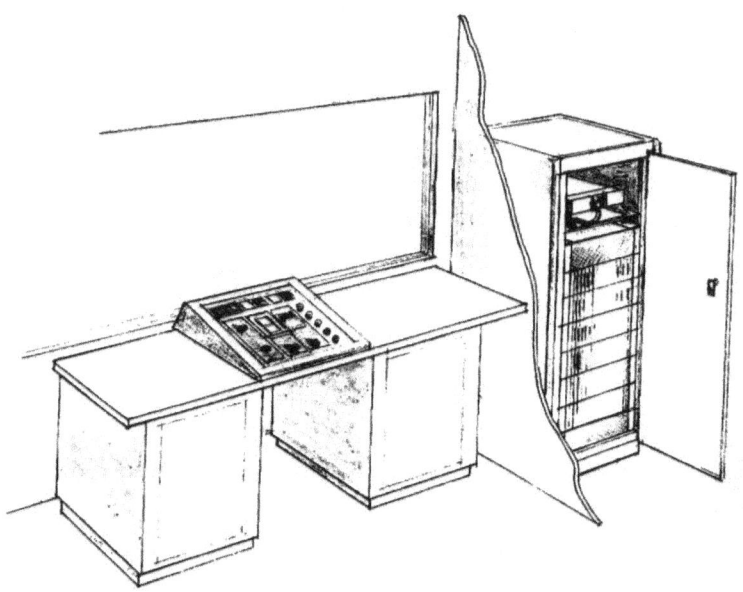

Figure 41: Sketch of proposed Tower Control Console installation. Radiation Corp. (Harris Controls & Composition), from the author's collection.

included a commitment to ship product by 1970. This imposed an immediate need for a coal port and terminal facilities on the southern BC coast. The proposed facility had to handle large bulk carriers, be equipped with efficient coal handling facilities to permit rapid train throughput in order to realise lower transportation costs, and be completed within 16 months.

Part of the Port of Vancouver, Roberts Bank is a twin-terminal port facility located on the mainland coastline of the Strait of Georgia within the Corporation of Delta[53] on the south side of the Fraser River estuary, approximately 35 km south of downtown Vancouver. Since opening in 1970, the terminal has been much expanded since, to become a major container port as well as the busiest single coal terminal in North America. The terminal is serviced by CP Rail, CN Rail, and BNSF Railway.

When its coal-train unloading facility was constructed, it was recognised that the intended throughput would require entire trains to be moved intact through a rotary car dumper. Originally, trains were moved through the dumper by a positioner arm called the 'sequencer' that travelled in a vertical position along a heavy steel beam and dropped between cars, whereupon it bore against the car structure to move the train ahead the desired distance to spot the next two cars for rotating. When this sequencer (known as 'the rabbit'[54]) broke after 1 week's service—without a back-up method—Ford, Bacon & Davis, the contractor responsible for constructing the car dumper, asked Don Selby to come to Roberts Bank and show them if it was possible to run the Remote consist to position the cars within an accuracy of 2 inches. Selby took the Lead locomotive off-line and, using the Remote units, was able to repeatedly position the cars within 1 inch of the chalk lines. Following this, Ron Gottbehuett and Dale Delaruelle were tasked with building a LOCOTROL master station in the dumper operator's cabin that became the Tower Control system.

Nowadays, CP Rail has about 24 unit coal trains in continuous service to-and-from Roberts Bank depending on demand. A 10 000-ton, 120-car train can be dumped in 2 hours.

<hr />

By late spring of 1970, all the road foremen of engines had been brought up to speed on the new equipment. They were busy because they had to ride every coal train to instruct the engineers and eventually—when the engineers were judged to have gained enough experience—to qualify them. Not everybody made it, and those who didn't were not permitted to operate these trains. The diesel inspectors were kept busy too, riding the

---

53   Delta is a city in British Columbia, and forms part of Greater Vancouver. In 1879, the area was incorporated as a municipality, named 'The Corporation of Delta', with the village of Ladner as its administrative centre.

54   The sequencer was winch-operated and connected to a wire rope for travel along its 'rail' (a heavy steel beam). The colloquial 'rabbit' nomenclature derives from its resemblance to the cable-operated lure that runs around a greyhound track.

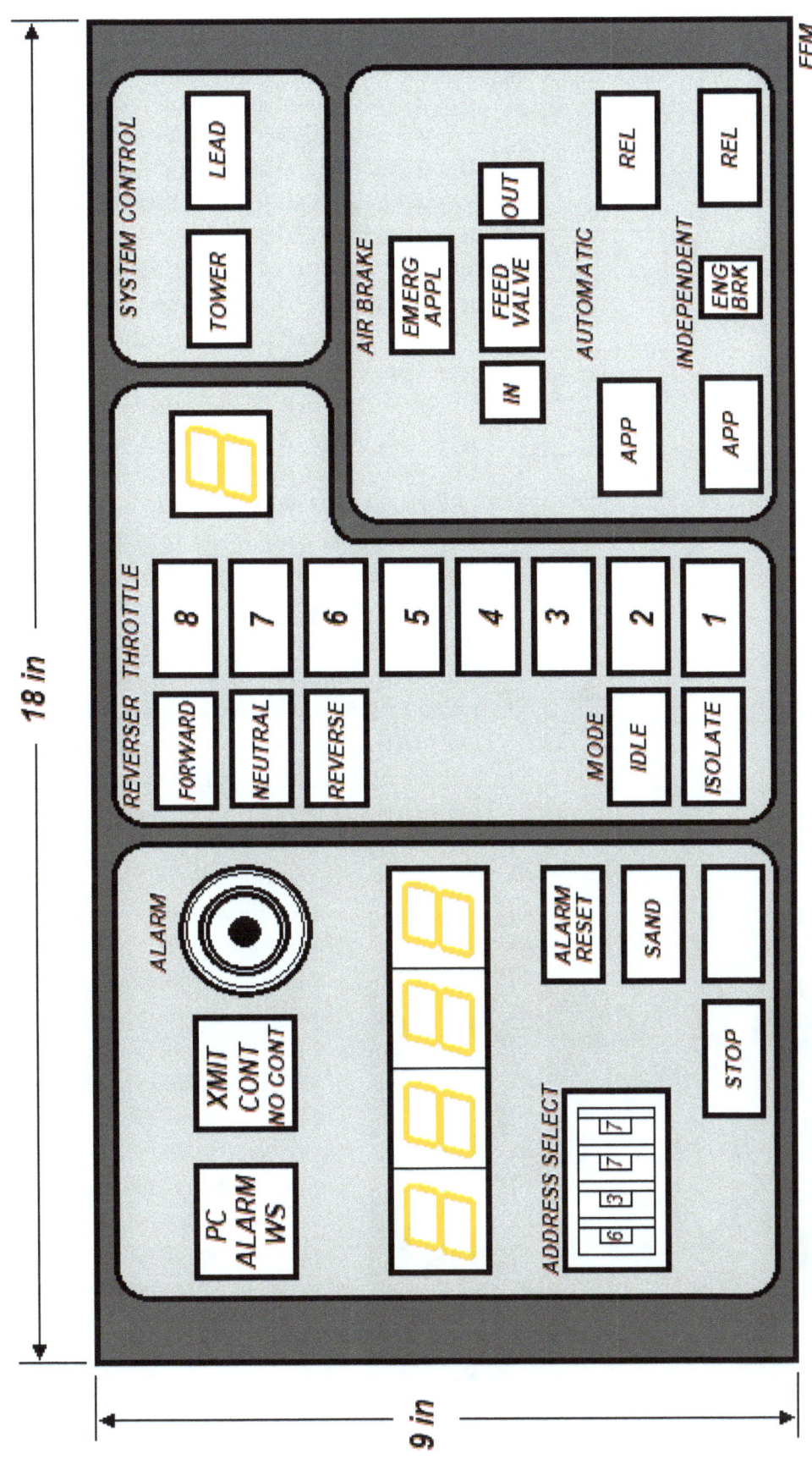

Figure 42: Sketch of proposed Tower Control Panel. Radiation Corp. (Harris Controls & Composition), from the author's collection.

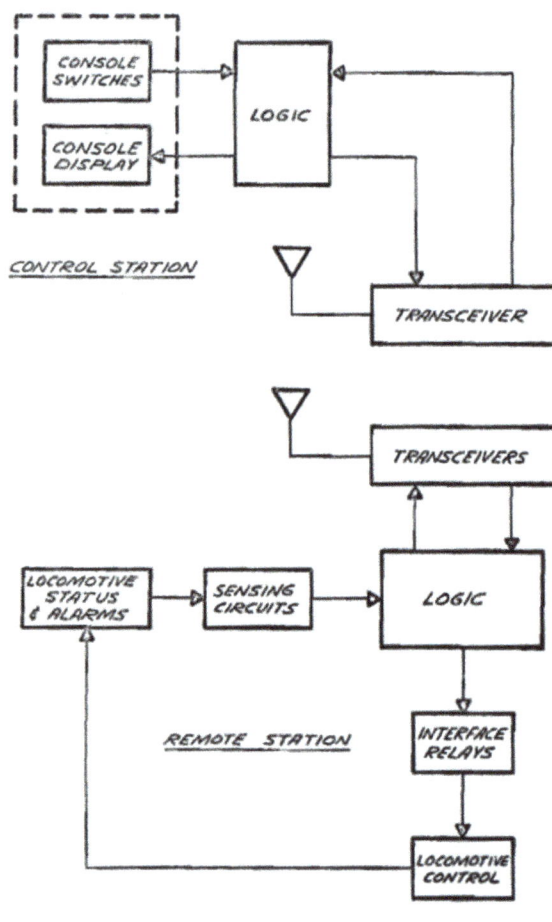

Figure 43: Tower Control block diagram. Radiation Corp. (Harris Controls & Composition), from the author's collection.

remote-controlled trains, sorting out problems caused by equipment failures and air brake problems. All-in-all, though, Deno felt things were going along well.

CPR's neighbour railroad, BC Rail, had purchased a few sets of LOCOTROL, and during the summer of 1970, Deno was asked to conduct a LOCOTROL Operations instruction class for some of their personnel. He decided to use it as a refresher course for his own diesel inspectors and road foremen of engines. He was looking forward to this school because his cousin, Bob Deno, had been promoted to Diesel Inspector on BC Rail and was going to be one of the participants.

The BC Rail and CP Rail personnel who did not live in Revelstoke stayed at McGregor's Motor Inn, which at the time was the best place in town, and therein lies a story. The school would end on a Friday, and with the divisional superintendent's concurrence, a conference room at the motel was rented and set up with a bar, bartender, and finger food. School would end at 4 o'clock and the get-together would begin at 5.00 pm. We were all looking forward to this shindig, including Superintendent Hill. Around 2.00 pm on Friday the shop clerk knocked on the classroom door and beckoned for me to come out. He had a message, that Jack Fraine would be arriving in town around 4.30 pm and wanted to meet with Mr Hill and me. The class was told to start without me.

Our meeting consisted mostly of a discussion about coal trains. Mr Fraine wanted to see the new servicing area for the Remotes, the crossover tracks, and the testing facilities near the roundhouse. We went back to Hill's office and Mr Fraine informed us he'd be staying at McGregor's Motor Inn that evening and invited us to have dinner with him. I was hoping that Mr Fraine's room would not be close to the party room; and thankfully, it wasn't. He didn't need to see my people in party mode. He'd never met any of the diesel inspectors or RFEs, but it was inevitable that one day he would, and I didn't want him to get a bad first impression.

Figure 44: Tower Control equipment at Roberts Bank in 2008. The control panel has changed a lot! Image by 'eminence grise', courtesy Trainorders.com.

There were two responsibilities I haven't yet mentioned that came under my jurisdiction as Master Mechanic at Revelstoke: [1] the maintenance conducted on all equipment associated with the Connaught Tunnel extraction fan system and the supervision of five employees involved, and [2] the operational supervision and maintenance of three large tugboats harboured at Penticton on Okanagan Lake.

At the fan house located at the west end of the tunnel, five houses were located at the base of Mount McDonald: the domiciles of the five maintenance men and their families. Up until the 1960s, and the arrival of the Trans-Canada Highway, this area was isolated and the only way in or out was by rail. Even with the highway, the only way in or out most of the time in the winter was by rail, making for a very lonely and isolated existence. I used to try to visit the families and the fan house as often as I could.

The original diesel engines that had operated the fans since their inception in December 1916 were dismantled during the early 1970s and replaced with fully-automated engines started up and shut down by track circuits dropped by the approach of trains. In the late 1980s, the Mount McDonald Tunnel was built, reducing the westbound grade to 1%. The Connaught Tunnel is still operational primarily for traffic in the eastbound direction.

♾

We'd been running these new heavy coal trains for several months and had learned a lot about how to make this operation a success. For instance, we learned that brake shoes would wear faster on one side of the train than the other, because Wabco hadn't centred the fulcrum point on the brake beams. Consequently, we assigned more carmen to that side of the train during inspection. We were mixing more oil into the (very fine) coal during loading at Sparwood, and this solved the problem of the loss of thousands of tons of coal that had been blown from the open-top bathtub cars during the run to Roberts Bank. We had to place speed limits on these trains while running through certain towns that sat on bedrock, because we were knocking the pictures off resident's walls and cracking some basement walls. We alerted the railroad police to check all locomotives before the train left Revelstoke for transients, particularly in the westward direction. These people—many in a state of inebriation—would get aboard and ride until they got to their desired destination, whereupon, if the train didn't stop, they would initiate an Emergency brake application, which you can do from any locomotive cab.

Sometime in mid-July of 1972, I was summoned to a meeting in Vancouver. Without divulging anything, Fred Booth took me to Jack Fraine's office. I have mentioned before that when these people get me in a situation that I don't know anything about, my bowels began to rumble. Mr Fraine greeted me with a big smile, shook my hand, and then said, 'It's good to see you again Milton…' I thought, hey, maybe everything is going okay.

Deno recalls they talked a little bit about last winter and the spring, and then Fraine looked at him, and asked, 'Do you know why you're here?' Deno: 'No sir, I do not.' Fraine told him, 'Fred and I want to know whom you'd recommend to replace you as master mechanic at Revelstoke, because you've been promoted to Director of Mechanical Engineering at System headquarters in Montréal and you're to report to the chief mechanical officer, Mr Raby, in his office at Windsor station, at 8.00 am on 7 August.'

I just stood there. I didn't know what to say. However, I do remember thinking, 'Jesus Milton, it looks like maybe you're only half a step away from getting a key to the executive washroom.' They congratulated me and asked who I'd recommend as my replacement. That was a no-brainer: diesel inspector Tom Johnson would be well-suited to the role. They concurred and the deal was done.

♾

A new project: an Arctic wind blows Deno some good

The Canadian government had become involved in a proposed study to investigate bringing oil and LNG gas out of the Arctic region by pipeline and/or rail[55]. Major players such as Queens University, University of Edmonton, Canadian Pacific Railway, and Canadian National Railway were to be involved in this study, as were federal government officials from the provinces of Manitoba and Alberta, along with senior government personnel from the Northwest Territories and the Yukon. The estimated duration of this study was one-and-a-half years. The leaders of these different entities were requested to provide the names of selected personnel to work on the various aspects of the study. Bill Stinson, Vice-President of Canadian Pacific Limited, and Charlie Pike, who was now System Chief Mechanical Officer, selected me to head up the study's mechanical team.

Seven other people from Canadian Pacific Railway were chosen, of whom I knew four. These individuals were to be involved with the operations and civil engineering teams. Eventually, the total cadre of people selected would number somewhere around 33: some from the Canadian Institute of Guided Ground Transport (Queens University, Kingston, ON) and the rest from Canadian National Railway.

When needed, 12 consultants from various parts of Canada were available. Also involved were 22 major energy companies from Canada and the United States. A block away from Windsor station—where my office presently was—a new skyscraper had been built, and the top three floors had been leased to house our study group, now called the Gas Arctic Northwest Project Study Group. My corner office was sandwiched very high up… high enough that when the wind blew strongly, the building swayed, which caused this prairie-dog some alarm! The big-city guys informed me everything was okay; the building was built to sway. Yeah, maybe.

Sterling Smith, Vice-President of Canadian Pacific Consulting Services, had been assigned as Deputy Director of the study and he was my immediate boss. A gentleman called Les McCagg from Canadian National was assigned to my team. He was a mountain of a man, but his temperament was just as calm as he was big. Les had been around the block a couple of times and was my type of guy. He became my number one man. Don't get me wrong; the other people on my team were very competent, they just didn't have Les' panache.

My team's job was to provide the study with proposals for appropriately equipped motive power and rollingstock to transport LNG and oil over a very long distance, through extreme weather conditions, in maximum-length trains. We were also charged with providing suitable maintenance facilities throughout the length of the proposed railway.

I suggested to Sterling Smith that because the Arctic presented so many operational and maintenance difficulties that were unparalleled in North America, we should seek out railroads operating under similar Arctic conditions and seek their counsel on how to combat the severe weather in which we'd be operating. Only one railroad in Canada, as far as I knew then, was operating in such similar conditions—the Québec, North Shore & Labrador Railroad, based in Sept-Îles, QC. Around Labrador City it was not uncommon for temperatures to plummet to minus 60°F in the wintertime (not including the wind-chill factor). Sterling agreed with me and said to make any arrangements needed.

For the next couple of weeks, Les McCagg and I were busy, arranging to meet with the engineering staff of various railway product manufacturers. We set up a trip to the Yukon and Northwest Territories so I could talk to various groups with a view to getting the manufacturers to build special equipment if required. It was my hope that I could accomplish this before the end of 1973, but I didn't make it because of other circumstances that prevailed.

---

55 *Arctic oil and gas by rail: a study of the technical feasibility and cost of transporting crude oil and liquefied natural gas from the Arctic southward by rail.* Prepared by Canalog Logistics Ltd and Canadian Pacific Consulting Services Ltd for Transportation Development Agency, Ministry of Transport, Commonwealth of Canada, 1974.

One interruption to my schedule occurred when Sterling summoned me to his office in late July and said, 'Milton, we have had a request from the Canadian government for two people from the study group to go to Cuba to look into the deteriorated condition of their sugar refinery railroads.' Cuba could not import railroad items and components from the United States because of the embargo placed on them after Castro had come to power there, but Canada was not a signatory to this arrangement and was willing to help. Because I was a locomotive troubleshooter, I was the chosen one. They also wanted an operations person, so a guy by the name of Dave Ferrell joined me for this unplanned experience, which was to be for no longer than a week. That didn't happen. Another interruption to my Arctic Study schedule occurred on 29 September 1973, after I had returned from Cuba. I had met someone I would eventually marry.

Communist Cuba had isolated itself from much of the world, and Deno found it was not the easiest place in the world to get to. He and Dave Ferrell eventually travelled via Mexico City and waited for Mexicali Airlines to operate one of their irregular flights to Havana, on a four-engine Ilyushin aircraft. It was 17 August 1973.

❦

With his exploration of the Arctic and Yukon completed, Deno returned to Montréal. For the next 7 months the team worked industriously to complete their written portions of the study. Time was also spent designing the control vehicle that was proposed to be in the lead position on the oil trains. Deno had formed the idea for this vehicle when it became apparent the round-trip was going to be approximately 1800 miles, and was going to require more than one train crew. The intent was to design a vehicle that would not be powered but would have all the control systems to operate a distributed-power train. This 'control vehicle' would be equipped with cooking and dining facilities, sleeping quarters, shower and bathroom facilities, and a snowcat to get around on if needed. The control vehicle would also convey fuel, which would be transferred to the lead locomotive consist by suitable piping and a transfer pump.

In early spring of 1974, the Arctic Rail Study reports were complete and under review by group leadership. The team were retained until the leaders and the Canadian Government were satisfied with the report. One day during this waiting period, a man who'd been a consultant on the study group, but with whom Deno was not acquainted, came into his office and introduced himself as Ray Corneil. He said that he had an interest in some of the work Deno had done, and if possible, whether they could have a one-on-one over a drink after work.

The two met and talked about the Control Car concept and some other things, before Corneil asked what Deno's education level was. Although Deno was nonplussed, Corneil reassured him and asked him not to be offended. Corneil said, 'I am the Dean of Engineering at Queens University in Kingston and I'm very interested in some of the work you've done but I've not had the opportunity to get acquainted with you.' Deno thought he was a pretty interesting guy, but chastened, informed Corneil that he had achieved only a Grade 10 education, and then elaborated on his career with the railroad. They chatted more, parted amicably, and Deno returned to the CPR in Montréal and to Charlie Pike, the System CMO.

I had only been back in Montréal for about a week when I got a call from the LOCOTROL people in Florida, asking if I'd come down to their plant, and do some consulting work for a week or so. I hadn't had a vacation for a while, and things were fairly slow on the railroad so I asked Charlie if I could have a fortnight's vacation. I called my wife and she jumped at the chance for a little downtime down in Florida. I called Harris Controls and told them that I would have to take vacation time to come down there so I wanted them to pay for my girlfriend's air ticket and mine, along with suitable remuneration for my time… to which they agreed.

We found a real nice motel right next to the ocean in Melbourne Beach. I only spent about 3 or 4 hours a day down at the plant, so there was quite a bit of time for us to scout out and enjoy the area. On one such occasion we decided to drive north on Highway A1A as far as Cape Canaveral. The guys at work had told me that a lot of fishing boats docked there, and there were restaurants that served fresh seafood, which we both really enjoyed. In 1974, you could drive up A1A from the Melbourne Causeway all the way to Patrick Air Force Base and have an unobstructed view of the sand dunes and ocean. Except on the oceanside of A1A in Satellite Beach, a new eight-storey condominium was nearing completion. Two- and three-bedroom apartments were being advertised at. We decided to have a peek.

Knowing something of the price of real estate next to water in Canada, I knew for sure a beautiful place like that, located on the ocean, with all the facilities it had was a way out of my price range, but I had to ask. The lady said, 'USD39 000, but they increase in price the higher up you go.' With a sad heart I told her a USD39 000 down payment was going be out of my price range. She laughed, and said, 'That's not the down payment Mr Deno, that's the total price!' After I'd recovered, I said, 'Look... we're Canadian and I don't have funds in the US.' I told her it'll take a little while to transfer down and asked whether she could hold the place until we found out some things, which we would do the next day. She said, 'If I hold it, I will need some money down. My boss will insist on that, so how much could you give me?' With rapidly diminishing expectations I said, 'One hundred bucks.' She said, 'It's a deal. I'll hold it for three days.' I had a home in Florida! What was I thinking? We flew home to Montréal a couple of days later.

ঔঠ

Sometime around the middle of June 1974, after getting off the commuter train and into the elevator to go to my office at Windsor Station, I was joined by Charlie Pike, who said, 'Don't bother getting off at the ninth floor Milt. Come to my office, I have something I need to discuss with you.' Charlie's office was elaborately decorated, with beautiful leather couches and lounge chairs, a large wooden coffee table, plush carpet, a magnificent desk and accompanying chair... and you never passed up an opportunity to go there. We sat in the lounge chairs and his secretary brought us coffee.

Charlie said, 'Milt, I believe you're acquainted with a Dr Ray Corneil, is that correct? You must've made quite an impression on him because he has been in touch with the President of Canadian Pacific Limited, requesting that you be sent under his care to Queens University to undertake studies in mechanical engineering. The president asked me to find out if you knew anything about this, and if not, what were your thoughts about going to university?' I said, 'Charlie, I don't know anything about this, but I think it would be a bad choice to send me, with a Grade 10 education, at 41-years-old, to the best engineering university in Canada. This must be a bad joke!'

It transpired that a few people knew about this, and they were not insignificant people. The boss of Canadian Pacific Consulting Services, Jack Fraine, Charlie Parker, Al Hill (my old boss at Revelstoke, now promoted to Vancouver as a Regional General Manager), all contacted me, and encouraged me to rethink my response. Thus, I told Charlie Pike that I'd give it a try. However—if at any point I thought it was an exercise in futility—I would have his assurance that I could cancel the university experience and return to Canadian Pacific. He agreed.

So, in July 1974 I was sent to McGill University in Montréal for 2 weeks of testing, and a report would be submitted to CPR and Queens University. In the end, it was obvious I needed to complete a lot of remedial work and I was told that if I completed that successfully, there was an excellent chance I'd be able to fulfil the requirements of an engineering degree.

I had approximately 6 weeks before I had to report for classes at Queens. Due to unforeseen circumstances occurring on the CPR I was only able to get 3 weeks of remedial work. My complaint was voiced to Dr Corneil, who replied, 'We will assist him when he gets here, and bring him up to speed.' That didn't happen either as it turned out.

My wife and I drove over to Kingston and with Ray Corneil's help I found a place to live. It was called Bellevue House, had been built in the 1800s, and had, at one time, been the home of Prime Minister Sir John Macdonald. This stately home had been divided up into apartments, of which I rented one. It was located conveniently near the university and downtown Kingston.

I met all the professors who'd be conducting my lectures and I also met a few of the kids with whom I'd be attending lectures: 41 men and one woman. We received an extensive list of books to buy. Calculators were a new item on the campus in 1974, and we were allowed to use these if we wanted, otherwise to stick with the slide rule. I didn't know what a calculator was, and I sure as hell had never used a slide rule. My new classmates told me to get a calculator if I could afford it. In 1974, the Texas Instruments scientific calculator cost CAD750. I didn't understand three-quarters of the things this marvellous little machine could do. Fortunately, a couple of the kids tutored me.

The lectures began and Deno struggled. He quit drinking and smoking. After he'd attended the last lecture of the day, he would try to find a professor who wasn't busy and was willing to help. If that worked, he'd stay with him for as long as possible, and would sometimes not get home until 7 or 8 o'clock, make a quick dinner, then hit the books and study for 3 to 5 hours.

Early on, it was obvious he was getting appreciably behind with new assignments. Having missed the 11th and 12th grades, and not having received the benefit of the remedial preparation he'd been promised had left him continually playing catch-up on subjects he should have had at his fingertips, particularly those vital for mechanical engineering, like math and physics. His wife was very supportive. When she could, she'd drive down from Montréal on weekends and prepare food for Deno for the coming week. On these occasions Deno often struggled to even make time for her; such was his study load.

The end-of-year exams were a miserable exercise in failure for Deno. He and his wife retreated to their new condo in Florida for the 2-week break: a 31-hour trek in a VW Beetle.

I worked my tail off during the last few months of my first year at school—to no avail. In some subjects, like programming, algebra, and drafting, I did well, but I was only mediocre in math and physics. I told Ray Corneil that I thought I should pack it up and go back to my job on the railway. He said, 'Let's give it another year, and see what happens.' I thought for sure somebody would line up some remedial tutoring for me during the summer break. No dice. Instead, they had me lined up to work with some of the professors on a study to determine the feasibility of electrifying Canada's railroads. That was very interesting work. I travelled to all the mainland provinces to meet with various government and railroad officials, but the highlight was when they sent me alone, down to the Black Mesa & Lake Powell Railroad, located on the Navajo reservation in northern Arizona. This railway was touted as the most modern automated, electrified (50 000 volts) railway in the world, and I was excited to get down there and see it. I flew into Flagstaff, and then drove to Page, where both the railroad centre and the coal-fired power plant it served were located.

I was accommodated by the superintendent of the railroad, who was a full-blood Navajo Indian. I told him I was pleased to have the opportunity to work with and observe operations on the BM&LP. He said, 'Let's you, and I have some coffee together, and I will bring you up-to-date on why the railroad is no longer automated, and a few other key items that you should be aware of.' Long story short, a vast deposit of thermal coal had been discovered on the reservation and the state government needed more energy across northern Arizona. They decided to build a mine, a power plant, and a 78-mile connecting railroad. But when they decided also to automate the train operation, they reneged on a deal they'd negotiated earlier with the Tribal Council to create jobs for Navajos. The *locals embarked on a strategy of well-structured civil disobedience, with the outcome that the railroad reverted to a manned operation and has been ever since, [although as this account is being completed in 2020, the mine and thus the railroad are about to close]. The superintendent informed me that I was welcome to stay. I had many discussions with local personnel and rode many trains. It was a delightful experience.*

I flew back to Montréal, then my wife and I drove to Kingston. I submitted my report on the BM&LP Railroad to Dr Corneil, taking that opportunity to discuss my coming year at Queens. I informed him once again that I had still not received the necessary remedial help I'd been promised.

I worked as hard as I could during the next school year, but I was getting further and further behind in the current math and physics requirements and consequently not attaining the required marks. Nobody was going to change my mind this time. I informed Dr Corneil, and Canadian Pacific that come June 1976, I was finished at Queens.

My time at Queens was not a total loss. I'm glad I took the challenge even though things didn't work out as some had expected. I learned a great deal: however, I came away from the experience unhappy with myself for not achieving what had been set for me. I completed my final semester at Queens, never did get math and physics under control, but met some really nice young people there. I returned to Montréal in a new role: Special Assistant to the Chief Mechanical Officer – System, Charlie Pike.

༺༻

The Québec, North Shore & Labrador Railroad that operates between Sept-Îles (Seven Islands) and Schefferville in south-eastern and north-eastern Québec respectively, had requested the assistance of Canadian Pacific Consulting Services in establishing a small production line for building locomotive power assemblies[56] in Sept-Iles, and to inspect their automated mine railroad near Schefferville with a view to suggesting improvements. Charlie Pike called Deno to his office and told him that CP Consulting had requested that he and retired CMO – System, Harold Hayward, be assigned. Deno was pleased; working closely with Harold would be a pleasure for him. He'd always admired Hayward when he'd been the boss. Pike told CP Consulting they could use Deno for 2 weeks, then he wanted him back. Deno thought it would be a nice change of pace.

Harold was going to fly down to Sept-Îles. I decided that I'd take Jana with me: she would be of assistance because she spoke fluent French, and that area was totally French Canadian.

Mr Hayward and I decided that to save time, he would remain in Sept-Îles to get the assembly-line underway, and I would head up to Schefferville. My wife and I had to travel by train: QNS&L corporate rules would not allow her to join me in flying to Schefferville in the company helicopter. That ride was more frightening than the one I had taken to a wreck in British Columbia, which had been in a nice big Sikorsky six-passenger job. This machine was one of those two-seaters with a bubble nose and a storage compartment.

The pilot was standing by the aircraft waiting for me. I was wearing a parka with ordinary dress clothes underneath, shoes and galoshes, and a briefcase in my gloved hand. When he saw me, his first words were, 'Where's your survival kit?' Taken aback, I looked at him. 'Survival kit?' 'Yes, you sure as hell can't go with me like that!' I said, 'Look, no-one told me anything about needing to have a survival kit to fly to Schefferville.' He mumbled something under his breath, told me to take a seat and wait while he made some phone calls, and returned to his company office. We flew: me, sans survival suit.

I was able to introduce some ideas to the mining people, and make them aware of Harris Corporation in Florida, who might be able to satisfy their requirements for upgrading their automated railroad.

༺༻

---

56   See Glossary.

On 26 November 1977, at 0035, LOCOTROL train 803 (Extra 5820 West) with 106 loaded coal cars for 15 292 gross tons, ran away, and de-railed at milepost 94.4 of the Mountain Subdivision, near a location known as Flat Creek. Three 3000 hp SD40 locomotives, a Robot car, 83 loaded bathtub coal cars and a bridge were destroyed, closing the CP main line for a week. I believe at the time, this mishap was North America's worst non-fatal railroad wreck.

Charlie Pike was informed of the wreck early on Sunday morning, 27 November. He contacted me around 7.00 am, related what he knew about the mishap, and told me to catch a flight as quickly as I could for Calgary where somebody would meet me, and take me out to the site. He told me that he had informed everyone on the Pacific Region and would also inform the Canadian Transport Commission that I was the CPR investigator for this mishap, reporting directly to him. He also asked me to keep the superintendent of the Revelstoke Division appraised of my findings.

I'd worked in various positions for 11 years on the Revelstoke Division, and had been present at and worked on numerous wrecks, but I had never seen anything even close to the devastation of this one. The train had been going so fast on the curves inside snow sheds and tunnels, the units on the head-end had been leaning over and scraping against and splintering the timber and removing rock material from the sides of the tunnels; the amazing thing being that when they came out of these structures onto tangent track, they fell back onto the rails.

The maximum permitted speed for a tonnage train descending a 2.2% gradient was 19 miles per hour. The head-end crew told me that at the point of derailment they were doing close to 80. One locomotive of the Lead consist plus two locomotives and Robot car of the Remote consist, and 83 coal cars were piled up at the bottom of the chasm, along with 6700 tons of coal. The remaining 28 cars and the caboose were not derailed or damaged and there were no casualties.

The engineer mishandled the train brake[57] but he claimed the Remote locomotives never went into dynamic braking, stayed in Motoring and pushed them down the hill. He also stated that the Emergency brake did not function. The whole crew had their story pat. The tail-end trainman was just a kid, the son of one of my road foremen when I had been Master Mechanic at Revelstoke.

I had the electronic boxes taken from both the Lead unit and Remote car and put on the simulator. They worked flawlessly. The engineman's story about dynamic braking not functioning was easy to discredit. I crawled down and looked at two of the units from the Remote consist that were still kinda together. The switchgear in the electrical cabinet was in the D/B position. That—of course—happens when the jumper cables are pulled out, such as often happens in a wreck. The switchgear would stay in the last commanded position before the circuit was broken.

The lead engine and the one behind it survived, and they only stopped because their brake beams were against the wheels providing some braking action. The brake shoes, brake shoe holders, all bolts and mechanisms, were burned off. On the train itself, there were no depleted brake shoes and no blue wheels.

The only good thing about this incident was the fact that they all had lived. You have never seen such a pileup in all your life. We made history in a stupid way. Right after the mishap, I issued orders for every coal train to be accompanied by an RFE. If the speed of the train exceeded 19 mph at the west switch of Glacier, they were to immediately put the train into Emergency. And if that happened, then upon their arrival at Revelstoke the entire train crew was to be pulled from service. We didn't have any more runaways on the Mountain Subdivision[58].

---

57 Accident Report: Canadian Transport Commission Inquiry under s226 of the *Railway Act*, 22 Jan 1980, Calgary, AB. File No. 31385.3845 (Library & Archives Canada 5147 8107 70). The investigation concluded that the cause of the derailment was loss-of-control of the speed of the train resulting from improper braking procedures by the crew.

58 Author's note: On 4 February 2019, westbound CPR train 301–349 ran out-of-control and derailed 99 cars and two locomotives at Mile 130.6 of the Laggan Subdivision, near Field, BC. The three crew members were fatally injured. See TSB rail transportation safety investigation R19C0015.

I can say that the subsequent CTC report does not completely describe the facts. As a matter of fact, I never did see anyone from that Commission out and around the wreck, or at the Revelstoke shops when I was testing the LOCOTROL equipment, and I was there for approximately 3 weeks.

A career change: paradise gained

In early July 1978, Charlie Pike called me to his office and without any preamble said, 'Milt, have you been making inquiries about getting a job?' I looked at him and said, 'What are you talking about Charlie?' He said, 'Bill told me to get you up to the office and find out what's going on. That's why you're here. What I'm talking about is a phone call I just received from Lou Goetz, General Manager Controls Division, Harris Corporation, asking if they could make you a job offer.' I said, 'Charlie, I don't know anything about this. Yes… jokingly, when I have been down to their plant or they have been up here, I have mentioned that if a job ever comes open for janitor or whatever, please let me know, but that's it.' Bill said, 'Okay Milt, leave me with it, I've got to get back to Bill and let him know what you said, then I'll get back to you.'

When he got back to me it was to say that Bill Stinson wished to see me… now. Bill reminded me that I had a good career going with Canadian Pacific, but if I wanted to pursue finding out what they were willing to offer, he would not stand in my way. When I said I would like to hear what they had to offer, Bill said he would call Lou Goetz and let him know that it would be okay to make me an offer.

The next day I got a call from Mr Goetz asking me to come down for an interview. I said I would but that I would have to take some vacation time, so I wanted them to pay for my time lost at CP and send prepaid first-class tickets for two. They did that and booked us into the brand new Ramada Inn on Highway US 1, in Melbourne. The interview would take place in my hotel room the day after we arrived in town, at 1.00 pm. They would be sending over the directors of personnel, program management, and engineering to conduct the interview.

I had never been interviewed before, so this was something new. I wasn't nervous, just curious as to what they had to offer. I certainly wasn't desperate for a job. I already had one, and like Mr Stinson had said, 'Your career is a good one, and will be in the future.'

Well, they didn't show up until 1.45 pm, and I was miffed, to say the least. Where I came from, time was of the essence, and if you're detained, you contact whomever you're meeting to let them know you're going to be late. Things got off to a bad start. At 1.10 pm, having not heard from anybody, I called room service to get some coke and ice because I had a nice bottle of rum in my grip, and as far as I was concerned I would have a couple of drinks, wait for my lady to return, take her on a sightseeing tour, and go out for dinner; and to hell with Harris.

I was just starting into my second drink when there was a knock at the door. When I opened it and before they could say anything; I said, 'You're late, and where I come from, that would require you to make a phone call to advise me.' The Director of Personnel glared at me. The Director of Engineering sought to defuse the situation by saying, 'Milt, may we sit down and get on with the interview?' He seemed like a decent sort. I said, 'Yes, but before we start would you all like to have a drink?' They declined. I said, 'Well, I'll just freshen mine up.' More dagger eyes from Mr Director of Personnel.

They asked many questions. Their language was not the railroad language I was used to, so on occasion I had to ask them what they meant. 'Are you capable of undertaking a turnkey operation to completion?' I told them I didn't know what a turnkey operation was (I'd never heard that phrase before). Someone said, 'Can you pick up a job right from the onset, and see it through to completion? Can you deliver a project where the product or service is ready to use, in a condition that allows for immediate operation?' I said, 'Gentlemen, I have been doing that for most of my railroad career.' We got through the nitty-gritty but when the Director of Personnel suggested employment terms that I was not prepared to agree to, there was some attempted remonstration. I said, 'Look, thanks for your job offer, but let's terminate the interview.'

My wife returned to the room right about this time and, of course, I introduced her to these people. The Personnel man left right away, but the other two remained and invited us to join them for dinner at Ashley's of Rockledge, which in fact sat right beside the Florida East Coast Railroad tracks. Maybe they thought that would soften me up some. We enjoyed the meal and their company. I told them we'd be leaving within a couple of days.

When we returned to our apartment in Montréal there was a message waiting for me to call Mr Goetz. He said, 'The interview panel have recommended you be hired, and we agree to your terms.' Surprised, I thanked him and told him that I'd discuss the offer with my wife and my boss and give him an answer in the next couple of days.

First thing Monday morning, after getting myself rearranged at work, I called Charlie and said I needed to talk with him about the job offer from Harris. He said, 'Fine, come right up, Milt.' I told him about having turned down the job when I was interviewed and then about the phone call from Mr Goetz and that I was now seriously considering accepting the offer. While I was sitting there, he called Bill Stinson and told him the story. Bill said, 'You tell Milt I want to see him in my office right after lunch: one o'clock will be suitable.'

Bill Stinson and I had known each other for quite a while, but I wasn't too sure how he was going to react to my wanting to leave Canadian Pacific. Hell, I wasn't too sure myself how I felt! He reiterated that I'd be leaving a good career that wasn't over yet, for a situation with this electronic company that I really didn't know much about. I said that I felt like I was in a rut, and that I needed to get into something that would broaden my horizons, and that Harris offered that opportunity.

Bill said that he understood and wouldn't stand in my way. He wished me well and told me he was going to put me on 18 months leave-of-absence so that if things didn't work out, I could return to CPR in my present role. That was extremely generous. I'd loved working for this company and had certainly been treated right by them. I said goodbye and phoned Lou Goetz to tell him I was accepting the offer, and asked when he wanted me there.

My career with Canadian Pacific Railway had spanned 28 years. In that time, I'd been fortunate enough to go from apprentice electrician in Regina, Saskatchewan to Special Assistant to the Chief Mechanical Officer – System, at head office in Montréal. As I saw it, the opportunity for further advancement would probably now be dictated by someone's retirement or death, and I was too impatient to wait for that. I was 45-years-old, still full of vim and vigour, and it seemed to me that it was the right time to start a new career with a dynamic company, so the die was cast. I had another 20 years to contribute before the retirement age of 65.

I was to report to Harris Controls at the beginning of the last week of June 1978 so we decided that I should leave right away and get organised in the Satellite Beach condo prior to starting work. My wife would stay in Montréal to sell up, and then join me.

☙❧

Deno's first undertaking at Harris was a series of meetings with a program manager, two engineers, and a design draftsman. The subject was the installation of LOCOTROL 105SS equipment for test purposes on the Iranian State Railway system in Tehran.

It soon became apparent at these meetings why the Harris Controls Division had hired him. They had top-notch people in many disciplines, 'But nobody knew the square root of zip about the intricacies of railroading, or railroad equipment.' This lack of knowledge on their part was clearly displayed when they brought out drawings of the Iranian locomotive depicting where the LOCOTROL equipment would be placed. To their dismay, Deno told them to scrap the drawings because where they intended to place most of the equipment would be disastrous. For instance, you don't put sensitive electronic equipment underneath locomotive radiators which tend to leak. Those boxes weren't waterproof, and even if they were, the additives in the cooling water system would eventually corrode the metal.

When the program manager said, 'Okay... what can we do?' Deno said, 'Nothing right now. We'll have to wait until I get there and inspect the locomotives. Only then can we make decisions on what we're going to do.' They said, 'But we're supposed to be starting work over there on 21 July. Deno said, 'Well maybe we better get over there a little earlier, and we'll need to take a design draftsman with us.

At this point Deno was curious as to why they'd not shown him drawings of the Remote equipment. When he asked for them, they said they didn't have anything because they hadn't got that far yet. Deno advised that was fortunate because there was no way the Remote electronic and air brake equipment could be put on those locomotives. It was too bulky and would have to go into some type of railroad car that he would choose when he got there, and that most certainly was going to add to the project timeframe. Next, Deno asked for the spare parts and installation equipment lists, which would now have to be modified due to having to construct a Remote car installation.

Deno was getting excited. Apart from Cuba, he'd never been to a foreign country before, and Iran certainly sounded exotic.

We travelled British Airways 'coach' (I got that changed to Business or First on all future long-distance flights) to New York, London, and then to Tehran. The Iranians had booked us into the Continental Hotel, which I later changed to the Hyatt Regency.

The next day, the Program Manager and I met with the Minister of Railways and the chief mechanical officer of the Iranian State Railways[59], I explained that in order to proceed I would need to see the locomotive they had chosen to put LOCOTROL on, and also some freight rollingstock that might be suitable to use as the remote-control car. They assured me they would have the locomotive ready to inspect in the morning, and then someone at the shop area would take me to look at some freight rollingstock.

One thing I had learned a long time ago was that if you were going to install new equipment onto a locomotive, the first order of business was to thoroughly inspect the locomotive and test all pertinent systems. You had to be satisfied that the locomotive was in top-notch shape prior to examining it for the installation of LOCOTROL electronics and air brake systems. You did not want to put brand-new equipment onto a defective locomotive. So, the next morning I thoroughly inspected their Class-60 EMD GT26CW locomotive.

It was in terrible shape. Its condition indicated to me that their maintenance program needed some work. I made up a work list that had to be completed before I would conduct an installation on this locomotive. The list produced a lot of commotion: lots of Farsi and lots of hand movement, and then they went off and eventually returned with a gentleman called Ali Ektadar. Speaking English, he introduced himself as the general locomotive foreman of all the workshops in Tehran. He told me that when these locomotives were purchased, adequate schooling on maintenance was not forthcoming from EMD; therefore, they were not in very good shape because the workmen didn't exactly know what to do. I was dumbfounded! I explained my railroad background with CPR in Canada and said I would help them where I could.

I needed a car and driver so I asked Ali if he could recommend someone. He said, 'Yes, I know of a very good agency. I will call them and arrange for someone to meet you at your hotel where you can interview them.' I also wanted an impartial interpreter: someone who did not work for the railroad. I wanted someone who would not

---

59 Now the Islamic Republic of Iran Railways.

be influenced by the railroad hierarchy when interpreting for me. I wanted somebody from outside the railroad industry. Ali said, 'I have a doctor friend who has a daughter—presently unemployed—who has a degree in business administration from UCLA in the US. She speaks and writes excellent English. I shall ask him if she's available.'

Ali arranged for me to meet and interview her at his house. I explained to her that she would have to be available on a moment's notice if meetings or something came up that I needed her for, and whether she was okay with that. She agreed, so we made a deal. Her name was Shorai.

☙❧

So, I began my career with Harris in Iran, training locomotive maintenance workers. Most of them were young, eager, and smart, so things mostly went along very well. It took about a week-and-a-half to get the unit that had been allocated for LOCOTROL installation to an acceptable condition. It required new wheels, new drawbars and knuckles, complete refitting of the air brake system, load-boxing to get the engine up to required horsepower, and miscellaneous other work. It finally tested out acceptably, so now we could begin installation.

In the meantime, I had chosen a fairly new German-made boxcar to be the recipient of the Remote LOCOTROL equipment. I had the workshop ballast it so that the combined weight of the LOCOTROL, electronics, and air brake equipment—along with ballast—would equal the carrying capacity of the vehicle. I knew how much our equipment weighed, so I had them advise me when they'd finished ballasting. Before our installation, I wanted the vehicle weighed to make sure the mass of added ballast was correct (they were using rails). In the event, it proved the car had been over-ballasted, and some of the weight had to be removed.

I had our design draftsman measure up and produce detailed drawings of where our equipment had to be placed plus details of where the Multiple Unit cable had to be run and terminated at both ends so the Remote locomotives could be hooked up to it and controlled. The technician would have his hands full running and connecting a myriad of wires plus pinning all the logic cables.

In the meantime, I was trying to figure out where in the hell we were going to put our equipment on this mini[60] locomotive. The air brake manifold and associated equipment I could fit in the short hood of the locomotive: however, the logic boxes, power supply, and relay interface box were not going to be so easy to locate.

When I returned to the hotel that evening, the concierge said there were two people waiting for me in the lobby. One was the rental car agent, and the other was the driver. We went up to my room where I interviewed them. The driver looked to be 18- or 19-years-old. He spoke enough English, and was extremely polite, so we negotiated a deal. His name was Ashad.

You don't realise until you go to some foreign countries how convenient everything is at home. For instance, without the help of Ali, Ashad, and Shorai, the job in Iran would not have gone along as well as it did. You had the distinct feeling of isolation, because all the TV and radio news reports were in Farsi (which none of us understood), and the hotel rooms did not have phone books so unless you had a list of phone numbers, you were not going to call anybody. Without Ashad or Shorai we wouldn't have been able to purchase anything. The list of ways in which their assistance facilitated the project was lengthy.

This was excellent on-the-job training for me. It became clear that any future foreign job should not be bid solely on the input of the marketing department. The bid should not be finalised until a technical team had undertaken an applications engineering visit to the customer's location. I ardently proposed this to Harris, and

---

60 The export models of American locomotive manufacturers were invariably built to a more modest loading-gauge than that of the US. To Deno, this locomotive apparently seemed to be small. His powers of perception would later be tested to their limits during installation on metre-gauge locomotives in India.

they immediately bought into it. After that, no other project—foreign or domestic—was undertaken until myself and my team had done an 'apps engineering' survey. This saved the company time, and effort, and increased the profit margin considerably.

Back to the job at hand. The next step was to get the proposed test train put together in the railyard, get the Lead and Remote consists connected to it, and do a series of timed air brake function tests, including Emergencys. That's when I learned, to my utter dismay, the significant differences between air brake systems designed and built in North America, and those designed and built in Europe. Brake pipe charging and vehicle brake release times took a great deal more time than I was accustomed to; the results of our tests led to a further delay. The test train was made up of Romanian-built oil tank cars, and unlike their North American equivalents, these had two inch-and-a-quarter brake pipes throughout their length; one of which was never used. I never found out why they had two. These pipes were interconnected at each car and had 18 different fittings throughout their length (I know because I counted them). When I attempted to do a standard air brake leakage test (the maximum leakage allowed on a train being 5 pounds-per-minute), I was blown away... excuse the pun. The brake pipe on this test train was leaking at the rate of 26 psi-per-minute! I met with Ali Ektadar and had him assemble all the car foremen, and carmen into the car foremen's office building along with himself, and Shorai, so I could hold an instructional class on brake pipe leakage limits.

[Author's note: As a North American with a relatively recent introduction to the vagaries of railway air braking, Deno was something of a neophyte. The second pipe of a twin-pipe system conveys main reservoir pressure from the locomotive to maintain auxiliary reservoir pressure on each car. Because the auxiliary reservoirs are recharged by the main reservoir pipe, the brake pipe pressure increase signal is required to only trigger the brake release on each car's 'operating valve' and not to recharge the auxiliary reservoir as well. Thus, a twin-pipe system better ensures adequate auxiliary reservoir pressure for use in brake cylinders and can facilitate a more rapid brake release. A consequence is a greater demand on the locomotive compressor.

Twin-pipe systems are usually found on European air brake systems and invariably have a 'distributor' as an operating valve in place of a triple or control valve. Since fitting a second pipe to every railway vehicle is costly, backwards compatibility is maintained so that vehicles fitted with twin-pipe systems and/or distributor valves can be operated in trains with single-pipe systems and triple valves.

Deno would again encounter this air brake set-up nine years later when he and the author worked together on the Indian Railways Broad-Gauge. Even then, we would ponder why go to the expense and (relative) complication of two pipes to only bother using one. We noticed that most of the Indian Railways South Eastern Division iron ore wagons did not have rubber gasket seals ('O' rings) in the main reservoir hosebag gladhand connectors, so were in no condition to be used anyway.]

While I was waiting for that to happen, I took my pail of suds, a paintbrush, and some chalk, and personally inspected and soaped every fitting on all 60 cars of the train. Every bubbling fitting, I marked with chalk. Of the 1080 fittings on those 60 cars, well over half were leaking in one way or another. No wonder this railroad was unable to assemble a train longer than 20 cars. Because of the excessive leakage, the train brakes would be partially applied all the time. Not to mention the increased cost of brake shoes and fuel.

My instructional class was well-received. Ali and Shorai ensured that everyone understood what I was talking about, and what I expected in the future. Then I instructed them to start work on the train immediately, fix all the leaks I'd identified, and report to Ali when they were finished, at which point I would come down, and test again.

It took the better part of the week to get the leaks fixed. I sent them back twice because the results were unsatisfactory: however, eventually they got leakage down to 1 psi-per-minute, which is quite good. I explained to everybody, including the top brass, that allowing leakage on trains the way I had found on this one was a costly and unsafe practice. I also suggested that a directive should be put out to all terminals that a train test—to include the leakage test—must be done on each departing train, and that if it should exceed the limit of 5 psi-per-minute, the train must be held until the leakage was repaired. This was done, so by the time I left the country all the trains that I'd been on were within the leakage limit. That, I think, was a significant achievement.

***

By mid-September 1978, Deno had completed his static testing and it was time to get some enginemen into the classroom to instruct them in the train handling techniques required to operate LOCOTROL. It was difficult to conduct training and instruction classes when everything you presented and spoke had to go through your translator, but Deno felt Shorai did a good job and he appreciated her ability. Some of the terminology caused 'moments' (for instance, 'angle cock', and even 'transducer') but eventually they were able to negotiate the basics.

Deno informed the Minister of Railways that he would be ready to run a preliminary test train during the first week of October and would require the services of a knowledgeable operational employee as an interpreter. Venturing out on the main line was a whole different story from testing in a railroad yard and Deno needed someone who could communicate his instructions promptly and correctly to the engineman and obtain a coherent response. Shorai certainly wouldn't be permitted in the cab on road trips.

The Ministry advised that the first road trip would be to Qom. This wasn't too taxing a run and it was uneventful, except during air brake releases. Because of the twin brake pipe set-up, this train *'sucked a lot of wind'* from the air compressors in order to recharge, which caused their output to drop dramatically. Air compressors normally cycle between 130 and 140 psi, but when the brakes were released on this double-piped train, *'We knocked her down to about 110 psi.'* Upon returning to Tehran, Deno arranged to have two extra locomotive air reservoirs added to the Remote car, which solved the problem.

The one prior thing Deno knew about Qom was that it had one of the largest and most beautiful mosques in the Middle East. He and the Harris team were keen to see it, but their hosts informed them that a function was taking place there, and it was not a good day to go. The team were not allowed to get off the train. At this particular time, Deno was unaware of the degree of political turmoil that was taking place within Iran concerning dissatisfaction with the Shah, but found later that the railroad people were afraid to allow them off the train in the holy city for fear of being set upon by Shiite enthusiasts. More about this turmoil later.

***

Towards the end of September 1978, I'd been told the railroad had arranged for a short trip with a small train—2000 metric tons—from Tehran to Karaj, just to determine how well everything performed on grades of 1.5%, and through tunnels of 100 metres. Though it was quite small, operation of this train would serve to demonstrate the system's functionality in a mountain and tunnel environment and, importantly, give the Iranians some confidence in it. They had included a stainless-steel, German-built dining car in the train for our gastronomical satisfaction, and they wanted it placed right behind the leading locomotives. I did not want this car destroyed. I said, 'No, it will go on the tail-end of the train.' The trip was uneventful; we made it in about 16 hours. The Iranians seemed pleased with the outcome and said they would begin planning for a longer trip through both desert and mountain terrain.

The dining car was beautiful, with a stainless-steel kitchen that was scarcely used. Instead the cooks had rigged up a 45-gallon drum that had been cut in half longitudinally and provided with legs and a grill top. This was erected in the vestibule at one end of the dining car and most of our hot food was cooked on it over charcoal. Everything was delicious.

෴

Contrasts and challenges in a foreign country: the good, the bad, and the ugly

For a couple of months, Ashad had driven us the same route to and from work, but for the previous couple of days had been taking a different route each day, so I asked why. It turned out there was growing political and religious discontent in the country with the Western-oriented monarchy of the Shah. A couple of Western businesses had been attacked and two American airmen had been killed by demonstrators. So he was taking us on different routes in case somebody tried to target us.

This information got my immediate attention, so on the way home that day, I had Ashad take us to the American Embassy. There, we were told not to be concerned, that these incidents were just the work of a small, radical group that the Shah would soon get under control using his secret police. I was wholly unconvinced and had Ashad take us to the Canadian Embassy, where their explanation was a similar: 'These are just isolated incidents.' Turned out that both embassies were wrong, or they knew what was going on but didn't want to tell us.

The railroad brass had decided they wanted two more qualification test runs through mountainous terrain with numerous tunnels, and through some desert territory. One of these trains would be what they termed a 'regular-sized' train (2000 metric tons) and another would be with tonnage at my discretion. The first test run would be from Tehran, through Daroud to Andimeshk and return, approximately 950 km. The second run would be from Karaj through Qom to Isfahan and return, approximately 730 km.

I didn't think this first train of 2000 tonnes provided a satisfactory opportunity to adequately display the capabilities of LOCOTROL, *so I decided the second train would be greater than twice the size of the first, somewhere around 5000 tonnes. That really got their attention: there had never been a train on the Iranian railway system as large as that. Some of them voiced the opinion that it couldn't be done and was dangerous. I informed them there was nothing dangerous about it, and if they wanted to really move tonnage on their system, I would demonstrate this as an everyday event.* [The author would encounter precisely the same nervousness from Indian Railways personnel some years later.]

We assembled the first train, that also included a dining car and two sleeping cars: one with an observation lounge at one end, so I placed these at the tail-end of the train. I was informed there would be 12–15 senior people accompanying us on this trip, estimated time-away-from-home to be 3 days. I don't know if any of these guys had ever been on a train trip before, but their estimate was way out of whack; we were gone for 6 days (largely attributable to train control issues). The run through the mountains was beautiful, although it was warm: 147°F [64°C]. No doubt the Iranian brass were quite comfortable in the air-conditioned sleeping cars and diner: however, the locomotives and Remote car were not air-conditioned, so we Harris people poured out buckets of sweat.

The return trip from Andimeshk to Tehran was interesting, and a little bit exciting. We were in the mountains and had to stop near a small village to wait for another train to meet us and go by in the opposite direction. Because the meets were all made by train order, sometimes you could wait up to half an hour for the other train to show up, so I got on my portable two-way radio and called up Mac on the Remote to tell him to get off and stretch his legs; I was going to do the same thing. I'd no sooner alighted than on the trail coming from the village appeared about six little girls in their chadors. All six pairs of these beautiful eyes were directed at me. I expect I was probably the first 'gringo' they'd ever seen.

In those days, I always carried a polaroid camera with me, so I asked my Iranian interpreter to hand it down and to tell these little girls I wanted to take a picture of them. I took a picture of the group, and then one picture of each of them separately, and had the interpreter tell them they could keep their individual pictures, but that I wanted to keep the group picture. When they saw their individual photos, their eyes became as big as soup-plates. There was a lot of giggling and they all took off down the trail towards the village. I would've given anything to be there when they showed their parents these pictures. Well, I was in for a surprise. About 10 minutes later, all six of them burst from the trail again, talking ten-to-the-dozen, until presently, the interpreter said to me, 'Their mothers are coming up to see you.' And soon, there they were. Clad, of course, in chador, and their eyes totally expressive.

Two of them were carrying trays, one with a few glasses and a large pitcher of ice water, and the other with some beautiful cut-glass tumblers and a large pitcher of pink liquid, which they indicated was for all of us on the train. They spoke at length with my interpreter who told me they wanted to thank me for taking pictures of their little girls, and if I had time, along with the others, to come back to the village later and share a meal with them. Our schedule was tight, and our interpreter regretfully declined on our behalf. We were, though, very happy to provide them the courtesy of partaking of their pink drink. Our interpreter informed us it was made only in this district and was a special infusion made from roses. It was delicious, and the episode remained a fond memory of this time in Iran.

At the beginning of Day Four it became clear that if our progress remained the same then we'd have a couple more days yet to go, and we'd only packed for 3 days. Then a frightening event occurred. LOCOTROL maintains a continuous security comms loop between the Lead locomotive and the Remote consist every few seconds. This is accomplished by UHF radios located at both these positions. The walkie-talkies we used for voice communications have a channel for this purpose and could also be set to the LOCOTROL frequency. We were climbing a 1.6%, grade travelling at approximately 14 mph, when quite suddenly—level with the right-hand side of the locomotive— appeared an Iranian Air Force helicopter gunship. They were equipped with loudspeakers, and—in Farsi—told us to stop the train. I was somewhat concerned, to say the least: however, the Iranian crewmembers on the train, and interpreter, were greatly alarmed. We stopped on the grade, and the helicopter landed on the rails about 50 feet in front of us. The loudspeakers came to life and told us to get out of the locomotive, and line up in front of it. An officer and three guys with weapons approached us, I took my walkie-talkie out of my pocket to tell Mac back on the Remote what was going on, which drew a burst of Farsi from the officer. My interpreter told me to put the radio on the ground… NOW.

Turned out the frequencies we were using on the LOCOTROL equipment, and the walkie-talkies were frequencies that were apparently assigned to the Iranian military. I suggested to the interpreter that someone be sent back to gather up the appropriate brass, and bring them to the front of the train to explain what the hell was going on, because months before we had told the Iranian Railroad the frequencies we'd be using and for them to get clearance. We had a piece of paper saying that had been done. Obviously, it hadn't.

Three of the big wheels from Iranian Railways spoke to the military officer and that seemed to calm things down a bit. We would be allowed to proceed to Tehran, but the equipment and walkie-talkies could not be used until we had clearance for the radio frequencies.

I had told Ashad, prior to leaving on this trip, that we would return to Tehran in 3 days and that he was to be at the railroad yard to pick us up and deliver us back to the hotel. Well that was all shot to hell, because it was around 7.00 pm on Day Six when we pulled to a stand in the Tehran railyard. There had been no way for me to get a hold of him to advise we'd be arriving later than 3 days, so this kid, being as faithful as he was, parked himself at the freight yard and waited day and night until we eventually arrived. I found out later from a car foreman that he had not left that spot except to go and get something to eat. When he saw me climb down from the locomotive, his eyes lit up, and he got a smile on his face that went from ear-to-ear. I made damn sure that he got compensated for what I considered outstanding faithfulness to duty.

◆

This saga wasn't over yet. Mac and I were exhausted, sweat soaked, unwashed, and I'm confident we smelled like polecats. We both looked forward to getting back to the Hyatt, having a shower, a few drinks and a meal, then a good rest. I told the railroad people we would not be in to work the following day.

When we arrived at the hotel, there seemed to be something special going on. There were a lot of limousines, ladies in long dresses, and men in tuxedos entering the hotel. Whatever it was, Mac, and I were neither sufficiently fragrant nor properly attired for the occasion, and we couldn't have cared less; we had more important things on our mind.

Upon entering the lobby, we had noted a long line of these dressed-up people standing in front of a desk with three SAVAK (secret police) in attendance. They were having everyone empty pockets and purses onto the desk and then be patted down. Well we certainly weren't part of this party, so we headed for the elevator. A loud voice said, 'Where are you going? Come back and get in line.' I had heard a little bit about the savagery of SAVAK from Ali Ecktadar and Shorai. Their advice had been along the lines of 'Don't mess with these guys. They have unrestricted horsepower.' Because several unpleasant incidents had taken place across the country, the Shah was using his secret police to quell those who were in favour of Ayatollah Khomeini taking over leadership of the country, hence those at this function were being checked out.

We finally got to the front of the line and, when queried about what was in my gym bag and briefcase, I opened the briefcase for them to check but mentioned they might not want me to unload the gym bag because it contained a lot of dirty clothing and underwear. It also had a zippered side pocket containing the walkie-talkies and charger that I hadn't had a chance to mention because Mr Efficiency said, 'Dump it on the table including your pockets.' So with exaggerated bravado, I dumped a really stinky pair of coveralls, three pairs of stinky socks, three snotty handkerchiefs, my sweat-soaked Budweiser hat that I wore on road trips for luck, and last-but-not-least, three pairs of unfresh underwear (not to overstate the matter, but the sleeping car had a style of toilet not usually seen in a Florida condo or a 5-star hotel. There was a piped water supply to clean your hand but toilet paper was not part of the equation).

Several of the tuxedo, and long-dress folks in line behind me, backed off a few feet to escape what they could see and their noses were detecting. Mr Efficiency told me to get the stuff back in the bag and to get out of the line. I gave him a big smile, and in no great rush gathered up my belongings and headed for my room.

Once there, I put my dirty clothes into a laundry bag, retrieved the four walkie-talkies and put them in the gang-charger, had my shower (did that ever feel good!), dressed, and headed down to meet the guys for drinks and dinner. Returning to my room later, I was beat and was in the process of getting undressed when there was a loud knock at my door. When I opened it, there were two big guys standing there, one with a gun in his hand pointed at me. They pushed me back into the room, went over and looked at the walkie-talkies, and one of them said, 'What are you doing with these radios in your possession?' With the gun still pointed at me I told them I was there at the request of Iranian Railways, that these walkie-talkies had been cleared for use by them, and that I had just gotten back from a trip and was recharging them for further use.

That didn't satisfy them. They told me to get dressed and that I would be escorted downtown to their headquarters. They hadn't told me who they were, but they sure acted with authority, and I fearfully suspected they were SAVAK. That was chilling. I can't be sure, but I reckon that when the housekeeper turned down my bed while I was downstairs at dinner, she saw the radios and reported that to management, who reported it to SAVAK.

Fortunately for me, the day we'd arrived back on the test train, the railroad's Chief Mechanical Officer had given me his business card, telling me that if anything came up and I needed assistance, to give him a call. I sure as hell didn't want to go down to the SAVAK HQ. I'd heard nothing but bad things about these guys—how people disappeared—so I made one more plea to them that I be allowed to call this senior railroad officer and he could explain everything. The two of them got into a somewhat heated discussion, and finally, one of them said, 'Make your phone call, but when it's answered, you give the phone immediately to me.' After a brief conversation which, of course, I didn't understand, he abruptly handed me the phone. The CMO apologised for what had occurred and said the SAVAK operatives were going to take the walkie-talkies and charger down to their headquarters. Tomorrow, he would send the head of the railroad's communications department down to the SAVAK headquarters, and hopefully resolve the frequency and usage issue concerning both the portables and the main LOCOTROL radios. I said, 'That's good, otherwise all testing will be dead-in-the-water.'

⁕

Resolution was indeed achieved, and it was time to get the final test trip organised. This would complete our contract, and after cleaning up some loose ends we could depart Iran. When the radios and charger had been returned to me, I was told emphatically, 'These radios and the charger must not leave railroad property, and when not in use must be secured by railroad personnel.' I followed those instructions. The radios are still there, though I disabled them before I left.

For the final trip, I assembled a train of around 5300 tonnes that included a sleeping car and the dining car. The Iranians were sceptical about running such a large train. This trip would take us from Karaj to Qom and thence to Isfahan, through desert and mountain terrain, with heavy grades and many tunnels. The round-trip took us four-and-a-half days and this time we took a week's worth of clothes. Everything went according to plan, and the Iranians were impressed and happy with the results.

The next day down at the shop, Ali Ektadar beckoned me to his office where he closed the door. He was not his usual cheerful self. He sat me down and asked when were we planning on leaving Iran. I said, 'Probably in a week, or a week-and-a-half because we still have some clean-up to do.' He said, 'No… you should leave no later than this coming Monday.' This was Thursday, so I said, 'We won't be finished, Ali, so why would we do that?' He replied, 'Mr Deno, please don't question me, just do it!' Because of the political unrest that was happening, and my latest go-around with SAVAK, I decided we were going to follow Ali's instructions; he'd proven to be a trusted friend and co-worker.

It didn't take long to determine something was going on because when we returned to the hotel and sat down with the concierge, I asked him to make flight reservations for us. He said, 'I will do my best, sir, but you may all have to take different flights out of Tehran as there is quite a crush of people trying to get out.' I asked him to book flights for Mac and Joe first, and then mine. They were going back to the States, but I was going to Athens, Greece, to meet my wife and take some R&R. Before I left on the trip, I had negotiated this to occur on Harris' ticket once the job was complete.

My wife was unaware of this trip. I'd wanted it to be a surprise, so I called and instructed her to contact Bob Hamilton, the program manager, and have him get tickets right away for her to fly to Athens, and when she got there to check into the Hyatt; I'd make arrangements for her. I also told her I didn't know exactly when I'd arrive, but for her to get into the hotel and I'd be there as soon as I could.

The concierge had some trouble getting flights, but he did get them, and we gave him a generous tip. Mac was the first to go, leaving in the morning on Lufthansa, flying to Frankfurt and then on to Melbourne, Florida. Joe would be leaving on Saturday on British Air, flying to Heathrow, and then on to Melbourne. Me… well I wasn't as lucky as those guys. I was going to have to wait until Monday morning, and then fly out on Iranian Air to Athens. To say the least I wasn't thrilled about that. I'd have much preferred some Western country's international airline, and to be leaving Tehran before Monday, which Ali had said was the deadline for us to get out. Adding to my concern about leaving was the fact that the concierge informed me I would not be flying out of the international terminal, with which I was familiar, but from the domestic terminal, because they did not consider Athens an international destination. I told Ashad about my departure plans and asked him to take me to the airport. I knew he was reliable and competent and could interpret for me at the check-in desk. Most staff at the international terminal spoke English, but I wasn't too sure how it might work at the domestic terminal.

Over the several months we'd been together, Ashad had mentioned on a few occasions his desire to leave Iran and go to the States. Mac had taken him down to the US Embassy and helped with the forms for a visa application. The kid had been very helpful to us, so we'd decided to slip him some funds to assist him to leave the country. When he finished helping me to get checked-in (and thank goodness he was there), he presented me with a picture of a Bedouin riding a camel carved out of ivory. I thanked him for the gift—which I still have—and then I gave him the money we'd saved up; it was almost 1000 bucks. Tears welled up in his eyes, we hugged each other, and then I had to leave. More about him later.

I was sitting in the departure lounge, and soon realised that I was the only light-skinned, Western-dressed person there. They were calling out flights in Farsi, so each time I saw people going to the gate I joined them, presented my ticket, and was asked to sit down again. I was really frustrated and a little afraid, because I sure as hell didn't want to miss this flight. I must have had a look of desperation on my face, because an Iranian lady came over and asked me if she could help. I could have hugged her. She sat down beside me and when my flight was called, she let me know.

As the aircraft lifted off, I breathed a sigh of relief, and settled in for the flight to Athens. Surprise, surprise: after only being in the air for about three-quarters of an hour or so, we started to descend, and landed, and now I was a very unhappy camper. I'd been told this was a direct flight to Athens, but it sure as hell wasn't Athens. One of the flight attendants told everybody they would have to depart the plane and enter the terminal, with no explanation as to why. An exhaustive inspection of all luggage and whatever was carried in your pockets was then undertaken, one person at a time. I tried to find out why this unscheduled stop and inspection was taking place and received some blank, not-so-friendly stares. When they inspected me, they took my Harris pen, my small Swiss-made Harris pocketknife, two packs of Marlboro cigarettes, and my lighter. As near as I can tell from different signs around the airport, we were in Tabriz. Tabriz was a few hundred miles north of Tehran. At least we were headed in the right direction.

We finally took off and eventually landed in Athens, with me a little the worse for wear. We'd been on the ground in Tabriz for over 4 hours. I had now reached my destination, and was anxious to get together with my *wife, but first I had to settle down a bit. The unscheduled stop in Tabriz, and the pillaging of some of my personal stuff had unnerved me, so when I exited the terminal, I saw a cafe with outside tables. I was going to get a drink. All the tables were occupied; however, there was one with just one older gentleman sitting there, so I asked if I might sit down with him. He replied in German something that I took to mean, 'sure'. I was so relieved to be out of Iran that I was babbling away to him about my trip, and he was saying something in German, and I don't think either one of us knew what the other was talking about. I finished my drink, bought him one, and left for the Hyatt Hotel to meet my wife.*

Thus ended both Milt Deno's and Harris Controls' first overseas LOCOTROL project. It had been a tremendous learning experience but there would be more projects and he'd be challenged again… many times over. I know, I was with him on one of them.

This first project was sufficiently successful that Iranian Railways purchased 40 sets of equipment. Deno was supposed to return to Iran around 1991 and install LOCOTROL II systems, and your author was to participate in the project. Fate, however, intervened. The General Electric Company purchased Harris Controls and decided immediately that, for political reasons, the project would not proceed. As of 2020, the Islamic Republic of Iran Railways has received no further LOCOTROL equipment.

Summary timeline of Deno's involvement with LOCOTROL

1964/65
Southern Railway operate the first ever *multiple remote-unit* train: a 300-car LOCOTROL 103 test train. LOCOTROL at this time was still a Radiation product, and a brilliant young electronics engineer, Dale Delaruelle, was involved. Dale would become a trusted friend of Deno.

1967
Deno commences his first involvement with LOCOTROL: 1½ years of testing on CPR. As Asst District Diesel Inspector,[61] Revelstoke, BC, Milt travels around 85 000 miles testing LOCOTROL 105 on coal trains between Field and Vancouver, living mostly in the remote-control car *Robot 1*—a converted boxcar—in which he had a camp bed and in which he ate most of his meals. Testing also occurred on some wheat trains between Lethbridge, AB and Vancouver.

1969
CP Rail buys LOCOTROL, becoming Radiation's second customer for the product. Milt is appointed Master Mechanic, Revelstoke, BC, to set up the first scheduled LOCOTROL operation.

1972
While located in Montréal as Supervisor of Motive Power – Mechanical, Milt—along with Harold Hayward, retired CPR Chief of Motive Power—is seconded as a contractor to the iron ore hauler, Quebec, North Shore & Labrador Railway to set up a locomotive diesel engine *power assembly* maintenance system at Sept-Iles, Quebec.

---

61 Even in 2012, long into retirement, Deno insisted to the author that CPR had pulled a 'fast one' on him, and that during this period he never knew he was Asst Diesel Inspector. He states he had not been advised of a promotion and thought he was still the Divisional Master Mechanic.

1973

Deno continues with the QNS&L at Sept-Iles, providing LOCOTROL training for operations supervisors.

(Spring) *Cuba:* Deno is sent to the Republic of Cuba on behalf of CP Consulting Services. At various locations he is engaged in repairing and in proposing a recommended repair and maintenance regime for EMD and Alco diesels for the state railway system.

(Summer) *Alaska:* Seconded by CPR to participate in the Canadian government's Arctic Rail Study to examine the feasibility of building and operating a railway to transport oil from the Alaska North Slope to Canada. He spends 1 year on this project, functioning as head of the constituent mechanical group specialising in looking at railroad operational aspects. On completion of this assignment, a senior engineering participant on the study team—Dr Corneil, Dean of Engineering at Queens University, Kingston, ON) strongly recommends to CPR that Deno be enrolled in mechanical engineering at McGill University, Montreal. Deno puts in a year of study but struggles to keep up with his younger student peers. He convinces his superiors to permit him to withdraw and return to railroading.

Figure 45: In 1983, Robot 1—now #1001—is laid up with other Locotrol 105 Robot cars and looks much the worse for wear. Date and photographer unknown - Bruce Chapman Collection. CP Rail Newsletter

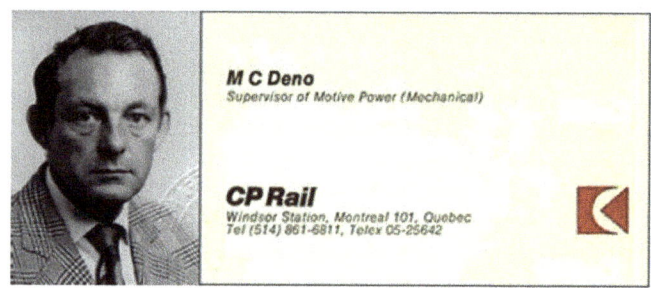

Figure 46: Deno, 1972. Author's collection

1975

(Summer) *USA:* Deno is dispatched by CPR to the Black Mesa & Lake Powell Railroad in Arizona, to produce a report on AC electrification. In 1984, CPR would become the first North American railroad to place (diesel) AC-traction locomotives into service and at the time of writing is believed to have the highest percentage of AC-traction locomotives in North American Class 1 service.

Later in 1975, Deno is appointed Special Assistant to the Chief Mechanical Officer of CPR, based in Montréal.

[Author's note: This book records that by the late 1960s LOCOTROL had been proven in regular operation on several North American railroads. In Australia, Queensland Railways (a 1067mm/3 ft 6in Narrow-Gauge network) had commenced using LOCOTROL 105 in 1972 installed on some of its new 2100-Class diesels (an Australian-built EMD export model) to haul coal on its Goonyella-to-Hay Point line. This equipment was installed by Harris Controls' Australian agent – Evans Deakin Industries, from Brisbane, QLD. The coal network, and the use of distributed power, has since expanded and the motive power is now mostly 25 kV electric. Queensland has several of the largest coal terminals in the world.

Concurrently, in the early 1970s, the Mount Newman Mining joint venture in Western Australia's Pilbara Region was looking at ways to increase iron ore production and railing capacity. In 1973, MNM personnel travelled to Canada to inspect the LOCOTROL operations on CPR's Mountain division and on the Iron Ore Company of Canada's Quebec, North Shore & Labrador Railroad. In 1974, MNM railroad personnel inspected the Queensland LOCOTROL operation. Their report concluded that while there was nothing to prevent the introduction of LOCOTROL onto the MNM railroad, it should not be considered viable until locomotive reliability could be considerably increased and until Remote unit locomotive faults were able to be reset automatically or a Remote throttle step-down feature incorporated.

Following this, a study undertaken for MNM by Canadian Pacific Consulting Services reported there were two basic ways by which MNM could expand its rail capacity to its required 70 million tonnes per year. These were to increase either train payload or the number of trains operated. CPCS advised there were two options to increase payload: by using distributed power or by regrading the railroad.

The first option was chosen and personnel from Evans Deakin Industries installed three LOCOTROL 105 systems. In May 1975, MNM operated its first LOCOTROL test run, with subsequent test runs configured with up to 270 cars and six Alco 636 locomotives. This early equipment was then used only intermittently due to a downturn in business and a diminished necessity for the planned extra capacity.

The burgeoning Chinese market for iron ore in the 1980s eventually saw LOCOTROL II equipment installed on three more Alco units and four new GE Dash 8s, and a regularly-scheduled LOCOTROL operation commenced on the railroad (by then BHP Billiton Iron Ore and now called BHP Iron Ore) early in 1986. The railroad continued to operate a full schedule, moving to the 3rd-generation LOCOTROL III (introducing the ability to individually control Remote groups at more than one distributed location in the train) and nowadays with 4th-generation LOCOTROL IDP equipment (where the system is installed when the locomotive is manufactured and is integrated into the on-board locomotive control electronics.)

1978

Florida: At the beginning of June, Deno finally accepts an offer from Harris Controls to work permanently on the development of LOCOTROL and its installation on client railroads. He figures that after being outside in too many freezing Canadian winters, he's earned some time among the orange groves of the balmy South. He moves to Melbourne, Florida where he and his wife have purchased an apartment on Satellite Beach.

Iran: In late June, Deno travels to the Imperial State of Iran on a 4-month project

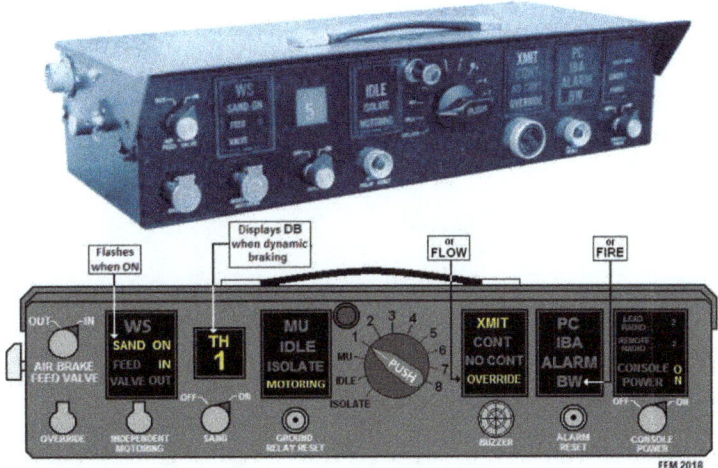

Figure 47: Locotrol 105 control console detail. Note: The upper image is from a Harris promotion brochure. It depicts an anomalous situation, with the Mode Selector at IDLE but the throttle indicator displaying TH

to carry out the installation of LOCOTROL 105 systems for Iranian State Railways. Work locations: Tehran, Andimeshk, Qom, and Tabriz. The overthrow of the Shah occurs while they are there, and Deno and his team only succeed in getting flights out of Tehran because of a timely warning from friendly Iranian Railways staff.

1979

Canada: LOCOTROL installations for CPR, QNS&L, and another iron ore miner, Quebec Cartier Mining.

1980

Japan: Deno spends 2 weeks in Kyoto on technical discussions with Hitachi Ltd.

Mexico: Installation of seven LOCOTROL 105 systems on (then) Ferrocarriles Nacionales de México (N de

M)⁶² GE C30-7 locomotives for operation over the Jalapa Division between the Gulf port of Veracruz (46 ft asl) and Mexico City (7374 ft asl), one of Mexico's first main railway routes and part of the original Interoceanic Railway of Mexico. The previous Helper area for which LOCOTROL is to substitute, is 81 miles of single track between Tamarindo (Municipio de Puente Nacional, Veracruz) and Las Vigas de Ramírez, that features 2.8%

Figure 48: Locotrol II airbrake rack. Harris Controls

uncompensated grades and 12° curves. Deno surveys the proposed route, which is especially contorted between Jalapa (4681 ft asl) and Las Vigas (7868 ft asl) where the railroad crosses the Sierra Madre Oriental range. During testing, Deno described it as "The toughest piece of railroad that I've ever been on in any country. There is an amount of curvature such as I've never seen anywhere else. During testing, the emphasis was on braking, with full stops on the steepest grades and sharpest curves. We were able to start the train from a stand in the 4th throttle position." The 5000-plus-ton trains were configured with 3 Lead-consist units and 4 Remotes. At the time of installation, N de M were expecting government funding for a reconfiguration of this tortuous route, and viewed LOCOTROL as a temporary solution. Milt is not convinced the permanent way is to an acceptable standard for regular LOCOTROL operation.

Work locations include Mexico City, Veracruz, Jalapa (also spelt Xalapa, it is the 'birthplace' of the jalapeño pepper), and Las Vigas de Ramírez.

1981

(Spring) *Algeria:* A 5-month installation project of LOCOTROL 105 for Algerian Railways (SNTF). Work location: Algiers.

1982

Morocco: A 5-month 'applications engineering' and installation project for LOCOTROL II systems for Moroccan National Railways (ONCFM). Work locations: Casablanca, Rabat, Khouribga. These were the first concept LOCOTROL II systems installed, and they included Deno's *A38* flow adapter invention—utilising two A19 adapters in series.

1982

Canada: Back-to-back testing of LOCOTROL II systems for CPR at Alyth Diesel Shop, Calgary, including over-the-road testing and instruction. CPR themselves carried out the installation of this equipment.

1983

Australia: Installation of LOCOTROL II systems for iron ore miner Mount Newman Mining (Note: MNM already had two LOCOTROL 105 systems, resulting from a consultancy visit during the late 1970s by Charlie Parker, a senior motive power engineer with CPR). Work locations: Port Hedland and Newman, Western Australia.

Canada: Testing of LOCOTROL II systems for CP Rail. Locations included Alyth Yard (Calgary) and the

---

62 Since privatisation of the Mexican freight railroad system in 1995, Kansas City Southern de México (KCSM, previously Grupo Transportación Ferroviaria Mexicana, TFM), is fully owned and operated by US Class 1 railroad company, Kansas City Southern, through a concession from the Mexican government. In 2005, KCSM assumed operation of the Veracruz-to-Mexico City line after absorbing the Ferrocarril del Sureste (FSRR, commonly known as Ferrosur).

main line (Note: Milt's recollection of this project is a little hazy. He believes he was testing the 'segmented brake pipe' system). In 1992, GE Harris and Knorr-Bremse developed a Segmented Brake Pipe LOCOTROL system for CP Rail (patent credited to Steve Heneka, Gene Smith and one other). The system was designed to operate with the brake pipe of the front section of the train disconnected from the Remote locomotive such that the train had two separate brake pipe sections (segments), rather than it being continuous. In operation, coordination of the two brake pipe sections was accomplished by Segmentation Units located on the Lead and Remote locomotives and on a detachable device—that could deplete the brake pipe but not charge it—mounted on the last car of the front section. The Segmentation Units communicated with each other via the LOCOTROL radio link so that their responsive actions emulated a continuous brake pipe. The design intent of the system was to eliminate the brake pipe connection between front and rear sections of a

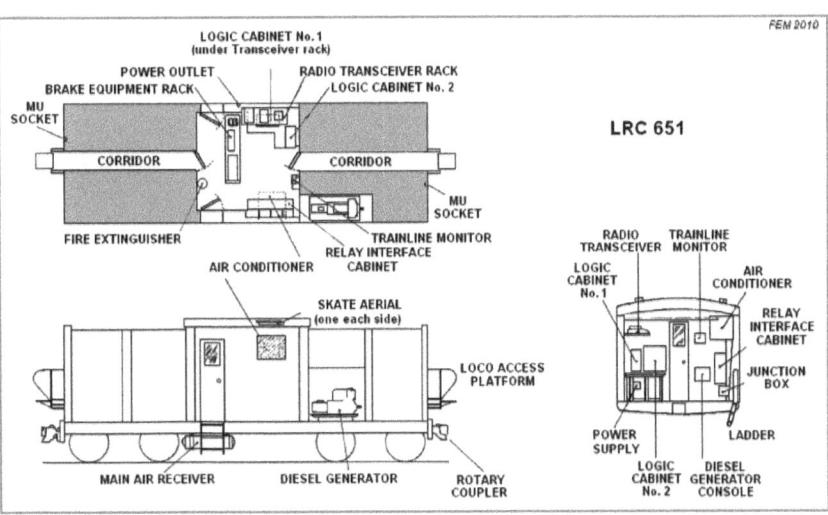

Figure 49: Mt Newman Mining Locotrol 105 Remote Control car LRC 651. Artwork by the author

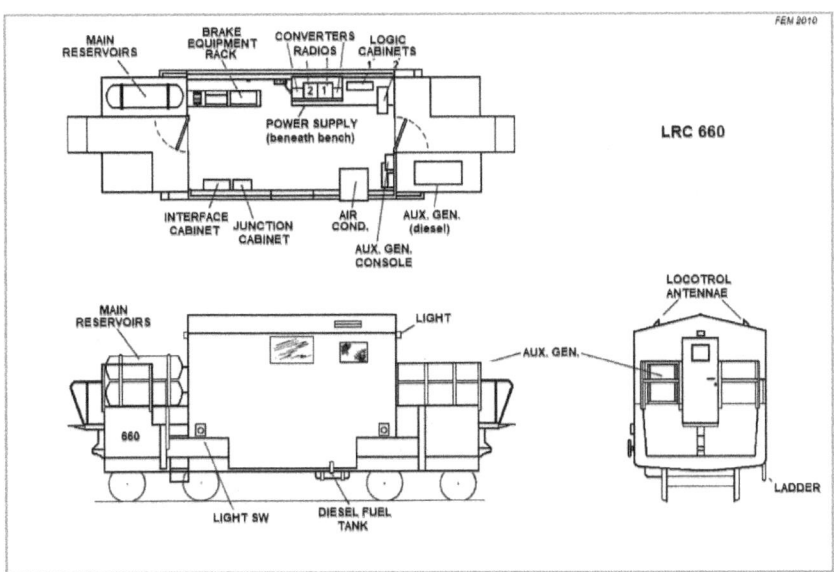

Figure 50: Mt Newman Mining Locotrol II Remote Control car LRC 660. Artwork by the author

LOCOTROL train, and thus also the need for railroad personnel to intervene physically in coupling sections together to form such a train. In turn, this would save time in building a LOCOTROL train. The testing failed to sufficiently prove the concept and Milt told CP they had to re-pipe the locomotives.

1984

Morocco: Further LOCOTROL II installation.

1984/85

Australia: Installation of LOCOTROL II systems—adapted for AC electric locomotives and the Davies & Metcalf P85 air braking system—for Queensland Rail. Work locations: Gladstone for ASEA locomotives, and Jilalan (Sarina) for Hitachi locomotives.

Figure 51: Mt Newman Mining Locotrol II installation, Port Hedland, Western Australia.

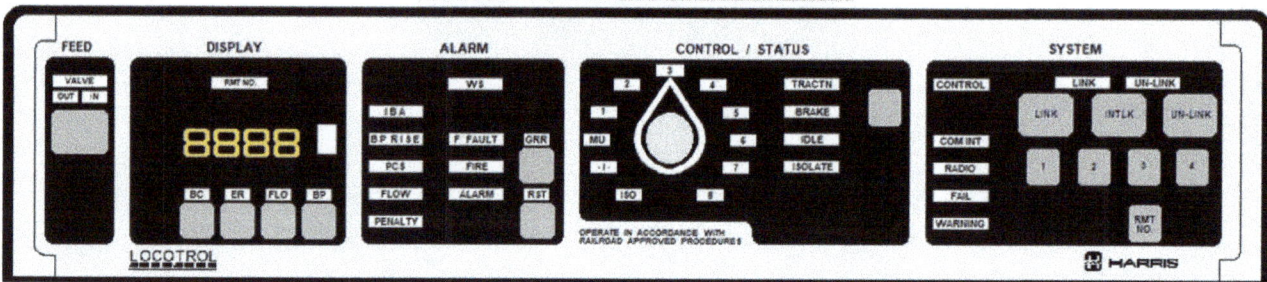

Figure 52: Locotrol II Control Console

Figure 53: Mt Newman Mining Locotrol units and LRC cars, Locomotive Service Shop, Port Hedland WA. Tom Winterbourne, author's collection

Figure 54 (left): Mount Newman Mining CM39-8 engineer's control station. Shows the Locotrol II Control Console integrated into the control stand, and Air Brake Console located adjacent to the conventional air brake controls. Author's photo.

Figure 55 (right): Mount Newman Mining CM39-8 engineer's control station. Shows the Locotrol II Air Brake Console located adjacent to the W30 air brake handles. The W30 is an Australian adaptation of the 30-CDW that splits and co-locates the Automatic and Independent portions. The pushbutton with a blue lens provides Quick Release (bail-off). A black rotary knob, almost invisible on the vertical face of the console, to the left of the ashtray, is the Hump [slow-speed] Control. Author

1986

India: Preliminary sales and technical discussions for LOCOTROL II with Indian Railways. Work locations: New Delhi (diesel) and Varanasi (electric).

China: Preliminary sales and technical discussions for LOCOTROL II with the Ministry of Railways of the People's Republic of China (now China Railway Corporation). Work locations: Beijing and Dalian.

1987

Australia: A 6-month installation project for LOCOTROL II systems applied to new GE Dash 8 locomotives, and crew training for BHP Iron Ore (ex-MNM). Work locations: Port Hedland and Newman, Western Australia. The author was a locomotive crew foreman at this time and was assigned by Barry Green, the acting railroad manager, to work closely with Milt in designing crew training material. Due to a bitter industrial dispute at the time, the author was banned by the union from riding on Lead units and providing instruction to crews. Before Milt departed following this project, he asked the author to join his team in India for a forthcoming

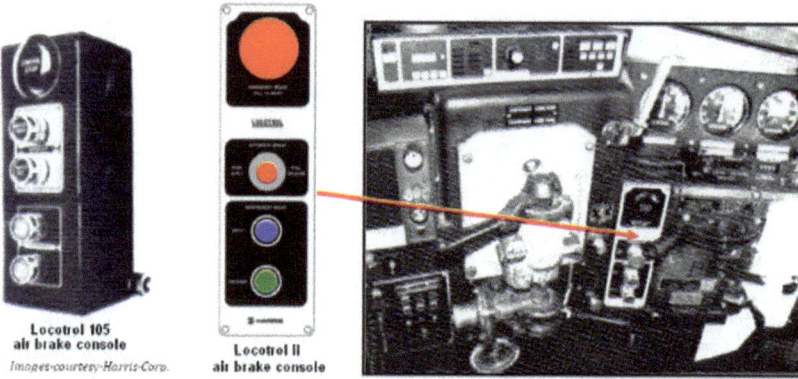

Figure 56: Comparison between Locotrol 105 and Locotrol II Air Brake Consoles. Note that Locotrol II has one button (Push/Pull) for Automatic brake functions. The image shows this latter console installed in an EMD locomotive control stand. Author

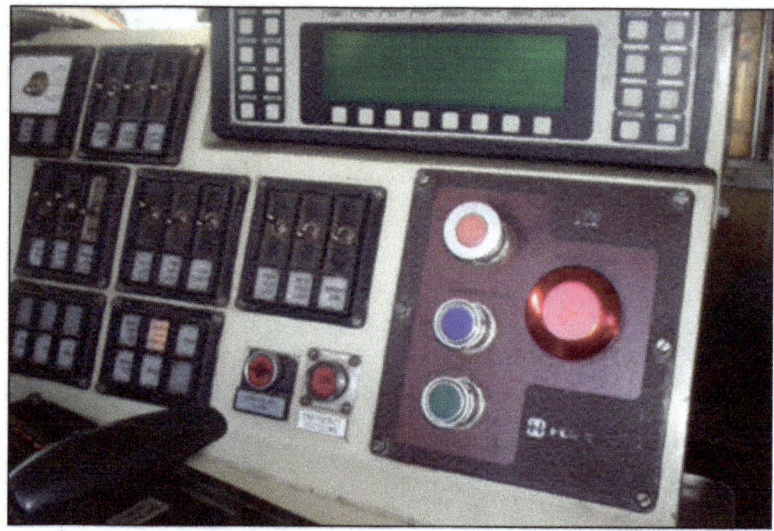

Figure 57: This photo depicts the 'space-saving' version of the later Locotrol Air Brake Console, in this case installed with the Locotrol III system on a narrow-gauge GE Dash 8 locomotive in Western Australia. Nowadays, with modern EPIC and CCB equipments, DP air brake control is incorporated into the locomotive brake controller (engineer's brake valve), making redundant the requirement for any separate Air Brake Console. Author

installation project. In fact, the official request was made to Andrew Neal, Railroad Manager, who released the author on 4 months' leave of absence.

1988

Canada: Installation of LOCOTROL III systems (including Tower Control) for CP Rail. Work locations: Calgary, AB, Roberts Bank, BC and en route. Testing of *fully-automatic* train operations (using Probe[63] and LOCOTROL) was also conducted between Airdrie and Didsbury, and between Calgary and Red Deer, AB. The Red Deer Subdivision—hard against the foothills of the Rocky Mountains—was chosen for these tests because of the undulating topography and because non-stop runs could be arranged. CP Rail participants included Don Johnson and VP Operations – Pacific Region, Jim Gheddis. Milt relates that the prototype system worked but was not completely successful due to problems with software algorithms. Development of the technology did not progress due to economic conditions then extant. Jim Gheddis, however, was reported to be extremely pleased with the outcome of this testing.

---

63 'Probe' was a computer designed by Harris Communications & Information and was tested by CPR as a data logging system installed in the cab of a locomotive. The device scanned various engine, alarm, and control parameters through a series of sensors located throughout the unit. Data concerning a locomotive's performance over a given time, distance, and stretch of track could be collected, stored, then transferred to a lab for analysis. The information reported the locomotive's previous performance and indicated faults or potential problems.

Figure 58: Milt and Gene Smith pose during Locotrol III back-to-back testing at Alyth Diesel Shop Calgary, in 1988. Author's collection, courtesy Dale Delaruelle

Figure 59: Milt conducting Locotrol III system testing aboard an SD40-2 at Alyth Yard, Calgary, in 1988. Author's collection, courtesy Dale Delaruelle

Figure 60: Westbound Locotrol III test train exits the Mt MacDonald tunnel, Revelstoke Div, 1988. Author's collection, courtesy Dale Delaruelle

Figure 61: Milt conducting Locotrol III system testing aboard an SD40-2 at Alyth Yard, Calgary, in 1988. Author's collection, courtesy Dale Delaruelle

Figure 62: Milt is photographed on the Lead unit (where he always was during over-the-road testing) of a westbound Locotrol III test train, having exited the Mt MacDonald tunnel. Author's collection, courtesy Dale Delaruelle

Figure 64: Locotrol III test train westbound on Notch Hill BC, in 1988. The east-bound descending track is adjacent. Author's collection, courtesy Dale Delaruelle

Figure 63: Gene Smith on the Remote unit (where he always was during over-the-road testing) and a colleague, outside the west portal of the Mt MacDonald tunnel, 1988. Author's collection, courtesy Dale Delaruelle

India: Wayne Barber (NYAB), and Gene Smith and Steve Heneka from Harris accompany Deno to New Delhi, Lucknow, (Uttar Pradesh), Rourkela (Odisha), and Guwahati (Assam) to commence applications engineering for LOCOTROL II on wide-gauge (air-braked) and metre-gauge (vacuum-braked) locomotives.

1989

India: A 6-month project installing and conducting back-to-back testing of LOCOTROL II systems for Indian Railways. Work locations: Calcutta (now Kolkata), Rourkela and Guwahati. The author was involved for the final 4 months, conducting classroom instruction of engineering staff, crew supervisors, and locomotive drivers. This role also included over-the-road instruction for crew supervisors and drivers. During the project, Milt sustains a minor broken-skin injury in the office at the Bondamunda Diesel Depot, Rourkela, and subsequently falls ill with a potentially life-threatening medical condition caused (the doctors said) by the deposition of mosquito-borne larvae into the wound. His treatment in Ispat General Hospital in Rourkela is ineffective and Milt is aero-evacuated to New Delhi, where he is hospitalised for several weeks. Following discharge, Milt flies to Scotland for a brief rest with his daughter before returning to the project. Gene Smith and I continue in Milt's absence.

1990

USA: Having held discussions with NYAB at Watertown, NY, Milt is in Syracuse to catch a flight to Montréal when he becomes ill with atherosclerosis and is taken to University Hospital where he spends 4 days undergoing angioplasty and recovery. Later, Milt flies to the UK for discussions with Davies & Metcalf regarding air brake manifolds for a forthcoming LOCOTROL installation project for the Chicago & North-Western Railroad. While at D&M, Milt suffers an unexpected partial paralysis due to a previous neck injury. He returns quickly to Florida to have a titanium plate surgically inserted and to undergo a second angioplasty procedure to correct complications arising from the previous one.

1991

Canada: A 2-month project testing the Adaptive Air Brake Control System[64] with BC Rail. Work locations: Vancouver and Prince George. This is not strictly a LOCOTROL project.

1992

Canada: Installation of LOCOTROL II systems for QNS&L and BC Rail.

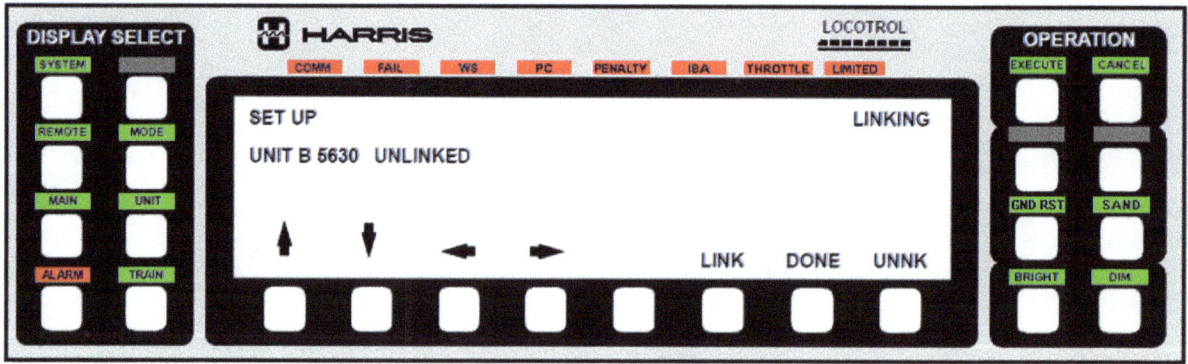

Figure 65: Locotrol III Control Console. Artwork by the author.

---

64 See Chapter....

1992

Australia: Applications engineering for LOCOTROL III systems (including Tower Control) for Robe River Iron Associates. Work locations: Karratha, Cape Lambert, and Pannawonica, Western Australia.

1992

(May) *United Kingdom:* Milt is asked by British Rail to trial the operation of manned distributed power of aggregate unit trains for Foster Yeoman Ltd (Mendip Rail) utilising voice radio between the Lead and Remote locomotive engineers. He warns against the proposal, but BR are resolute. A 115-car train weighing just under 12 000 tonnes is built at the Merehead Quarry, with two EMD Class-59 locomotives—one on the head-end and the other 58 cars back. Precise coordination of control actions by the drivers on the two units is not achieved—as Deno has predicted—and the train parts ahead of the Remote unit only a few miles into the run, pulling out the complete coupler assembly. Deno is furious and abandons the test run, climbs off the locomotive, and hitchhikes 15 miles back to his hotel. A few days later he flies out for Florida. The trials are abandoned.

1992

USA: Installation of LOCOTROL II systems for Conrail. Work location: Altoona, PA.

1993

Iran: Applications engineering and tests for a tunnel repeater system. An evaluation of Islamic Republic of Iran Railways following the war with Iraq is undertaken and a test run is made from Tehran to Andimeshk. Iran has some extremely rugged railroading territory. IRIR has recently taken delivery of a fleet of EMD and GE locomotives that have been furnished with LOCOTROL II equipment. Due to the threatening nature of contemporary Iranian politics, GE—the new owners of Harris's LOCOTROL business—elect to cancel the project. The locomotives will never operate as LOCOTROL units, and the author—about to sign a contract with GE-Harris for a 12-month participation in this project—is blindsided. He eventually secures contract railroad work in Australia with a labour hire company.

1993

Mexico: Deno returns to Veracruz; N de M wish to operate LOCOTROL trains over the 3% mountain grades (see '1980' on pages 97 and 98). Milt notices the remains of numerous train wrecks and the inferior permanent way condition, and rejects it for heavy, long-train operations. The project does not proceed.

1993/94

USA: The Burlington Northern borrows some LOCOTROL II units from another railroad and Milt is asked to officiate on over-the-road trials and instruction in mountain grade territory. There is a clash with operating supervision who have their own ideas. At one point, the railroad's private office car—marshalled within the train (against Milt's advice, as he doesn't believe it is of sufficiently robust construction to be cut into the body of the train)—is subject to excessive in-train draft forces and is 'stretched'[65]. Milt judges the current intentions of BN to be 'unsafe' and the trials go no further.

---

65   Deno states this was immediately apparent from the appearance of disjointed internal panelling and from flooding caused by the separation of pipework in ceiling-mounted plumbing installations.

Chicago Northwestern RR. Installation of LOCOTROL II systems for Powder River Basin coal trains and over-the-road testing and instruction.

Union Pacific RR, Eugene, OR. New GE Dash 8 locomotives have an air brake fault that is incompatible with LOCOTROL. Naturally, GE are sceptical, and some time is expended by Deno and others in order to prove their point.

1994

USA: General Electric Transportation. Milt undertakes 'integration' testing of two new-build locomotives equipped with Harris LOCOTROL. Work location: Erie, PA. The purchase by GE of Harris Controls is imminent; GE has decided to produce all new-build locomotives as 'LOCOTROL-ready'.

(May) Union Pacific RR. Milt says he was 'all over the property', instructing staff and running over-the-road trips. Work locations: Pocatella and Nampa, ID and Hinkle, OR.

(June) *Canada:* QNS&LRR.

(July) *USA:* GE/Harris integration. Work location: Salt Lake City, UT.

(August) Southern Pacific RR, Carbarton, ID. Participants: Mike Wall, Butch Vermilion, G Primmer, T Wells. [Author's note: As he did with Eugene Green from the N&W railroad, Milt considers Vermilion to be an outstanding Road Foreman of Engines. In conversation with the author, he will speak often of these two railroad officials.]

1995

Australia: Installation, over-the-road testing, and crew instruction for LOCOTROL III systems (including Tower Control) for Robe River Iron Associates. Work locations: Karratha, Cape Lambert, and Pannawonica, Western Australia. The author—who was a contract locomotive driver for Robe River at the time—participated in the instruction of mine load-out personnel (who were not railroad staff) in the operation of the Tower Control system for loading trains.

The application of LOCOTROL by Robe River was notable for several reasons. Firstly, it was used for rear-end remote-control only—to control the banker consist (pushers) that had until then been crewed—and secondly, it included the first (and thus far the only) application of an automatic set-off system for the rear-end Remote unit once the train had reached the summit en route and rear-end assistance was no longer needed. The project also included the first (and at the time, the only) application of Tower Control for mine load-out operations.[66]

Robe's 200-car ore trains required assistance to climb out of the Mesa-J mine site and train-loading yard in the Robe River valley and run the 100 or so kilometres to the summit of the Chichester Range at Siding Two. At this point, the loaded train met an empty returning to the mine, so Robe River railroad officials conceived the idea of cutting off the unmanned banker consist as the loaded train ran through on the main line. This would require the creation of some new software logic and protocols to effect control of the whole sequence. Robe questioned Harris as to the viability of this scheme and Harris said, 'Yes… you get us a self-closing, pull-apart air hose set-up and we'll make it happen.'

This cut-off of the unmanned trailing locomotive group was made at the summit of the Chichester Range, after which the train completed its mostly-downhill run to the port at Cape Lambert. The cut-off was made as the loaded train ran down the main track at Siding Two, and the trailing locomotive group (usually a double-unit)—with a self-sealing brake pipe connection—came to a stand, not in Emergency, but with an

---

66   In more recent years, Tower Control has been installed at several of BHPIO's mine facilities to facilitate train load-out.

Independent brake application and was automatically secured. An opposing empty train was in the passing track, and its engineer walked across to the detached locomotives, moved them back and then in through the mainline switch to couple onto the head-end of his train. The empty train now continued to the mine where its two leading units—formerly the detached unmanned bankers—formed the head-end power for the next loaded to depart and the second two units became the bankers.

The 'detach' sequence was initiated by the engineer on the loaded train. A button on the Control Console was pushed and the coupler lifting lever automatically activated, the trailing Remote locomotive group throttled back to Idle, and—as the loaded train pulled away—the Remote banker consist came to a stand under an Independent brake application.

Upon completion of the installation project, Deno was asked by the railroad if he could arrange a run that would showcase the functionality of full LOCOTROL III operation and perhaps set a new world record for a long/heavy train. Ever happy to help, he builds a 350-car, 45 500-gross-tonne test train powered by seven GE Dash 8 locomotives. This enormous train (3.8 km/2.36 miles long) was configured as 2 + 120 + 2 + 115 + 2 + 115 + 1, and on 18 March, RRIA engineer Glen Davies—with Deno and a railroad supervisor on board—ran it from Mesa J to Cape Lambert, taking 4 hr 31 min at an average speed of 39.74 km/h for the roughly 200 km (124 miles). Milt believed this train, as of 2012, was the longest *successful* distributed power train in the world, where the 'success' of a test—in Deno's view—was always defined by the train completing its entire journey without parting en route.

In 2001, the RRIA railroad was merged into the Hamersley Iron system to become part of Rio Tinto's Pilbara Iron network. Hamersley Iron—who had always sought to do things their own way in heavy-haul railroading, and whose corporate railroad management had maintained a refined culture of rivalry with their neighbour up the coast—had never followed Mount Newman Mining by embracing distributed power and weren't about to start. Robe's DP operation and its associated automatic banker detachment was soon terminated and dispensed with. Pilbara Iron is a conventional head-end power operation, and—as of July 2018—they have commenced the automatic operation of unmanned ore trains[67], with oversight from the railroad operations control centre 1500 km distant, in Perth, WA.

(April) *USA:* At Wabco, Wilmerding, PA. Testing of EPIC brake with LOCOTROL III to resolve interface problems with GE's Integrated Function Control.

(June) As above, General Electric at Erie, PA.

(July) The Harris Controls' LOCOTROL business is purchased by GE, who install an idealistic young engineer, Don Herndon—who likes to play his guitar in the office—as VP Railcar Systems. Evidently deciding that the fresh viewpoint of a new crop of GE engineers will be of greater benefit to the organisation than the collective corporate knowledge of long-serving Harris staff, Herndon proceeds to purge the Melbourne plant of most of those with a long-term involvement with the product. Many resign or retire. This utterly offends Deno's long-held perceptions of company loyalty. He is not prepared for such brutal, new-age reality and considers it an act of corporate bastardry. Concerned for the welfare of many of his friends and respected colleagues, he is aghast and confronts Herndon. Milt is advised his job title is to be changed from Railroad Product Line Manager to Field Service Engineer and he is to have no further contact with customers. The practicalities of the business soon result in this edict being moderated.

(November) Union Pacific RR. Commissioning GE locomotives in Denver CO. Resolving EPIC brake problems.

---

67   Rio's technology is called 'AutoHaul'.

1996
(March) Commissioning GE locomotives as above.

Canada: CP Rail. Testing LOCOTROL system air flow sensitivities to make comparisons for Southern Pacific RR.

1997
(September) Tired of watching talented and effective people with whom he has long worked being discarded, and contemptuous of Herndon's manner—which he considers unprofessional, unprincipled, and incompetent—Deno retires in utter disgust.

1997
(October) Settling into retirement, Deno is contacted by a colleague at the Melbourne plant and prevailed-upon to go to Brazil as part of a current installation project. He is not disposed to assist, but after some serious negotiation, relents. Deno drives a hard bargain, which includes that he will not occupy the same room or corridor in the plant at any time with Herndon. From now on he will only interact with this colleague. Milt goes down to Rio de Janeiro to 'have a look'. LOCOTROL has been installed on the Metre-Gauge Estrada De Ferro Vitória a Minas (EFVM[68]) but is not operative.

1997/98
Brazil: Technical discussions with EFVM. Work locations: Rio de Janeiro and Vitória. Two months.

1998
(January) *USA:* Milt meets with GE to discuss what they are doing in Brazil.

(February) *Brazil:* Installation of LOCOTROL III systems for EFVM and testing the interface with GE locomotives. Location, Rio de Janeiro. EVFM operate 17 000-ton Metre-Gauge iron ore trains and intend to utilise LOCOTROL to increase train mass to 25 000 tons in a 2 + 2 configuration with 240 cars.

Canada: CP Rail. Assisting GE with back-to-back testing.

Germany: Site tour and technical discussions with Deutsch Bundesbahn and Knorr-Bremse. Work location: Munich.

Figure 66: GE Dash 8 Remote units on an EFVM (Estrada De Ferro Vitória a Minas) iron ore train. RRPictureArchives.NET, Leandro Ribeiro Barbosa

1999
Germany: Preliminary tests for operations and testing of the Knorr-Bremse airbrake manifold. Work location: Munich.

---

68 EFVM is administered by Vale [*Vah*-lay], SA, formerly CVRD (Companhia Vale do Rio Doce).

Brazil: Check on status of operations.

Germany: Installation of LOCOTROL III systems for Deutsch Bundesbahn. Work locations: Munich and Cottbus.

Following this project, Milt decides that he has had enough travelling, and retires for good. In 2013, I was very pleased that my nomination of Milt for the Canadian Railway Hall of Fame was accepted, and in 2014 he was inducted into the 'Industry Trailblazers' division[69]. It was a great pity that Milt was not well enough to travel to Ottawa for his induction ceremony.

Milt resided at his tidy home in Croton Woods, a residential area of Brevard County, Melbourne, Florida. Over the ensuing years, his health slowly deteriorated, and he passed away on 7 April 2019, aged 86. The author is grateful that he was able to visit him in Florida, three times during this period. I am indebted to his daughter Barbara, who lives near Calgary, AB, for the invitation to accompany her and her husband into the mountains to Field, BC—Milt's first supervisory location—to release his ashes to the rails in September 2019. (see Dedication page at the start of the book).

---

69   Visit the Canadian Railway Hall of Fame website (http://www.railfame.ca).

3

OPERATING Locotrol TRAINS ON CP RAIL

Further development

In December 1971, Charles W Parker, CP Rail's Assistant Chief of Motive Power & Rollingstock, presented a paper at the AAR Conference on *Track/Train Dynamics Interaction* in Chicago, discussing CPR's experiences operating LOCOTROL unit trains. In a January 1974 article published in *Railway Engineering Journal* (Institution of Mechanical Engineers – Railway Division), Parker further described numerous aspects of the operation of remote-controlled locomotives on CP Rail freight trains. He stated that whatever the reasons for a railroad using remote-controlled distributed power, the attributable cost had to be paid for in some practical way, perhaps with faster scheduling, perhaps through the better utilisation of equipment, or perhaps by crew wage savings resulting from the increase in train size.

From its use of the technology through British Columbia, CP Rail realised gains by being able to increase their maximum coal train tonnage over the ruling grade from 7500 to 12 300, and the time for the 1400-mile (2250 km) round trip to-and-from the Pacific Coast halved from 6 to 3 days. The main operating challenge faced by CP Rail in the mountains of Western Canada was the limiting factor of drawgear strength and the resultant restriction on the size of the head-end locomotive group. To meet this challenge, their practice was to operate manned Pushers on grades such as Beaver Hill on the eastern slope of the Selkirk Mountains. Here, there were a succession of adverse grades spaced just far enough apart that the same Pusher consist could not be retained in the train to assist on all of them. This reality revealed the economic advantage of using an unmanned remote-controlled locomotive in the train instead of a sequence of manned Pushers and the infrastructure they required, and the 'dead' running that resulted.

In his conference presentation, Parker delivered a case study on CP Rail's unit coal trains operating from south-eastern British Colombia to the Pacific Coast export coal terminal at Roberts Bank near Vancouver. These trains—operating through the heavily-graded Rocky and Selkirk Mountains—ran the 1400-mile round trip over 72 hours, utilised thirteen 3000 hp units, and weighed just under 14 000 gross tons with 105 cars. The motive power was separated into three power groups: a four-unit Lead consist (head-end), a five-unit Remote consist (mid-train), and a four-unit manned Pusher consist (hind-end).

Over 4 years of experience with LOCOTROL, CP Rail had developed significant expertise in heavy distributed power unit train operation using remote-control. Throughout this period, the railroad experienced a variety of train handling problems that were treated analytically to accumulate data on how best to marshal such trains and handle them over-the-road. This chapter draws on the comment of the late Mr Parker and the recollections of Milt Deno arising from his experience with these trains and the testing and experimentation that occurred.

If it can be acknowledged that the Southern Railway and other south-east US railroads played the lead role in developing prototypical distributed power systems—culminating in the emergence of LOCOTROL as the pre-eminent product—then it must be acknowledged that, as the first major production customer to intensively use it, CP Rail can be credited with having played the principal role in the ongoing development of successful operating strategies for the everyday use of LOCOTROL over challenging, heavy-grade terrain.

The operation

CP Rail had been operating regularly scheduled unit trains[70] and experimenting with LOCOTROL since 1967. In mid-1970, they operated their first unit coal train (88 cars/9000 tons) from Sparwood, BC to Roberts Bank, BC. A typical such coal train would depart the mine at Sparwood hauled by four 6-axle units operating conventionally[71]. The train then descended the western slope of the Rockies to the Kootenay River valley and proceeded north-west along the Kootenay and Colombia rivers to Golden where it would join the CP Rail main line.

At Golden, a five-unit Remote consist was inserted about 50 cars into the train, and the now nine-unit train continued to Beavermouth where a four-unit manned Helper consist was cut in 25 cars from the hind-end to assist with the ascent of the 2.2% (1-in-45) Beaver Hill to the summit of the Selkirk Mountains. Parker stated that in the ascent of Beaver Hill, it would not have mattered if the positions of the Remote consist and Helper units were interchanged; indeed, this would have positioned the Remote consist in a more conventional location within the train. However, once the train reached Stoney Creek at the summit of Beaver Hill, the characteristics of the track profile (including the location of the east switch relative to the grade) made it desirable to have the Helper marshalled behind the Remote consist to facilitate its safe removal. Although this configuration resulted in the Remote consist being located within the forward half of the train it was found to have the advantage in the subsequent journey of promoting improved radio communications between Lead and Remote consists.

With the steep descent to Revelstoke disposed-of, two of the Remote units were removed from the train, leaving seven units to handle it over two more 1.4% (1-in-70) grades to Chase, where two Lead units were set out, leaving a five-unit train to continue to Roberts Bank. These trains were mostly powered by 6-axle Alco-MLW 630 units equipped with a unique Dofasco truck and Barco-Summit wheelslip control. In territory where adverse weather and rail conditions tended to promote wheelslip, these units were very sure-footed. Parker stated that in 1½ years of coal-train operation there had not been a single recorded occurrence of a broken coupler knuckle on these trains. This had not been the case during the test program of 1968/69 when solid trains of grain, sulphur, potash and coal were handled by SD40 units in trains of progressive mass, eventually exceeding 14 000 tons on the 2.2% (1-in-45) grades and 18 000 tons over more benign topography. During this test period, about 20 train separations occurred under various circumstances, the critical analysis of which led to the formulation of a marshalling theory and train-handling strategies that materially assisted in the later successful operation of these trains.

Positioning the Remote consist in the train

Since the earliest days of testing and experimenting with LOCOTROL, the North Electric Company, Radiation Corporation, and then Harris Controls had examined innumerable aspects of the system. These included—for Harris Corp—the cooperation of the conveniently located Florida East Coast Railroad, whose assistance with

---

70 3700-ton trains of sulphuric acid from Sudbury to Sarnia, ON.
71 As conventional (non-DP) trains.

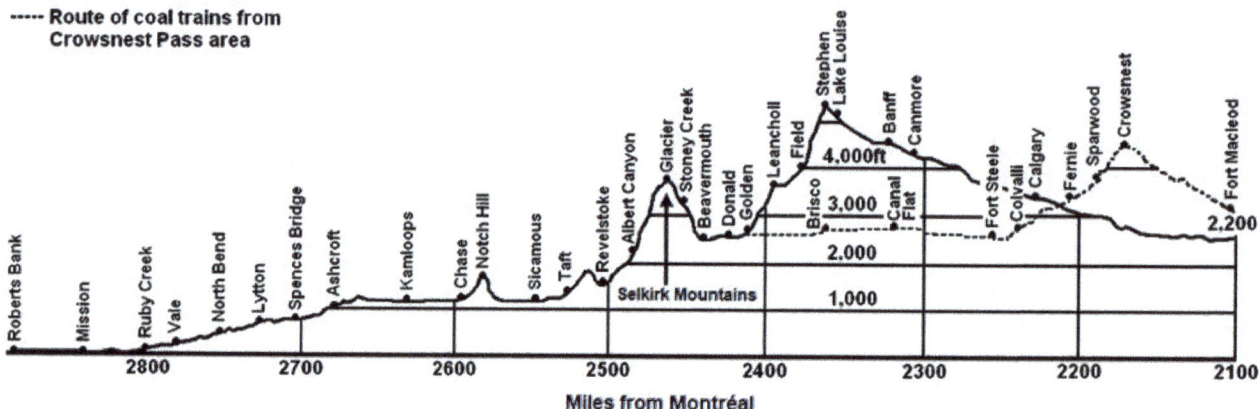

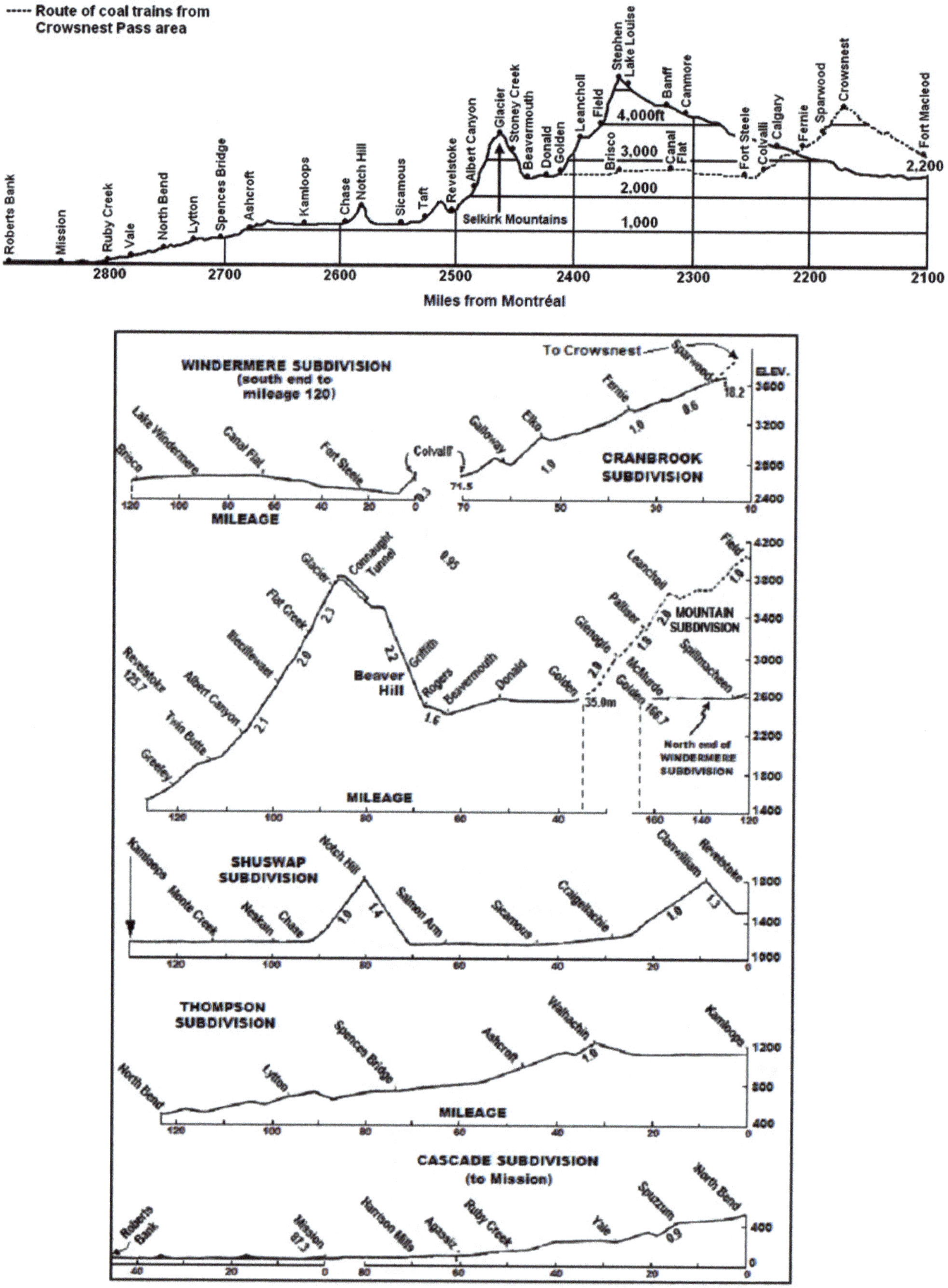

Figure 67: Route profile for CP freight trains with distributed power. Coal trains originate from Sparwood and Fording in the Crowsnest area. Sulphur, grain and potash trains from prairie regions will have remote power inserted at Calgary.

one-off requests from Harris to facilitate on-track tests of systems and equipment greatly assisted with the evolution of LOCOTROL over the years, and encompassed both the intricacies of esoteric technical detail plus more practical considerations such as where within the body of the train to locate the remote power group(s). CP Rail continued this work and were able to contribute profoundly to the sum of knowledge on the subject of how, precisely, to operate distributed power trains.

In using three power groups (the third being a manned Pusher cut into the train between the Remote consist and the hind-end), CP Rail were managing multiple remote power groups from an early stage in their heavy-haul long-train operations. The technology to be able to separately control more than one Remote consist was yet 30 years away as they experimented with various train sizes and remote-control configurations, such as 4-4-4, 4-5-3, 4-6-4, 4-2-4, and 3-2-5. To assist their train dispatchers with assigning the location of the in-train power groups, CP Rail created the formula shown below, using the following assumptions:

- All cars are either loaded or empty
- All locomotive units have the same tractive rating
- Since contemporary operating circumstances required that the third power group be a manned Pusher consist operating over just part of the total run, its removal must leave the Remote consist located such that satisfactory train handling would not be compromised
- On adverse ruling grades, the Lead consist will be operated in full throttle—occasional use of the lead unit *power reduction* feature excepted—in which case, train handling must not be unfavourably affected.

Let: l = number of units in the Lead consist

r = number of units in the Remote consist

p = number of units in the Pusher consist

A = tonnage between Lead and Remote consists

B = tonnage between Remote and Pusher consists

C = tonnage behind Pusher consist

T = total train tonnage $(A + B + C)$

(Note: A, B, C, and T can also represent the number of cars if all are of uniform weight)

Assuming that the Remote consist is to be located such that its output will achieve 35% pushing and 65% pulling, then:

$$A = \frac{(l + 0.35r)T}{l + s + p}$$

Assuming that the Pusher consist is located such that its output achieves 40% pushing and 60% pulling, then:

$$B = \frac{(0.65r + 0.40p)T}{l + s + p}$$

and

$$C = \frac{(0.6p)}{(l+s+p)}T \text{ or } C = T - (A \div B)$$

Example: A 130-car unit train is to be handled in a 4-5-4 power configuration. Since all cars are of relatively equal weight, the calculated results will be sought by number of cars rather than tonnage.

$$A = \frac{4 + (0.35)(5) \times 130}{13} = 58 \text{ cars}$$

$$B = \frac{(0.65)(5) + (0.40)}{13} \times 130 = 48 \text{ cars}$$

$$C = (0.6)(4) \times 130 = 24 \text{ cars}$$

or

$$C = 130 - (58 + 48) = 24 \text{ cars}$$

If all cars are not of uniform gross weight, then A, B, C, and T cannot be represented as numbers of cars but must appear as tonnage blocks.

In this example, the distribution of Remote consist haulage capacity is divided 35/65. This could be varied from 25/75 to 40/60 to suit local conditions. Similarly, the Pusher consist haulage capacity could vary between 30/70 and 50/50. Experimentation determined the optimum arrangement for operation over a given topography. The Pusher should not be located too far back, especially if it comprised three or more units. If special conditions required that the Pusher be located on the hind-end of the train, it could not exceed two units, and would preferably be just one.

Using the principle of the 'floating node' to locate the in-train power groups, CP Rail commenced the scheduled operation of unit coal trains in 1969 and—4 years and many millions of miles later—could count only one instance of a coupler failure in normal operation. This exceptional record stood in sharp contrast to that experienced in the preceding test phase and prior to the development of train configuration principles.

In the illustrations that follow, locomotives are assumed—for ease of depiction—to equate to three car-lengths, and each vertical square equates to a known unit of tractive capacity. Cars are depicted such that each horizontal square on the graph is accepted as being equal to one car-length, and for simplicity each car is assumed to have a 100-ton mass. The graphical relationship between the numbers of locomotives (assumed in these diagrams to be identical units) and therefore total drawbar force, and the length of the various portions of the train provides a reasonably precise approximation of where the 'node' for drawbar stress will be. This knowledge is critical to determining the optimum position at which to insert the remote motive power group(s) for the kind of territory through which the train will operate.

The red sections in these diagrams represent relative draft drawbar force throughout the train while blue represents relative buff force from the resultant 'pushing' effort of the remote power group(s). The draft force is depicted as being at its maximum value immediately behind each power group, reducing to zero immediately ahead of the Remote group as well as at the last car of the train. At the point of zero drawbar stress—for which CP Rail coined the term 'node'—the couplers are neither in tension (draft) nor compression (buff).

Parker reported that—in addition to predicting that train separations could be expected in the early months of operation—the manufacturer had also offered advice on locating the Remote power group within the train: advice that served as a starting point for CP Rail's experimentation in their take-up of the system. Depending upon the relative train consist and nature of the territory to be negotiated, conventional wisdom proposed that Remote power should be inserted somewhere between the middle of the train and three-quarters back towards the rear (see the following figure).

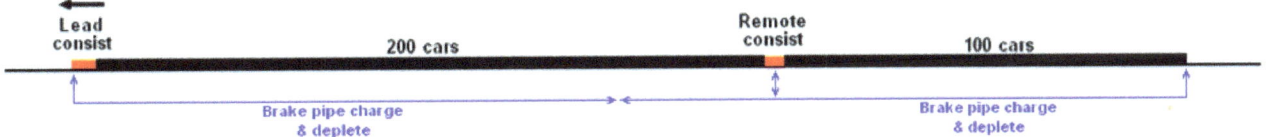

Figure 68: Conventional location of Remote consist in a long train.

If, for example, four Lead and four Remote units were to be assigned to operate a 300-car train over relatively level territory where air braking response was of primary importance, it would be customary to place the Remote units within the area two-thirds to three-quarters of the train length (somewhere between the 200th and 225th cars). Conversely, if the same motive power were to operate a 10 000-ton train up a 2.0% (1-in-50) grade, the need for each power group to handle its share of the tonnage assumed greater importance, the theoretical location for the Remote units then being behind 5000 tons rather than a particular brake pipe distance.

This latter scenario—addressing as it did, operation in mountain territory—resulted in configuring the LOCOTROL train effectively as two separate, connected trains coupled nose-to-tail and is depicted diagrammatically in the next figure.

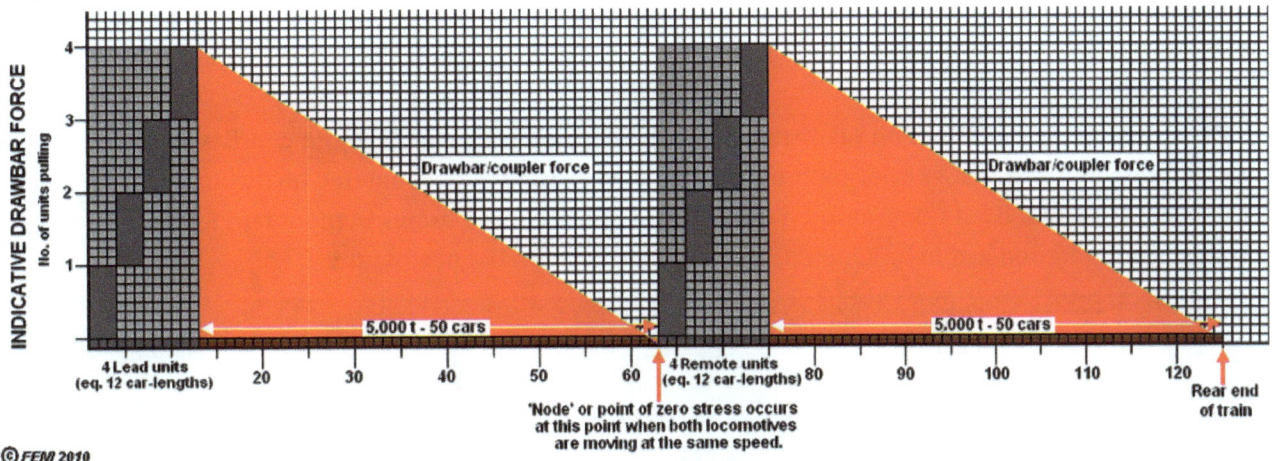

Figure 69: Drawbar force diagram for an 8-unit, 10,000-ton train in balanced, synchronous operation on ascending grade.

This configuration is acceptable for operation up sustained heavy grades of relatively uniform profile, where the tractive force exerted by the Lead consist is fully expended pulling that portion of the train lying ahead of the Remote consist such that a theoretical condition of zero drawbar force exists immediately at the front coupler of the leading Remote unit. However, if the sustained gradient profile should vary from time-to-time to any significant degree, then the potential exists to create a run-of-slack—however seemingly modest—that might cause a train separation; the dynamics of this circumstance being as follows.

The two portions of a train containing a Remote consist will tend to assume different speeds at different times due to the (sometimes transitory) influence of variations in gradient and track curvature. Should the head-end of a train configured in this manner slow marginally due to one of these influences but then advance as this influence was negotiated and resistance diminished, the slack created by the momentary slowdown is picked up again car-by-car until the process reaches back to the Remote consist, which is too heavy to be suddenly accelerated as the draft tension arrives there. This quick 'snap' can easily result in a broken knuckle or a pulled drawbar immediately ahead of the Remotes (or close-by if there is a car with a hidden, pre-existing weakness). Accelerations of as little as 1 mph (1.6 km/h) at this point could develop an instantaneous peak force of sufficient magnitude to part the train.

With their unit coal train, CP Rail protected against this possibility by inserting the Remote consist ahead of the theoretical halfway point (see next figure). By this means, a small buffer of unexpended head-end drawbar pull ensured couplers immediately ahead of the Remote consist remained in tension, preventing the development of slack action at this location.

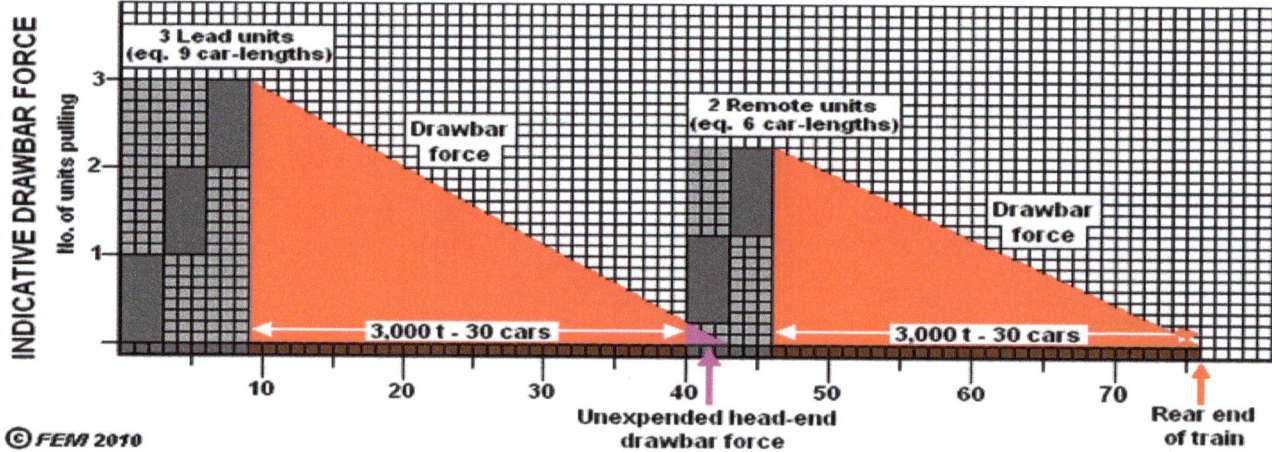

Figure 70: Remote consist placed ahead of the point at which the node would normally develop. Entire train normally in tension.

Parker made the point that on other railroads where four or five-unit Remote consists are not used, the configuration described above does not produce the adverse results as encountered on the Revelstoke Division. To manage the severe conditions imposed upon CP Rail's LOCOTROL trains in mountain territory operation, a revised configuration had to be adopted, and this is depicted in the next figure.

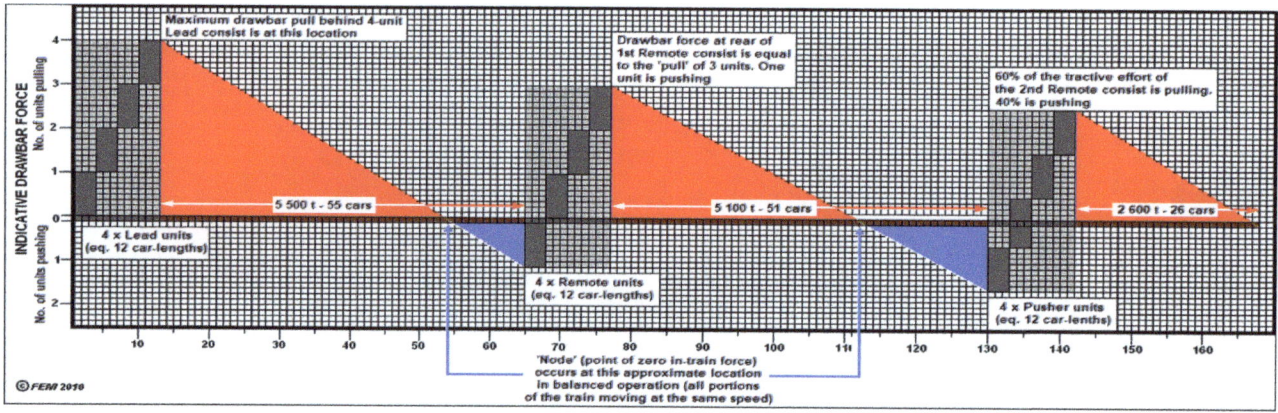

Figure 71: Drawbar force diagram for a 12-unit, 13 000-ton train configured for mountain grade operation.

In this scenario, the distributed power groups include a Remote consist and a manned Pusher consist. These groups were located within the train such that in a situation of steady, balanced 'pulling', the node existed not immediately ahead of the Remote consist as in the first of these three graphics, but at a point significantly ahead of it and the first Pusher unit. The slack ahead of the node was stretched, and behind it was bunched. If the Lead consist should encounter a temporary easing of the grade or curve resistance relative to the Remote consist, or the Remote consist encounters the same condition relative to the Pushers and begin to increase speed, they would gradually and progressively pick up more of that portion of the train that was in buff between the node and the following power group without having to suddenly encounter the massive and unresponsive weight of that trailing group of locomotives. This gradual, rearward migration of the node back towards the power group behind—and the resultant deceleration of the Lead consist as it began to lift more of the load behind—tended to retard its rate of speed increase, while the trailing power group—now relieved of the load to shove against—tended to accelerate so that a new condition of equilibrium for the train was arrived-at and the node returned to its original area.

In recurrent tests with trains of 13 200 to 14 200 tons using 4-4-4, 4-5-4, and 4-6-3 motive power configurations, it was established that the Remote consist should expend 25–40% of its tractive force capacity in pushing, and the remainder in pulling, while a Pusher consist should be located so as to expend 40% in pushing and 60% in pulling. The purpose of restricting the tonnage behind the Pusher consist to 60% of its combined rating was to reduce the possibility of damage should a severe run-in or run-out of slack occur to that otherwise unrestrained portion of the train. CP Rail found that when very long trains operating over flat or moderate cresting grade territory were configured for optimum brake pipe support (that is, with the Remote consist located two-thirds to three-quarters back), the result accorded with the method described above. The node tended to float in the middle of the train—well ahead of the Remote consist—and instances of drawgear failure were relatively infrequent.

Braking under power

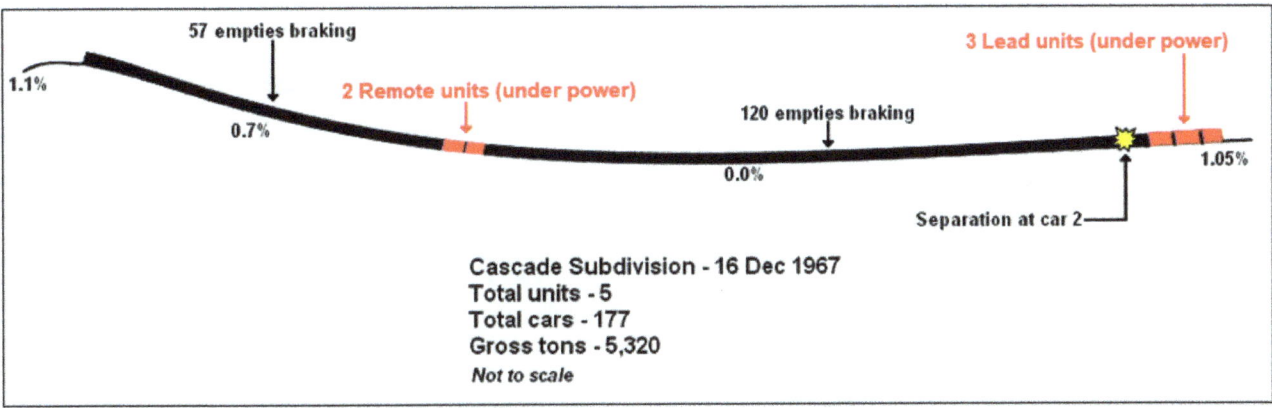

Figure 72: Locotrol train braking under power.

CP Rail's first indication that the handling of Remote consists could be problematic occurred in December 1967 when a train of 17 empties separated near the head-end while stretch-braking to a stand on fairly level track. This technique involves leaving the throttle open as the train is slowed by air braking, thus stretching slack to avoid a run-in, and progressively reducing the throttle to Idle approaching the point of stopping. The front portion of the train was on level track but the rear portion, including the Remote consist, was on a slight descending grade. The use of power on the Remote consist combined with the influence of the descending grade under the rear portion of the train bunched the slack ahead of the Remote consist and—although the throttle was closed just before the train stopped—the buff force drove the head-end forward so that it ran out with a jerk and snapped the rear coupler knuckle on the car immediately behind the Lead consist.

This incident demonstrated the vital requirement for the locomotive engineer to 'think' for the Remote units as well as the Lead, and to give due consideration to the effect of the Remote consist on train dynamics.

Starting trains

The inclusion of Remote units in a train that is being started on level track provides no great handling challenge: however, when the train is stretched over a hump or crest, there is the potential for a discouraging outcome. This was exemplified by the case of a long train stopped at Leanchoil (17 miles west of Field) for a brake test prior to commencing the descent of a 2% grade as shown below. In starting from this location, when the hind-end started to move, the slack ahead of the Remote units bunched, propelling the front portion forward so that it momentarily moved at a higher speed than the rear portion. The result was a succession of separations behind the Remote consist caused by the brief runout of slack as both Lead and Remote consists moved the train away from the portion still draped over the crest.

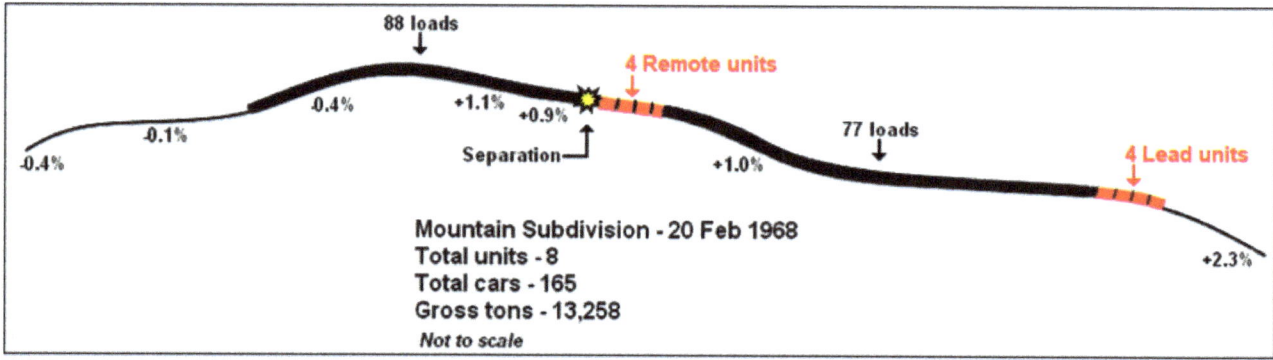

Figure 73: Starting a Locotrol train that has a portion stretched over a crest.

Should a sufficient component of the rear portion of the train stand on the upgrade at a crest, as depicted in the next illustration of a train at Donald, 51 miles west of Field, the tractive effort developed at the head-end is more than sufficient to move the front portion of the train away from the Remote consist such that the unexpended drawbar force reaches back beyond the Remote consist. This force is added to that created by the Remote consist itself and is sufficient to break a knuckle immediately behind the Remote units.

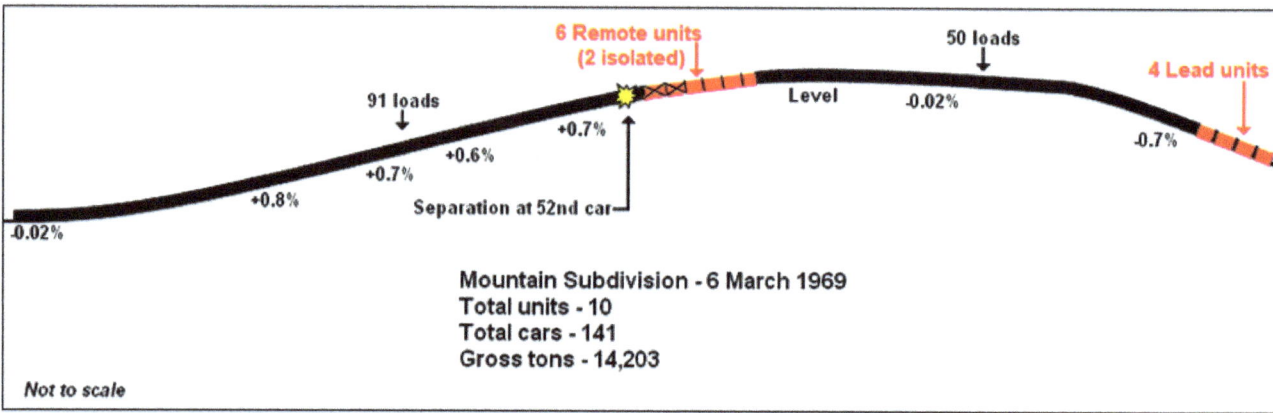

Figure 74: Starting a Locotrol train that extends over a crest but has most of its length on the ascending grade.

One solution to starting under the conditions depicted above was to apply all or most of the starting tractive effort at the Remote units: the challenge then being to avoid the head-end running out and causing a separation in the front portion of the train. Judicious use of Lead consist Independent brake as the train gets underway can be used to restrain the head-end and achieve this outcome.

Operation on heavy grades

When trains containing Remote power—and particularly when assisted by Pushers cut into the rear portion—are operated on heavy grades at full power, loss of rail adhesion at any of the power groups can be disastrous. If the Lead units slip or lose power momentarily (and perhaps make backward transition), the other units can bunch enough slack that when the slipping units 'regain their feet' and stretch the slack, a break-in-two results. It is important to understand that the heavy mass that is represented by a portion of a unit train cannot have its velocity changed both quickly and gently at the same time. Something will give!

Such an incident occurred at Griffith on the Beaver Hill (71 miles west of Field) with the train depicted in the next illustration. As the Lead units negotiated a reverse curve at which a trackside rail lubricator was installed, a wheelslip occurred that caused the units to lose momentum and bunch some slack behind them. When the units regained adhesion, the negative drawbar force permitted them to surge forward, stretching the head-end and snapping a knuckle on the fifth car. The resultant train separation and Emergency brake application caused a heavier braking effort on the motive power groups than on the loaded cars, and the hind-end of the train ran in on the Pusher group, crushing the tenth car behind them.

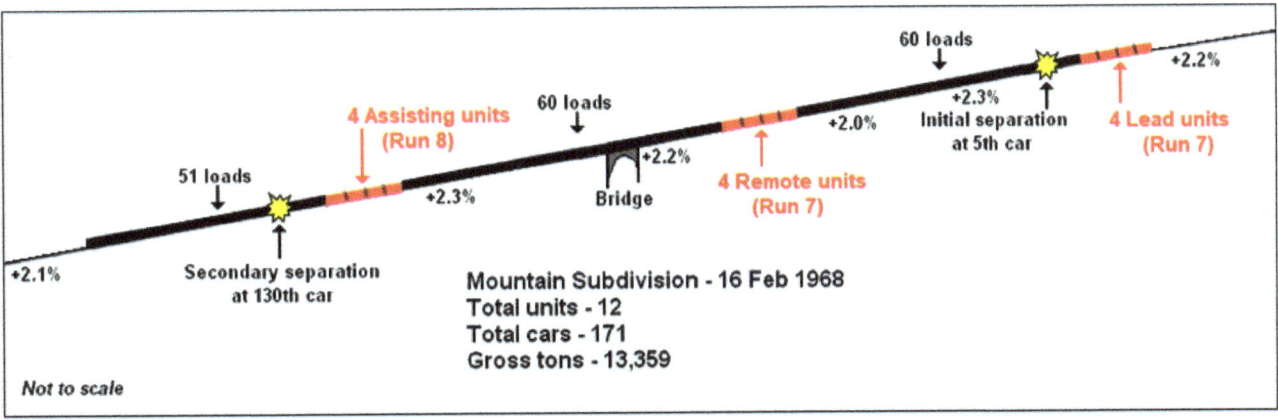

Figure 75: The effect of wheelslip on a Locotrol train on a heavy ascending grade.

The effect of change-of-grade

When a conventional train is ascending a grade, the entire drawgear system is in draft tension except momentarily when locomotive wheelslip might occur, the possible effects having been described above. If the grade intensifies or moderates, the slack still tends to remain stretched so long as the throttle position is not varied dramatically. Slack reversals are the most prevalent cause of train separations and engine service personnel are required to employ advanced train handling kills in order to avoid excessive slack action.

With remote-controlled distributed power operation, negotiating varying grades calls for even more skilful handling than with a conventional train due to the Remote consist tending to reverse the slack condition in the forward portion of the train at every significant variation in grade unless the distributed motive power groups are controlled individually through the affected area. The next illustration depicts three specific locations on the Revelstoke Division where westbound LOCOTROL trains incurred separations from this cause.

At each of the locations shown, the ascending grade not only moderates or levels out, but where it levels out there is a short descent before it does. If the engineer leaves the head-end units in full throttle the drawbar force thereat is not fully utilised in the forward portion of the train and the unexpended balance can augment the drawbar force at the Remote units to an extent sufficient to break a knuckle immediately behind the

Remote consist. To avoid this, the engineer must revert to Independent Control, leaving the Remote consist in full throttle while reducing the Lead consist throttle notch-by-notch after reaching the initial crest of the grade.

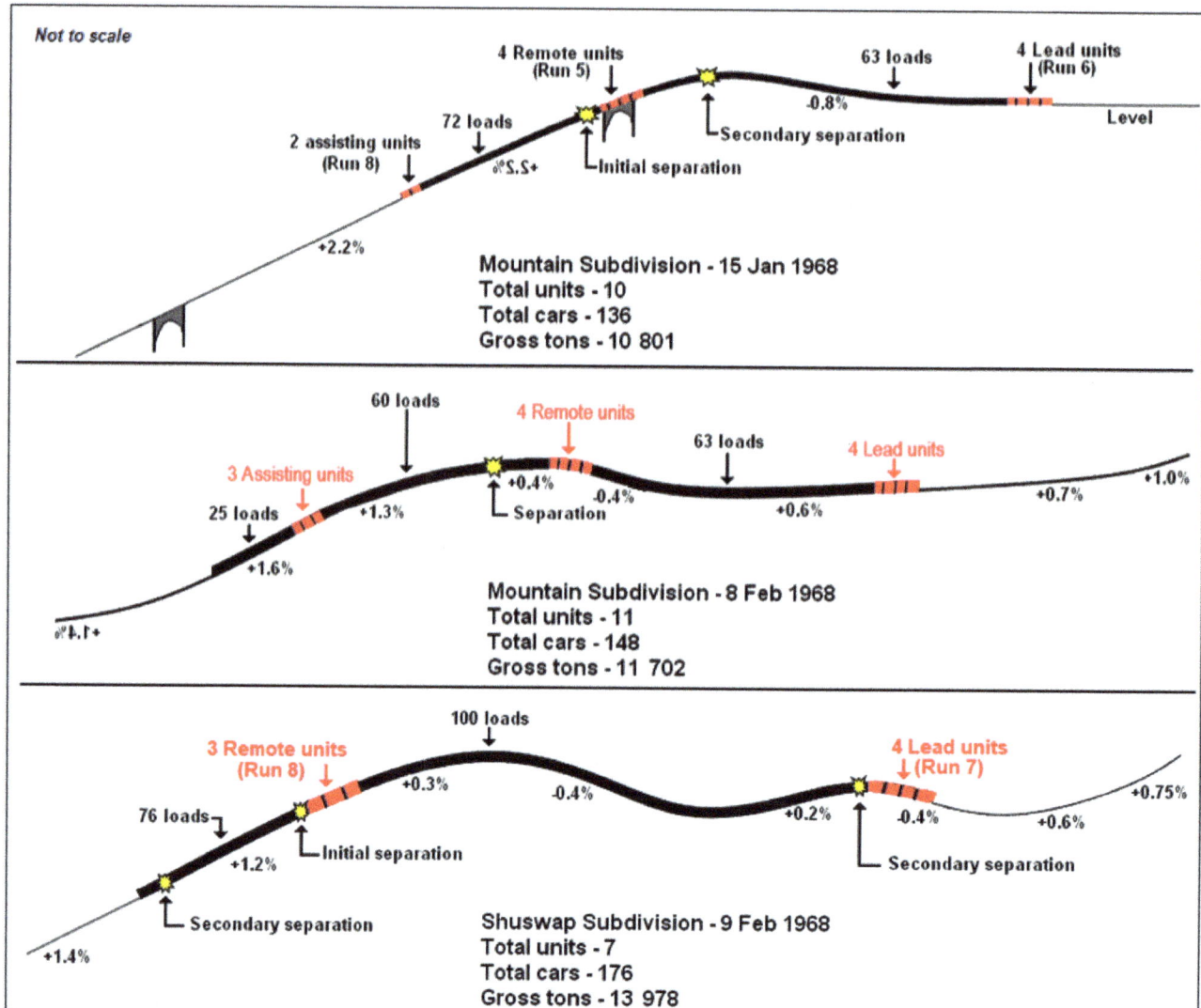

Figure 76: *The effect of gradient moderations on a Locotrol train ascending a grade.*

Should power be reduced too much on the Lead consist, the slack tends to close up excessively, and in extreme cases this may produce a sufficiently excessive load on the Remote consist as to cause it to stall. The engineer's objective, therefore, must be to endeavour to balance the throttle settings on both power groups to maintain the same speed at each group without causing excessive drawbar force behind the Remote units.

On both the centre and lower profiles of the previous illustration, a further complication is presented by the increase in gradient after the Lead units pass through the level section. Now the engineer must progressively increase the throttle setting on the Lead consist at just the correct rate to avoid too rapidly drawing out any slack that might have developed ahead of the Remote consist. If the Lead consist is powered up too quickly, a break-in-two can result.

Pusher engineers are fully aware of the requirements for successful operation through locations such as these. The trains depicted in both the upper and centre profiles contain a Pusher group. When the train has advanced through the level section to the point where the Pusher is approaching the crest, the Pusher engineer observes the condition of the couplers head of the lead Pusher unit. He knows they must remain in compression and that should they start to loosen up in advance of going into draft tension before the Pusher consist is well onto the level section, he will notify the head-end engineer by radio to reduce power to maintain the desired distribution of force within the train.

Operation over cresting grade territory

At the time of writing his paper, Parker contended that analysts of train handling—while having written extensively on the art of handling long conventional trains over the humps and sags of undulating territory— had been silent on the subject of handling trains with remote-controlled distributed power across rolling terrain. The inclusion of remote power in the train, Parker argued, was supposed to facilitate better control of slack through the provision of independent operation of the Remote units but there were so many factors involved that any engineer insufficiently skilled in the exercise of Independent Control of the Remote consist was actually more likely to break the train apart than if they were operating the same train without Remote motive power.

To successfully handle long trains with remote-controlled distributed motive power over rolling terrain, the engineer must possess an accurate knowledge of the track profile, the make-up of the train (if it is not a unit-train), and a situational awareness of the relative location of the Remote consist at any time. To be optimally effective, an engineer also needed to have an awareness of the effects of curvature, gradient, throttle, brake, and train resistance on the various portions of the train.

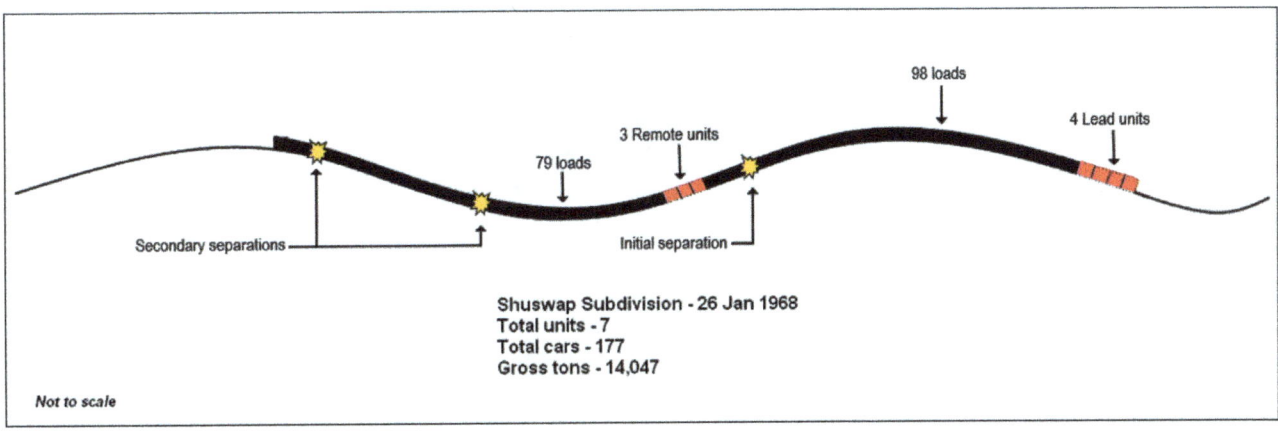

Figure 77: Negotiating cresting grade territory.

The illustration above is representative of a stretch of comparatively level track along the South Thompson River east of Kamloops where the train depicted, operating with SD40 units, suddenly broke into four sections while negotiating a series of slight humps (which—in the illustration—are exaggerated for effect) at 30 mph (50 km/h). The position of the train relative to the topography at the time of the incident shows the reason for the separations. The Remote consist—which was being operated in the same throttle setting as the Lead consist—closed up the slack as the head-end, having passed over the second crest, started to run with the slack—the sudden take-up then parting the train 15 cars ahead of the Remote consist. From the resultant Emergency air brake application, the hind-end of the train bunched up tight then 'cracked the whip' under the influence of the draftgear recoil, causing two secondary separations.

A second incident relating to running through a sag (the Galloway Dip) is depicted in the next illustration. On the previous day, an almost identical (conventional) train operated through this location at about the same speed with no difficulty. When the LOCOTROL train ran the same course, a break-in-two occurred in the rear portion after the hind-end cleared the bottom of the sag. It is surmised that the slowing of the front portion of the train on the brief adverse grade caused a bunching of slack ahead of the Remote consist. This slowed the rear portion but eased the drawbar force on the Lead consist permitting it to accelerate as the rear of the train decelerated, resulting in a separation behind the single Remote unit. In a subsequent run of this train, the Remote unit was operated at a lower throttle setting after negotiating the sag, permitting the train to stretch out gently and avoiding trouble.

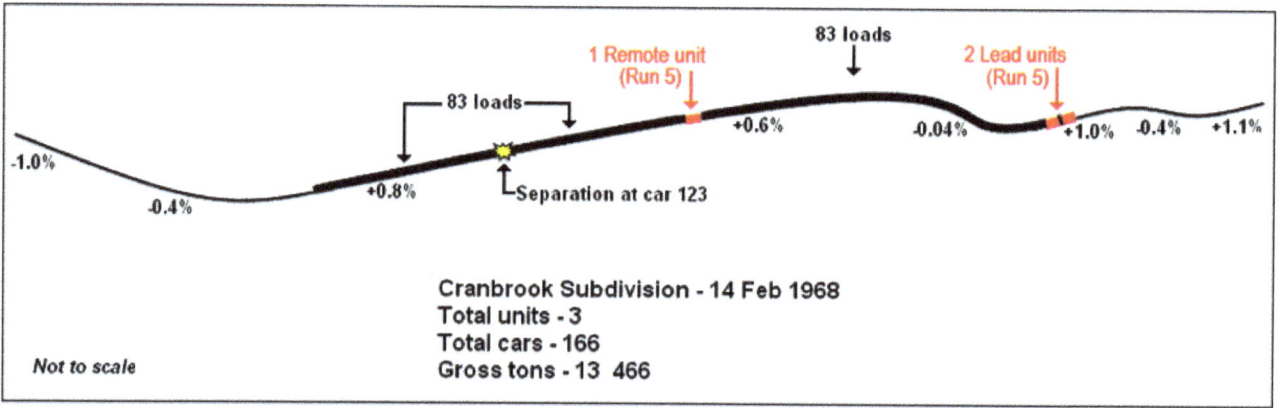

Figure 78: Negotiating a sag.

Operation through tunnels and cuts

When LOCOTROL trains are required to operate through long tunnels or deep, curved cuttings, Lead-to-Remote radio contact will inevitably be lost for periods of time. In this case, the locomotive engineer can see from Control Console indications that this 'No-Continuity' condition has occurred. Under these conditions the radio system enters a default state in which the established control settings (throttle or dynamic brake) at the Remote group are automatically maintained for a set period, after which the Remote group will throttle back—one notch at a time—to Idle.

This programmed setting can often work perfectly well for the locomotive engineer's requirements but, inevitably, there are times when the engineer will need to make an 'in-tunnel' throttle or D/B control adjustment on the Remote group. Therefore, a means had to be provided to enable the Lead and Remote groups to remain in radio communication under these restrictive conditions, and this was achieved by stringing a 'leaky feeder cable'[72] through the tunnel (or part of the tunnel) and a radio propagation tower on a high point near the cut.

Tunnels of less than one-quarter of a mile in length pose no problem to remote operation unless two or more tunnels are located close enough that the Lead unit is inside one while the Remotes are in the other. When radio continuity is lost, the Lead and Remote transceivers will make repeated attempts to re-establish contact, failing which—unless the engineer has placed the system into Override—the Remote consist will throttle back to Idle after 45 seconds' loss-of-continuity, and the feed valve will drop out. If continuity is restored quickly, the Remote consist will resume the mode of operation set by the engineer under these conditions. Should the

---

72 A coaxial cable that can emit and receive radio signals along its entire length and which functions as an extended antenna, providing two-way radio communication throughout the length of a tunnel.

train be subject to significant brake pipe leakage, the brake pipe pressure throughout the rearward portion of the train may reduce sufficiently to result in a light brake application that can cause a train separation.

Automatic (or Manual) Override is a means of delaying the Remote run-down for up to 30 minutes during an interval of discontinuity in radio communication. Override retains the Remote consist in whatever dynamic braking or throttle position it was operating in when continuity was lost, for that period. Lead consist throttle changes can be made during this interval that will not affect the Remote consist. During the Override interval, however, a brake pipe reduction will terminate the Override state and cause the Remote consist to revert to Idle. While this can have awkward consequences, it is a vital provision since the brake pipe is now the only means of communication between the two power groups, and a brake pipe pressure reduction is the only way the engineer can signal the Remote consist to reduce power (to Idle) should this become necessary.

CP Rail's longest tunnel at the time, the 5-mile/8-km Connaught Tunnel[73] under the Selkirk Mountains near Revelstoke, BC, required the use of Override for both east and westbound LOCOTROL trains. In the westbound direction, the train entered the Connaught Tunnel at about 30 mph but needed to emerge at a speed of less than 20 mph (32 km/h) because of the requirement to stop at Glacier for a brake test before descending the severe grade beyond. Careful choosing of the Lead and Remote throttle positions was required as the Remote consist throttle could not be varied while radio continuity was lost, and although the speed needed to have been reduced prior to the train emerging from the tunnel, an air brake application—that would result in the Remotes throttling back to Idle—was not desirable. To ensure radio communication for at least a train-length within the west end of the tunnel, an antenna was strung for 8000 ft eastward along the bore from the west portal, with a repeater station located at the portal. This permitted the restoration of radio communication before the Lead consist reached the west portal. When this occurred, Override was automatically cancelled and the Remote consist throttle setting could be reduced, or an air brake application effected as desired.

The installation of this antenna materially eased a train handling problem affecting eastbound LOCOTROL trains as depicted in the next illustration. Operating eastbound, the Remote consist could not be throttled back by any significant degree before reaching the summit or the train would stall, but could not be retained at full power either, after entering the tunnel, or speed would quickly have become excessive. Before the antenna had been installed, it was necessary to reduce the Remote consist throttle setting to as low as possible immediately before the loss of continuity, then to control the descent through the tunnel using dynamic brake on the head-end only. Once the antenna was available, maintaining radio continuity within the west end, the Remote consist could enter the tunnel prior to having its throttle setting reduced to Idle, providing greater precision in speed control.

Conclusions

In his paper, Parker drew the following conclusions:
- Of the several advantages claimed for the inclusion in a train of radio-controlled remote power, there was no question regarding those of improved charging times, faster brake applications, quicker releases, and the reduced stopping distances deriving from the more rapid brake response. In more temperate climates, train lengths and tonnages could be markedly increased – a factor that could produce gratifying economics in mountain territory operations.

---

73 The 9-mile (14.7-km) Mount Macdonald Tunnel—also through the Selkirk Mtns and close to (in fact crossing beneath) the Connaught—was completed in 1988.

- Under frigid weather conditions that promoted brake pipe leakage and forced the railroad to limit train lengths, the regulation mandating that the Remote feed valve be isolated when performing the brake pipe leakage test prevented the benefits of LOCOTROL trains being realised. Considering that these trains were operated with the Remote feed valve cut *in*, he felt this regulation was unduly harsh and restrictive.

- The claimed advantage for LOCOTROL of improved train dynamics would not automatically accrue, and this was evidenced by the various incidents discussed. On certain kinds of territory, only experienced and skilful handling of a train incorporating remote-controlled motive power would avoid worse slack action than might occur with the average conventional train.

- Remote control operations increased maintenance costs and the equipment required special servicing and maintenance facilities and the frequent attention of skilled technicians. The radio components were the most prone to failure and their dependability required improvement.

- Overall—when used strictly to obtain the operating advantages for which it was best suited, and when applied to operations where these advantages could be exploited to maximum effect— radio control of Remote locomotives could provide excellent benefits and produce an operation unexcelled by any other means[74].

Questioned by conference participants on the subject of the static holding ability of the Independent brake when these trains were stopped on a heavy grade and the train brakes released to recharge the system, Parker explained that with 80 psi (550 kPa) maximum Independent brake cylinder pressure, the nine 3000 hp units used to control a LOCOTROL train on the 2.2% descent west of Glacier had insufficient holding power for this task. For this reason, all coal cars were being equipped to receive a #20 pipe Independent brake supply from the locomotive, and it was expected that about 20 cars on each train would be so-configured – perhaps using 10 cars behind each consist to reduce the application and release time for what was, in effect, a straight-air process. This Independent Application & Release extension pipe was connected to the locomotive brake cylinder branch pipes via double-check valves and a cock located near the Independent brake valve that the engineer would open as required to admit the #20 pipe supply to the trailing cars. In later years, Queensland Rail in Australia used this arrangement on coal trains.

Parker was asked if he thought it possible to have a one-dimensional computer model that might predict and solve problems related to train separations and the strategies for configuring the Remote power in a LOCOTROL train. He thought there were currently many obstacles hindering effective application of this idea—not least predicting internal forces and being able to take account of the variables of train handling—but also that '... *clever Canadian engineers* [were] *researching the subject and we wish them every success.*'

---

74 This was an entirely reasonable position for the era. The later development of ECP braking (which would provide hardwired DP capability through the electric braking medium) could, at this point, have only been contemplated.

4

A Locotrol INSTALLATION PROJECT—INDIA

July to October 1989—an irregular diary

In the October 1988 issue of *International Railway Journal*, this item appeared on the World Market page:

INDIA: The Controls & Composition Division of Harris Corporation, United States, has won a multi-million-dollar contract for LOCOTROL II *on-board locomotive control and monitoring systems for the South Eastern Railway and North East Frontier Railway. Unique aspects of the equipment involve interfacing with vacuum-braked trains in the north-eastern region and a new radio development for communication between the Lead and Remote locomotives.*

Reading the article, little did this author know that he was to become intimately involved in this project. The following account has been produced from comprehensive notes kept throughout these 4 months. It is offered with many of the warts still attached, to provide the reader a nuanced sense of one's experience of participating in an expatriate project such as this. It was, indeed, an intensive experience.

☙❦☙

I originally learnt of the proposed Indian Railways LOCOTROL installation project in late 1988 when Milt Deno and Gene Smith from Harris Controls, plus Ian (Squizzy) Taylor and Kerry Kline from Harris Corporation's then Australian agents—Brisbane-based Evans Deakin Industries[75]—were on site in Port Hedland, Western Australia, to install some new LOCOTROL II systems. I had been assigned by the acting railroad manager, Barry Green, to work with Milt on crew training and to help in the provision of Harris' first specialised LOCOTROL Operator's Manual for locomotive engineers. Although LOCOTROL customers had often produced their own for their engine-service staff, and Harris had a generic document, the provision of a specific operator's manual for a customer later became a standard part of the service provided by Harris and then GE.

After completing a training run on an empty train from Port Hedland to Newman, we were relaxing in the bar at The Walkabout Hotel when a colleague asked Milt where his next job was, and he replied that after this one they'd be off to India. Later, Railroad Manager, Andrew Neal asked Milt to extend his visit at Port Hedland to review and report on aspects of the supervision and control of locomotive crews at Mount Newman Mining, which he did... on one page. At one stage, Neal asked him if he'd be prepared to remain in Port Hedland and run the Railroad Operations department for a period—an offer that Milt declined with grace. I would say he was over-qualified.

---

75 Now known as Downer EDI. EDI's LOCOTROL agency with Harris Corp was later ended due to a disagreement over product pricing.

During this period, he happened to mention that he was looking for a road foreman of engines for the Indian project and asked whether I would consider the job. Milt subsequently made me a formal job offer, stating what he needed me for and that he would be able to arrange a secondment if I was agreeable. My task would be to provide both classroom and practical training to enginemen and to rotate with Milt on the head-end when running commissioning and training trips. The project would be operating in two diverse locations: the regional steel-making centre of Rourkela in the state of Odisha (in Indian Railways' South Eastern region, the state being previously known as Orissa) and at Guwahati, the main centre for the state of Assam (Indian Railways' North East Frontier region).

That portion of the project in Rourkela would involve two complete systems to be installed on diesel locomotives of Indian Railways' Wide-Gauge (1676 mm/5 ft 6 in) for iron ore trains on the South Eastern Railway. The SER caters to the states of West Bengal, Jharkhand and Odisha. SER also runs regular electric multiple unit commuter services from Kolkata's (previously Calcutta) main station of Howrah to adjacent outer suburbs. SER also handles major freight traffic to Kolkata and Haldia. The Assam portion of the project was for two systems to be installed on Metre-Gauge diesels for use on vacuum-braked general freight trains over mountainous territory on portions of the North East Frontier Railway. The NEFR is responsible for rail operations in the entire north-east and parts of West Bengal and Bihar.

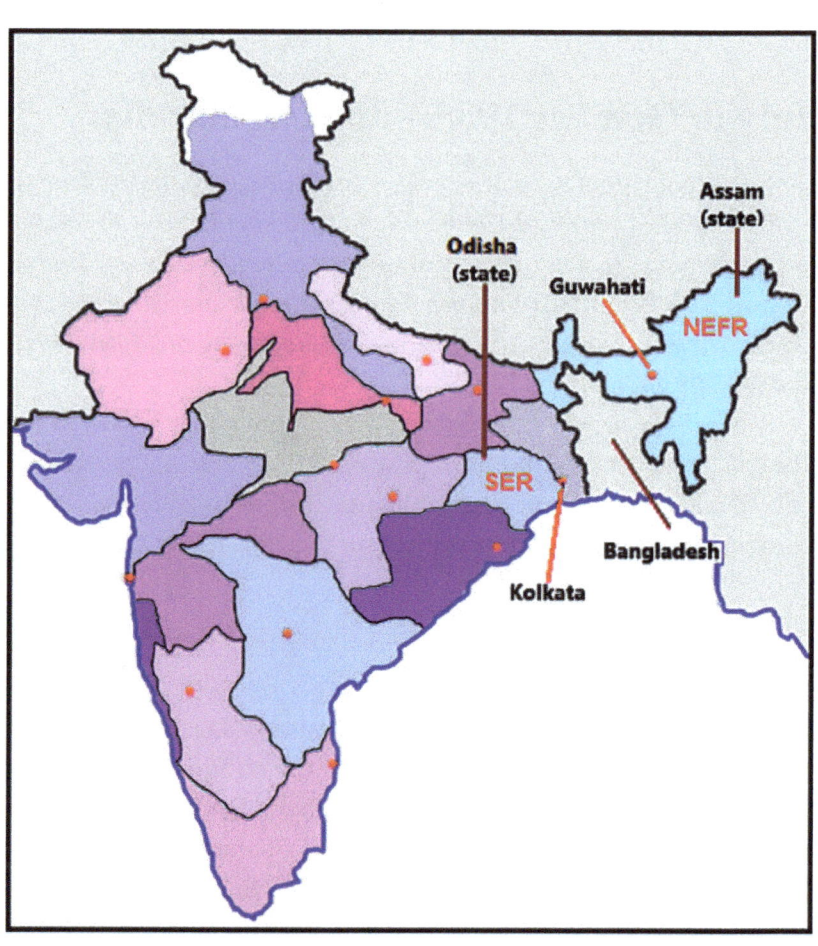

Figure 80: Map, showing locations of Locotrol project and Indian Railways regional HQ offices.

By now I was very interested, and, after some months, word arrived that I was to fly to Kolkata, book into the Oberoi Grand Hotel and contact Mr G C Mukherjee from Harris' Indian agents, the grandly-named Scientific Timesharing Systems Ltd of Salt Lake City, a Kolkata suburb. He would arrange for me to catch an overnight train to Rourkela where Milt and Gene were already on the job.

[Author's Note: The following narrative is presented, as originally diarised, in a more 'present' tense.]

Tuesday 4 July

I presume my colleagues in Rourkela will not be taking time off to celebrate Independence Day as I depart Australia to join them. Qantas flight QF71 from Perth to Singapore is a B767 and I find myself sitting beside Dennis Wholley—a Trade Commissioner with Austrade in Shanghai—who has some insights into doing business in India. It is fascinating to fly up the Western Australian coast and to look down from 35 000 ft upon familiar towns and points of interest. I can see the familiar North West Coastal Highway, scrawled like a pencil line across this huge map. Our progress is charted on the forward bulkhead of the cabin on a huge video screen (this was pre-seatback video days)—showing the WA and Indonesian coasts and our track over Geraldton, abeam Carnarvon and Monkey Mia and across the Indian Ocean to the Sunda Straits before turning sharply left and heading up the Java Sea to Singapore. As darkness falls, I peer through my window and see the lights of Jakarta twinkling 35 000 feet below before the wing abruptly rises as we bank left to head north along the Sumatran coast.

This is my first visit to Singapore, and I get lost getting out of the terminal building at Changi Airport (it's easy when you know how—follow Qantas Jetabout passengers as they exit at luggage Belt 19 for transfer into the city). I caught the airport minibus with other passengers and toured the central city, dropping people off at various hotels. Mine was the Glass Hotel (nowadays the Intercontinental Group's Holiday Inn Atrium), which has a sort of mini version of the atrium roof of the Burswood Resort Hotel in Perth.

Wednesday 5 July
I taxi to the Asian food centre at Newton Circus for lunch (per a recommendation from Mr Wholley), choose a live crab and have it cooked with ginger and garlic. It takes me 45 minutes to eat it! I now visit Newton Circus whenever I return to Singapore. Thanks, Mr Wholley! I make the day-long coach tour of the city of Singapore (Mt Faber, Selangor Pewter factory, Botanical Gardens, etc.).

Thursday 6 July
This morning it's back out to Changi Airport to catch Singapore Airlines flight SQ414 to Kolkata and Kathmandu. (I wish I could remain on board for the second leg of this flight as it will approach the Himalayas near Mt Everest. This would be a stunning sight in fine weather and a frightening ride in bad weather I imagine.) Our aircraft is an Airbus A310—a modern wide-body twin-engined aircraft built in 1983. There is no in-flight movie on this 3 hr 50 min flight. SIA only shows movies on its B747 flights [this is 1989!]. I'm sitting totally alone in business class. The chief purser asks if there's anything I'd like, and being a private pilot I ask if I can see a map showing our route from Singapore to Kolkata as we track up the Strait of Malacca and out across the Andaman Islands and Bay of Bengal. He returns with an invitation from our (British) captain to visit the flight deck. The captain introduces himself and hands me an En-Route Chart showing in detail our route SIN–CCU. A flight attendant brings me coffee—all is well with the world. I am impressed by the competent and professional manner of the captain and his young Singaporean first officer and by the fact that even this far out from Kolkata they are perusing charts and planning their descent and approach—to be flown by the first officer. I remain here for 2 hours—including the approach and landing into Dum Dum International Airport (now the Netaji Subhash Chandra Bose International Airport) at Kolkata, a very special privilege even at that time [and one utterly unavailable today].

The flight deck is configured for three crew, but SIA don't use a flight engineer on the A310. Thus, I am ensconced in this lambswool-covered armchair, situated behind and between the flight crew. I have cranked up the height adjustment and rest my feet on two fold-down footrests—rather like motorcycle foot-pegs—one either side of the centre console. I sit higher than the flight crew and have 180º of vision through the cockpit windows. We make our descent in clear weather over the Ganges delta with the Bangladesh border to our right and I look down upon seemingly limitless flooded fields and rice paddies. The aircraft has the ultra-modern 'glass cockpit'—three flat LCD screens in front of the pilots instead of the previous electromechanical instruments I am familiar with as a private pilot. It is fascinating to watch the tiny aeroplane icon in the centre of the copilot's screen as the autopilot 'flies' it along the magenta line which is the track to be followed en route to Kolkata and the approach to the airport.

Any doubts I may have about this wonderful technology are put to rest as we roll out of a 180° turn and there, dead in front, and 10 miles ahead is Runway 19 of Kolkata's airport. The huge vultures wheeling lazily around cause me more than mild concern – the consequences of one

Figure 81: Map of Kolkata showing the location of the Oberoi Grand Hotel and Howrah (Haora) Railway Station. Only major streets are shown.

coming through the windscreen at 160 knots don't bear thinking about! However, the captain is calmly calling out altitude numbers and the first officer, left hand on thrust levers and right hand on control yoke, flares the aircraft at what seems to me to be an absurd height above the runway (more familiar as I am to the view out of the front of a Cessna 172). We touch down smoothly and roll out on a rather bumpy runway surface with tufts of grass growing through the many cracks in the bitumen. Our transition of a time zone on this flight has lost us 2 hours. The crew will continue to Nepal – but I have arrived in India. How long have I been pondering this and what preconceptions will soon be shattered?

Two hundred and seventy kilometres to the north of Kolkata, at a place called Farakka, just before the mighty River Ganges crosses into Bangladesh and becomes the Padma, a barrage constructed across it diverts some of the flow into a canal that becomes the Hooghly River, and flows south to the Bay of Bengal, keeping to the Indian side of that great delta. On the way, it bisects the waterlogged city of Kolkata.

What can one say about Kolkata? You can see all the TV programs and documentaries in the world, and you still must be here to really understand it. The city in 1989 is scungy, squalid and sordid. The streets are paved with cracked, bumpy bitumen. The most incredibly rickety old trams absolutely packed to the rafters, sway and totter along a sinuous and very non-permanent-looking way, sometimes ducking mysteriously behind high-rise buildings. Hundreds of cars—nearly all of them taxis (an Indian-made version of the 1958 Morris Oxford, called an 'Ambassador')—jerk and weave their way through the mayhem, engines revving. The drivers are crazy, and the only road rule seems to be, '… if it's bigger than you, give way to it'. The vehicles duck and weave, missing each other by the thickness of a coat of punka Dulux. Horns honk and toot and squawk non-stop. Nobody seems to get permanently upset.

The Grand Hotel—one of a chain of such 5-star hotels throughout India and several overseas countries (there was one in Melbourne, Australia; now reverted to its previous life as The Windsor) owned by the Oberoi family, scions of a rich, old Indian Aristocrat—is a sea of opulent tranquillity in this third world madness. It is very much in the style of the British Raj; I'm sure Singapore's Raffles Hotel must have been like this in 1930. However, I won't get much of a chance to sample it, as arrangements have been made for me to depart by train at 10.00 pm tonight for Rourkela, where I am due to arrive at 6.00 am.

Mr Kundu from Scientific Timesharing Systems meets me at the hotel at 9.00 pm and escorts me to Howrah Railway Terminal. More mayhem! Unlike Australian city railway stations, those in India are in a constant and total state of dynamism. Abject poverty and middle-

Figure 82: Staff at the Oberoi Grand Hotel, 1989.

class wealth rub shoulders (with the 'poverty' doing the lifting and carrying and the 'wealth' walking serenely behind) as lower-caste families make their home in and around the station precincts while upper-crust families search for their rooms in air-conditioned sleeping coaches. All the while, porters swarm around like so many busy ants, pompous railway officials sit behind desks and talk animatedly into old-fashioned black Bakelite telephones, and locomotives sit trapped at stop blocks while their dark-skinned crews grin at the scurrying souls passing along the platforms. Here, a youngster squats over a platform edge and defecates onto the tracks a metre below. There an old man lies dead on a filthy blanket while flies walk around on his face. An Indian railway station is a stage and the teeming humanity therein is the show. This is what I love about travelling!

I find myself sharing a First Class sleeper with a German who either does not or doesn't want to speak *Englisch*. The train departs on time smoothly and without fanfare, and glides—electric-hauled—out into the darkness of suburban Kolkata. There is little for it but to sack in. It's a humid night and the train stops often. When it does, so does the air-conditioning, belt-driven from the axles. I don't sleep well, although the bunk is comfortable, and the linen is crisp and clean.

Friday 7 July

Figure 83: Radhika Hotel, Rourkela, Odisha State.

The train pulls into Rourkela station at 6.20 am. It is a smaller version of Howrah—a bustling human throng. Delicious food-smells mix with putrid organic smells. The car driver Milt has hired for us comes aboard the train and meets me at my compartment door holding a sign above his head that reads, 'G'DAY MOFFAT': I wonder where that idea came from? He takes my luggage and we proceed out of the station to his car – the ubiquitous Ambassador taxi.

Our hotel is one block from the station. I check in and go up to my room on the fourth floor to shower and change into work clothes. By this time, breakfast is being served in the

Figure 84: Road entrance to the Bondamunda Diesel Shop in 1989.

dining room and I proceed down to meet Milt and design draughtsman Steve Heneka at 7.15 am. Milt says, *'How ya doin' mate?'*

We drive out to the diesel running shop at Bondamunda – a 15-minute ride. The whole rail precinct is a massive, sprawling complex of railyards (with hump shunting being handled by a steam locomotive!), diesel and electric shops, a steam depot seeing out its last years, a modern wheel lathe shop, and a network of interconnecting tracks. A large railway staff settlement lies adjacent with an associated 'railway market'.

Milt shows me around and introduces me to the various foremen and the depot manager with whom we'll be working closely. I spend most of the day reading: getting myself up to speed on the air brake system and the Class WDM-2 locomotives (**W**ide-gauge, **D**iesel, **M**ulti-use, version #**2**). Built in India at the giant Chittaranjan Locomotive Works, these units are based on an ALCO RSD-29, this design having become Canadian builder MLW-Worthington's C-Series model for export. The WDM-2 is a C-26 model of 2400 hp with 63 000 lb of continuous tractive effort and an Alco 16-251C engine. Similar units—the Goodwin-Alco DL531 model—have been built in Australia, and include the New South Wales Government Railways 48-Class, the South Australian Railways 800-Class, three for the Silverton Tramway at Broken Hill, as well as the NSWGR 45-Class (known as the 600-Class in South Australia).

The units have a high short hood and dual control stands. The seats have circular padded vinyl squabs (unless the squab has worn off and is missing, in which case the seat merely consists of the circular timber base—well-polished!) and no backrest of any sort. The thing is basically a stool. Neither do the units have an operable vigilance device.

The four units we have here (two to be operated as Lead and two as Remote units) were all recently reconditioned and are thus quite tidy. The LOCOTROL II system to be installed is the new 'Universal' model that places all equipment on the locomotives and thus requires no remote-control car to be coupled to the Remote consist. Locomotives can be configured as Lead or Remote units at the flick of a switch.

The air brake schedule is 28L-AV (operating air-over-vacuum: the units being used by SER on air- as well as vacuum-braked trains) using the A-9 Automatic and SA-9 Independent brake valves. Not familiar to we Aussies or Kiwis, the A-9 is basically a rather low-capacity valve originally designed by Westinghouse for use in switching locomotives. Brake pipe pressure is 70 psi (500

Figure 85: Steve Heneka (L) and Milt confer in our office at the BND Shops.

Figure 86: Two Locotrol units undergo back-to-back testing outside the BND Shop

kPa), however the locomotive air pressure gauges are calibrated in kilograms-per-square-centimetre, so we read the pressure as 5 kg/cm^2. This means that a minimum equalising reservoir reduction is 0.5 kg/cm^2 (50 kPa); nothing too strange there.

For the air/vacuum brake system on the locomotive, the WDM-2 units have a combined exhauster/compressor. One interesting facet of this air/vacuum brake system—to this railroader anyway—is the process on the locomotive by which a brake application produces locomotive brake cylinder pressure. There are five steps in the process versus the three an engineman might ordinarily be familiar with: ER reduction > BP reduction > control valve admits MR pressure to the brake cylinders.

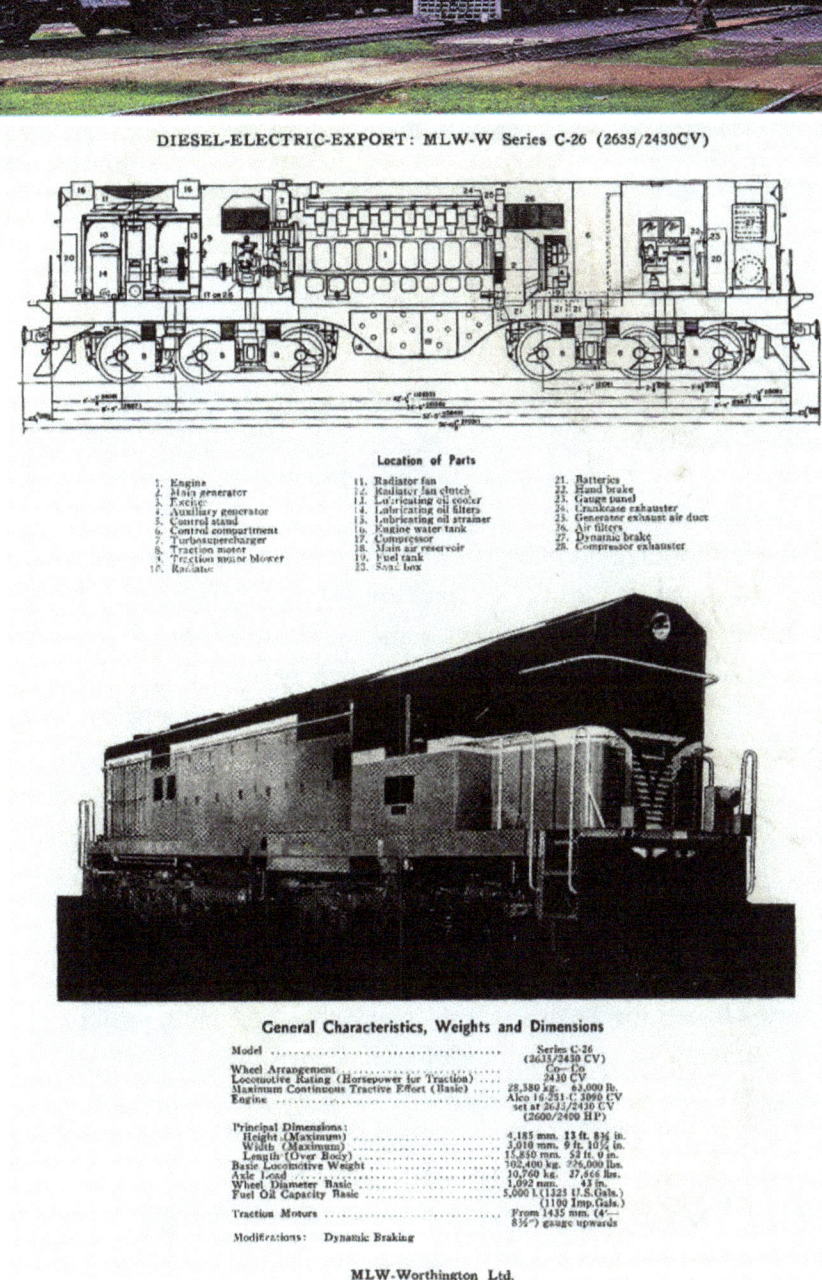

Figure 87: MLW-Worthington C-26 locomotive - General Characteristics.s.

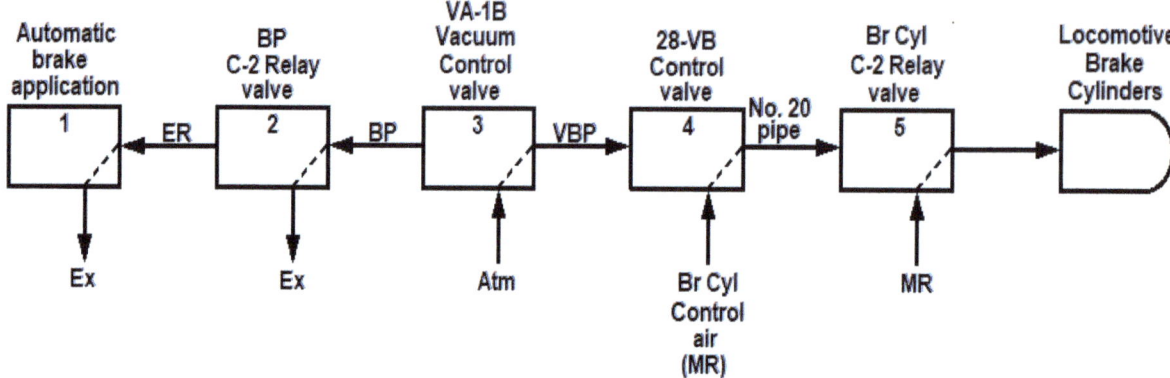

Step 1: Automatic brake application reduces Equalising Reservoir pressure
Step 2: Brake Pipe relay valve responds to ER reduction by reducing BP to the same value
Step 3: Vacuum control valve responds to BP reduction by admitting air into the Vacuum Brake Pipe (partial destruction of vacuum)
Step 4: 28-VB control valve responds to increased VBP pressure by admitting a control pressure to the Brake Cylinder relay valve
Step 5: Brake Cylinder relay valve admits Main reservoir pressure to the locomotive brake cylinders

Figure 88: WDM-2 locomotive air-over-vacuum braking system – development of brake cylinder pressure.

There is a steam shed about a kilometre away, from which are dispatched yard shunt, local passenger and 'inferior' goods locomotives. We have a good, roomy office that is air-conditioned and has a Western-style toilet. These facilities are equal to those enjoyed by the more senior Indian Railways regional managers. Milt has seen to that.

This railway region is on the broad 5 ft 6 in gauge, known here as 'Wide-Gauge' (and also as 'Indian Standard-Gauge') and the main lines are electrified. The diesel shop is very Dickensian: old, filthy, and very labour intensive. Sari-clad peasant women walk around barefoot in oil and grease, carrying cement and bricks on their heads or doing other menial labouring jobs. There are many men working in positions that could best be described as tradesmen's assistants. Their usual footwear is rubber thongs. No-one wears any sort of personal protective apparel whatsoever. There are guys working on our units who are hacksawing steel (for example, cutting the holes in the cab control stands into which the LOCOTROL Air Brake Consoles will be installed) using just a blade held in a piece of cloth! Obviously, this is a very laborious task so there are usually two men who alternate between sawing and holding the work, or resting.

Saturday 8 July

No rest for the wicked as we go back to the diesel shop today. I busy myself with more reading and prepare some test sheets for the drivers I shall be instructing. I am given to believe they do not have a great technical grasp of their job and that some are illiterate. I wish myself luck! My classes are going to be videotaped, I heard today. (*'Oh, did I not tell you about that…?'* exclaims Milt, wide-eyed).

The regional Chief Mechanical Engineer from Kolkata arrives, complete with armed escort, to inspect progress on the project thus far. Apparently, he expected to see a demonstration – he is disappointed. His entourage tag along at a respectful distance as Milt waxes eloquent on Interface Modules, equalising reservoir leakage and software programs. Around here the CME is G-O-D. This gentleman is a well-travelled engineer and seems to understand quite easily what is being explained to him.

Sunday 9 July

Eugene Smith (the 'bearded wonder', as Milt likes to call him), Harris' electronics engineer[76] on LOCOTROL and Milt's right-hand technician has flown out from the US to join the team and is supposed to arrive from Kolkata by train, as did I, at 6.00 am. However, we hear that his train is stuck somewhere on the wrong side of a freight derailment. God knows when he'll make it. Rest day today.

The Hotel Radhika is okay. Each of the four floors has what they call a 'suite', and we four *gringos* occupy them. Milt has also seen to that. My fourth-floor room contains a wide double bed (two single mattresses), colour TV (video movies twice a week: I saw the ever-hilarious John Cleese in *A Fish Called Wanda* last night), a writing desk, a coffee table and two easychairs, a large robe, and a noisy wall-mounted air-conditioner.

Figure 89: Eugene Smith, 2007

There are two adjoining rooms. One is a roomy bathroom with both western and Indian style toilets (side-by-side, take your pick—sit or squat), shower and basin. The other is a small lounge room with another coffee table, two vinyl armchairs and settee and (most valuable) a refrigerator with padlock. The refrigerators are not standard items for these rooms; Milt has insisted upon them. The floor coverings, the decor and the electrical wiring are nothing to write home about, but it's all much better than I'd expected. The hotel dining room is quite good but there's no bar. The telephone service appears reasonable—I shall ring home once a week.

Monday 10 July

Gene made it into town Sunday evening. Today, he gets his first look at the Bondamunda shop; he will start running his technical school next week. Work is continuing slowly on the units. Milt says the Indians always want to see your plan but are not good at planning ahead themselves. The tradesmen will work, but you must be there telling them what to do. Our Indian agent has supplied two people—one technical and one operating—who are supposed to remain here after we have left to assist the locals to keep this equipment operating. They are to become the local gurus on LOCOTROL. Well, I think that the technical guy is going to be useful, but the operating guy—an ex-locomotive driver—is not. I gave him a test on basic knowledge of a WDM-2 locomotive (he had considerable experience on this class) and he scored about 30%. Also, it appears that he has

Figure 90: View of the main street, Bisra Rd, from my 4th floor bedroom window above the front entrance to the Hotel Radhika.

---

76 In 2007, Gene, a South Carolinian, was awarded the prestigious General Electric Edison Award in the 'Transportation' division. This award—named after one of GE's founders (and perhaps history's most prolific innovator)—is presented to an individual for recent technical contributions that have made a significant impact on the current and future vitality of their business. Gene's unique combined expertise in the remote control of locomotives and electronic and pneumatic air brake systems was acknowledged. In association with the award, Gene chose a university to receive a $25 000 research grant. Gene is co-holder of numerous patents.

never driven an air-braked train in his life. I think we're going to have to let him go. Not his fault, he's a pleasant-enough fellow. Bad choice by Scientific Timesharing Systems.

Tuesday 11 July

Milt has been hobbling around all day and this evening a local doctor diagnosed some blood poisoning. It apparently came about through an incident a couple of days ago when he barked a shin on the sharp edge of a heavy wooden rubbish bin in our office. The doc offered Milt a shot, which—not being convinced of the quality of Indian penicillin—he declined. So, he has settled for some oral medication.

Figure 91: Bisra Rd from street level, close to the hotel front entrance (note the 'working' elephant in middle distance).

Friday 14 July

I've completed my third day in the classroom. It is not easy. The room is upstairs in the diesel shop and I must compete with locomotives being throttled in Notch 8 and air horns being tested with long, drawn-out blasts. The facilities are very primitive (I have a blackboard and coloured chalk to draw diagrams) and there are language difficulties. The drivers do not understand much of what I say. I have an Assistant Divisional Mechanical Engineer, Mr Gupta, who translates into Hindi for me when required; God bless him! There are several locomotive inspectors in each school (Mr Gupta informs me that only drivers who have received a high school education can be promoted to Loco Inspector). They are a little more knowledgeable, but I am not overly impressed by these peoples' basic knowledge of locomotives and the air brake, or their powers of comprehension of the subject. I suppose LOCOTROL is a challenge to grasp whatever country you're in, and I don't mean to demean these folks… I get to know many of them and enjoy their company.

Milt and Gene had a win today. They got the first set into LINK mode! This was a major achievement but getting to the next step—CONTROL mode—will be harder: the locomotives have so many equalising reservoir and brake pipe leaks. We celebrated at dinner by lashing out and ordering some different dishes from the menu. Let's hope we all survive!

Sunday 16 July

And on the seventh day we rested. We have breakfast late (I have a head cold and can't taste anything) and spend the morning individually bringing our expense sheets up to date. Later, we meet to review progress on the project. Gene is having trouble with the radios. He thinks he may have to get someone from the manufacturers, Aerotron, to come over and look at them. (It'll take at least a week to get someone here from the USA). Also, some of the equipment got wet in the monsoon when it was left outside at the Kolkata airport customs yard. This is also causing headaches. Speaking of headaches, Milt's leg isn't getting any better (but no worse) despite a tetanus shot and some penicillin. He will go to Kolkata by train if it doesn't improve tomorrow. Apparently, there's a better doctor there. I lay on my bed and watched Agatha Christie's *Evil Under the Sun* on TV this afternoon.

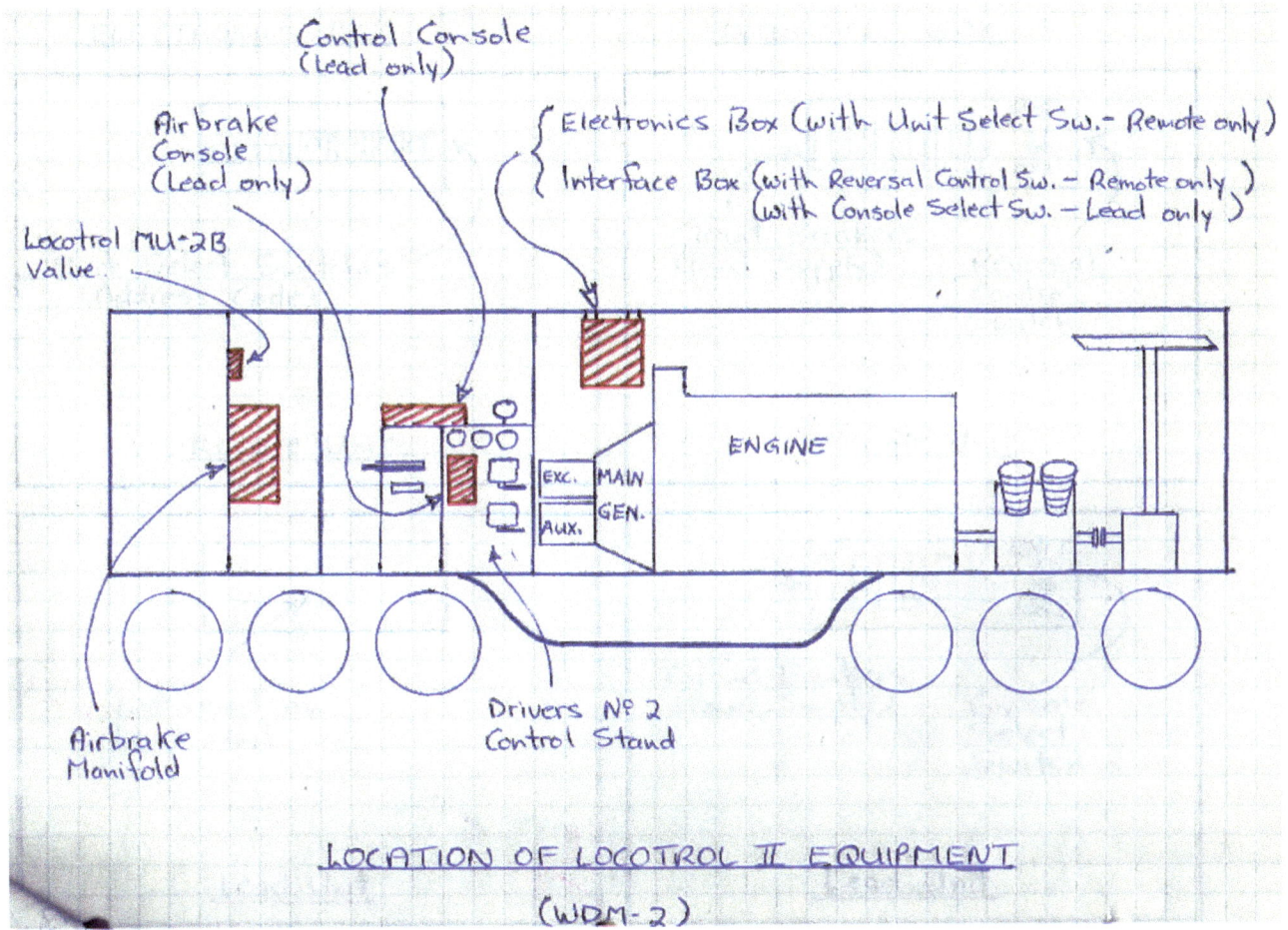

Figure 92: Sketched in a notebook, this is my plan for an instruction slide depicting the location of Locotrol II equipment installed on WDM-2 locomotives.

Tuesday 18 July

Forget Kolkata—Milt went into hospital here yesterday. He has a good case of blood poisoning and a very sore vein running right up the inside of his leg to his groin. We visited after work today. He's pretty drugged up – I think they took a lot of blood for tests. Being a *gringo*, he's something of a celebrity. Everybody wants to see him, using any pretext to come into his room. The nurses all wear white saris with a white or red leather belt around the waist. The hospital is one of the many facilities built by the Germans in the mid-50s when they built the steel works here. It's quite good. Milt has his own room with bathroom, so he's got the top facilities available, I think. He says the nurses, and everybody are just great and eager to please. Driving out to the hospital I got my first glimpse of a whole new part of town. The Germans have built a huge suburb out here with dual-carriageways, tree-lined streets, multi-storey apartment blocks, shopping and recreation facilities—and trees everywhere. The omni-present cattle still wander around at will and lie on the median strip, chewing their cud.

The Indians have been pressing us for a road trip on Sunday, but Gene said, 'Forget it!' Their units are still in a mess with ER and BP leaks a-plenty and Gene has been trying to tell them the system won't tolerate that. Anyhow, Milt has already warned us that if we don't get both sets of units finished before we run any road trips, they'll just get so carried away playing with the first set completed, they'll never get around to completing the second set. We're running out of time on this contract and I think some sort of extension is going to have to be

Figure 93: Ispat General Hospital, Rourkela.

negotiated. I asked for a group of locomotive inspectors to train for 4 days, and happily, I begin with them tomorrow.

Wednesday 19 July

Milt has been diagnosed with cellulitis and filariasis (get your medical dictionary out!). His symptoms are inflammation and infection of the soft tissue in the affected leg, blood poisoning, and thrombophlebitis. So, he's receiving dual medication and hopes to be out by Friday.

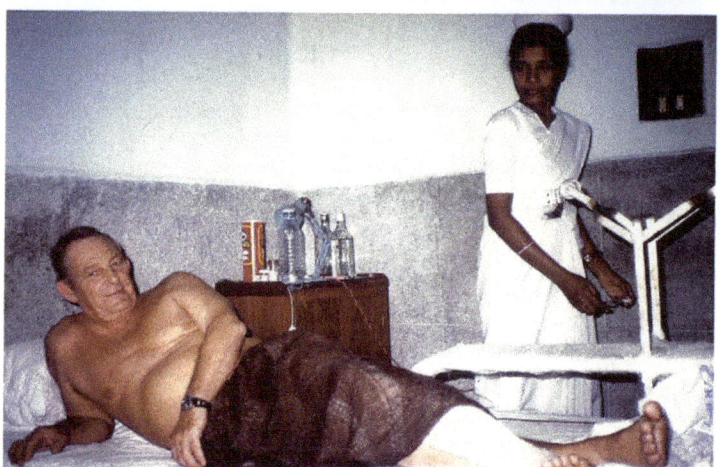

Figure 94: Milt receives celebrity attention at the Ispat General.

Our driver, Malkit Singh, is a Sikh who doesn't wear a turban. He tells us that once, when he was driving in the country, he was stopped by thieves who hung him in a tree. Fortunately, a passer-by cut him down before he choked but he has been left with a medical problem that causes discomfort if he wears a turban, so he has a permanent medical dispensation from what is otherwise a religious requirement. He's 23-years-old, speaks good English and is a real character. He's devoted to all of us but is not the least bit servile and wants to know everything about

Figure 95: Our project office, technician instruction classroom, and equipment store at Bondamunda diesel shop. Entirely adequate.

Australia and the USA. He is invaluable and will always ensure that we don't get cheated. I heard today that both Milt and I have received formal authorisation from the regional head office in Kolkata to drive the Indian Railways trains (so long as there is an operations inspector in the cab).

Sunday 30 July

This is our day off, so I'm catching up on some paperwork. I have completed my weekly expense sheet and am going to send some postcards home to see if the Indian postal system works. Malkit has come by and he's going to show me how to wrap a turban. Regarding Milt, the thrombophlebitis has placed him in danger of developing a blood clot that might travel to the heart, and he has been evacuated by Harris' medical representatives to New Delhi on a corporate charter flight. He will probably not return to Rourkela. Gene wants him to go back to the States. Milt says that if we go up to Guwahati he'll come up when he's discharged. Gene is not happy about that: there is presently bad flooding of the Brahmaputra River up in Assam as well as some political turmoil. There is news of people being killed up there. Milt and Gene are advocating caution and feel that perhaps we should all return home until the troubles die down, then return and finish the job. The Harris people in Florida, however, say they have made enquiries and that it is 'okay' for us to go up there. We'll see what happens.

This past week, I have made two road trips south of here on their standard 56-car ore trains to check the road to the mines (let's call it 'learning the road'). It is rugged jungle country, but the track profile isn't too bad. There's only one stretch of 2% (1:50): however, it includes two tunnels (one with a reverse curve) so it'll be interesting to see how LOCOTROL handles it. There is now only Gene and I here in Rourkela to finish the system testing and to do the road test trips. Then Gene is to hold a week-long maintenance school and I will complete the road trips by myself.

As it has turned out, my presence here has prevented the project from being postponed until Milt can get back, since Gene cannot run road trips by himself. He is an electronics engineer, not a railroad operating man. The road test and commissioning trips will be of 3 days' duration each. We will leave here on the empty train and travel south to the mine – about 60 km distant. They will load the train while we stay overnight. Next day we are scheduled to depart, and return loaded to-and-through Bondamunda (the railyard complex for Rourkela) and continue north to the steel mill at Bokaro – about a 13-hour trip they say. Here we are supposed to overnight again while the train is unloaded, then return with the empty to Bondamunda by a different route (thus completing a circle) and then do it all again, hopefully a couple of days later. I have requested two loco inspectors to be the engine-crew and they have agreed to this.

I have been accompanied on the road trips thus far by Senior Locomotive Inspector Mr Anwar Ahmed who is an extremely personable and helpful gentleman. On my second 'learn-the-road' road trip to the mine, we returned on the local passenger service—steam-hauled by a dilapidated 2-8-2. They have two firemen on board and those boys really get into their job! This locomotive has a 46 sq-ft hand-fired grate and a tender that holds 19 tonnes of coal that requires a lot of shovelling forward. They're good with the banjo though! On each side of the huge cab is the remains of a small swing-out seat that once had a wooden squab on it. Nowadays only the rusty metal base remains and is unused—both driver and firemen stand! The cab deck is wide and deep and extends right back in under the shovelling tray, obviating any need for an apron plate. It was a great trip and Mr Ahmed was clearly in charge—speaking often with the driver and motioning the boys to shut off the injector or put another fire on. Not that they couldn't have done it without him. But here, the system is very old-school hierarchical, and when someone senior is around, he receives due deference. It reminds me of the 'old' NZ Government Railways, and if Milt were here it would no doubt remind him of his CP Rail.

Wednesday 2 August

Today Gene and I (with me pretending to be Milt) completed back-to-back testing of the second set (Lead and Remote). LOCOTROL-wise it is fine: however, there are still air leaks for the Indians to attend to. I should explain here that Harris had originally asked for—and been expecting—new units on which to install these systems. What they got when they arrived was 'refurbished' units out of the shop after an overhaul. There is a major difference. Because they've never run very long trains in India, they're not used to having to keep air connections tight. Although an iron ore train weighs over 4500 tonnes, it is only 56 cars long and vacuum-braked ore trains are limited to 26 cars. Milt says there are similarities with Iran. There are indeed… Indian Railways runs the same two-pipe air brake system on their Wide-Gauge freight trains here that Milt encountered for the first time in Iran. As in Iran, the SER only bothers using the brake pipe, and the main reservoir pipe is never connected.

The units they have supplied are an excellent example of an 'overhauled' locomotive. Almost all joints in the locomotive piping leak. The Lead unit has 11 psi/min leakage on the brake pipe and 7 psi/min from the equalising reservoir. The Remote unit is a little better. The LOCOTROL system cannot tolerate much leakage from the ER circuit. These four units we are working with are pathetic considering they have just received A-grade overhauls. One had 40 psi/min leakage from the BP when we first tested it! It has been a herculean task to get that down to 11 psi/min, including much soaping of pipe joints. As I've mentioned above, the Indians have lived with this sort of thing for years and consider it normal. They just plain didn't believe Milt when he told them that at Mt Newman Mining we have 240-car trains with no BP leakage. We are pessimistic about our chances when we finally get into the yard to test a full-sized 112-car consist. It is believed that leakage from the cars will pose grave problems and may even prevent the system from operating. There is no reason to believe that the cars are maintained any better than the locomotives.

Presently, they are rushing to fit high-tensile 120-tonne drawgear to the 112-car consist we will have on our commissioning trips. This was supposed to have been done months ago. Tomorrow I shall take a group of locomotive inspectors and some drivers and we will take the first Lead/Remote set and run it up and down a track in the yard so they can get some 'hands on' experience. Gene starts his maintenance school today. We hope to do our first road trip late next week. Milt will be in hospital in New Delhi for another 7 to 10 days. Then he intends to fly to Scotland to visit his married daughter and renew his visa so that he can return to India and go up to Guwahati if necessary. Guwahati is looking more doubtful every day. Flooding is critical up there and there is no way of getting our equipment from here by road or rail. It's beginning to look as if we'll have to leave India and return later. Steve Heneka should be going up to Guwahati today. If he gets there (Air India B737 flight) he'll hopefully be able to contact us and let us know the situation regarding floods and political problems.

Figure 96: Gene graduates his first Indian Railways Locotrol Technicians course.

Figure 97: Gene Smith's techie's classroom at BND.

Sunday 6 August

On Thursday, Friday and Saturday I had groups of drivers in the yard with one of the LOCOTROL sets, back-to-back. This is valuable for them as they can begin to understand and practise the operating sequences. It's valuable for me also, as I can begin to gauge the value of my instruction and identify where any changes may be required. Regarding the second portion of the project, it now appears that our equipment is to be flown to Guwahati. We have also heard from Steve that flooding there is receding and that the Hotel Brahmaputra is better than the Radhika here at Rourkela. It is beginning to look as if we'll be continuing from here to Guwahati as planned.

Tuesday 8 August

Today I made a road-learning trip to check the route north from here to Hatia. Much of it is graded at 1% and there are numerous cuts and curves that should prove interesting as possible Communication Interrupt areas ('Comm Int' is the system state when Lead and Remote radios are unable to maintain their radio link). However, there are no tunnels on this route.

The scenery is spectacular – verdant, rugged, and hilly, with most areas wooded. Valleys and gullies are in rice paddy. It's amazing where they can put a paddy field and equally amazing as to where the water comes from for those at the higher levels. Any land neither wooded nor in paddy is badly eroded and the rivers are permanently muddy. Along the way, there are small country stations - many nowadays open only for passengers. Some are extremely picturesque with the main line curving through town beneath large, leafy trees.

The Distant signal gantry lets us know if we've got a run through town and—if the signals are 'off'—we do indeed come charging through, kicking up dust, with the horn blaring lustily. Our exhaust blasts through the luxuriant canopy overhead, whipping branches and sending leaves flying. The stationmaster comes out in his white uniform and waves. People stare with interest, dogs peer with boredom and scrawny chickens run for their lives. Our brief, intrusive presence may not excite everyone, but it cannot be ignored.

I travel out on a loaded ore train bound for Hatia and beyond to Bokaro Steel City. We are a standard consist of three units and 56 cars (class 'BOXN'[77]). The cars are gondolas and are also used for carrying coal and limestone. They are 25 tonnes tare and hold 58.5 tonnes of iron ore, which fills them about one-third full. When loaded with coal or limestone, they are up to the gunwales.

I returned on the local passenger—making all stops. Between stations the driver wasn't averse to some fast running—sometimes hitting 95 km/h. I called him 'Casey Jones', and when I explained the legend behind the name, everyone thought it a great joke. His name was Abdul and he had 2 years to go until retirement.

Every station has a signal cabin at each end and the most magnificent traditional semaphore gantries. Half the signal lights don't work but that doesn't seem to faze these boys. They know where they are and at night they simply slow down until they can see whether the stick is off or not! Unless we have a 'meet', we pick up the safeworking ball-token on the fly with a hand exchange. Sometimes they become short of cane slings, so they use a length of sapling bent into a rough circle and tied where the ends meet.

Figure 98: Indian Railways BOXN ore wagon.

---

[77] IR rollingstock classification: **B**ogie/**O**pen (Gondola)/'**X**' = high-sided pneumatic (air-braked). They're good for 80 km/h empty and 60 km/h loaded.

The brick or stone signal cabins of those stations that are closed still stand at each end of their overgrown yard locations, looking for all the world like pillboxes on the Siegfried Line. Yard trackage has been lifted and these places exist only as passenger stops. The local villagers make frequent use of the train. They will often jump a freight or ore train and ride from one station to the next. Sometimes they climb onto the locomotive consist and ride the rear platform. Sen. Inspector Ahmed tells me that sometimes drivers who eject these 'passengers' are threatened with physical harm should they come through the area again. Elsewhere, locals have become adept at removing brake shoes from wagons of stationary trains. They will then offer them back to the railways at Rs10 each.

Friday 11 August

I attended a meeting this morning with Devendra Rudola, the Senior DME (Divisional Mechanical Engineer) and overall manager of this diesel shop. Gene informed him that due to delays by Indian Railways we were behind schedule and outside our contract time. Gene intimated that we might have to pull stakes at the end of next week if we didn't get the 112 cars for testing in the yard on a full-sized LOCOTROL train – a standard commissioning procedure. This made a difference. Later that afternoon we were invited to Rudola's office and informed that officials in Kolkata were getting things moving. We now expect the cars early next week. Unfortunately, Tuesday is their Independence Day holiday. If we get the cars next week, we may be able to run two commissioning trips before we must leave for Guwahati. This would not leave these people the slightest bit equipped to continue running LOCOTROL trains. We have advised Harris of this situation and I believe that George Vorhees from Harris Controls in Florida is intending to come to Kolkata to discuss another contract for more training with SE Railway officials. What this would mean for me (if anything) I don't know. Also, Lewis Cox, a lead engineer with Harris, is coming over next week with equipment to make some software changes to the system here.

Figure 99: Learning the road on a local passenger train. L-R, the author, driver Abdul, driver's assistant, Sen Insp Ahmad.

Dev Rudola is hoping he can have the office word processor/PC (which is down) fixed. Technical support for the computer in their office apparently comes from Bombay[78] and takes a month. I have just completed the compilation—writing longhand—of a driver's operating manual for these folks and this delay means that I cannot get it printed and checked. I am also presently trying to get a short training video made. The major daily problem is the rain, which is torrential and is accompanied by the loudest thunder I've ever heard. The resultant power cuts which are stymieing our video production are a way of life here.

Sunday 13 August

A good day today. I got calls through to friends in Port Hedland, and to my wife who is holidaying with her parents in New Zealand. I stayed in my room most of the day and relaxed. I need to get my strength up for another week of frustrations beginning tomorrow. We heard from Milt. He's expecting to leave New Delhi for London and then Scotland on Friday and then back in India by the 25th. Heard, too, from Steve Heneka

---

78 Now Mumbai.

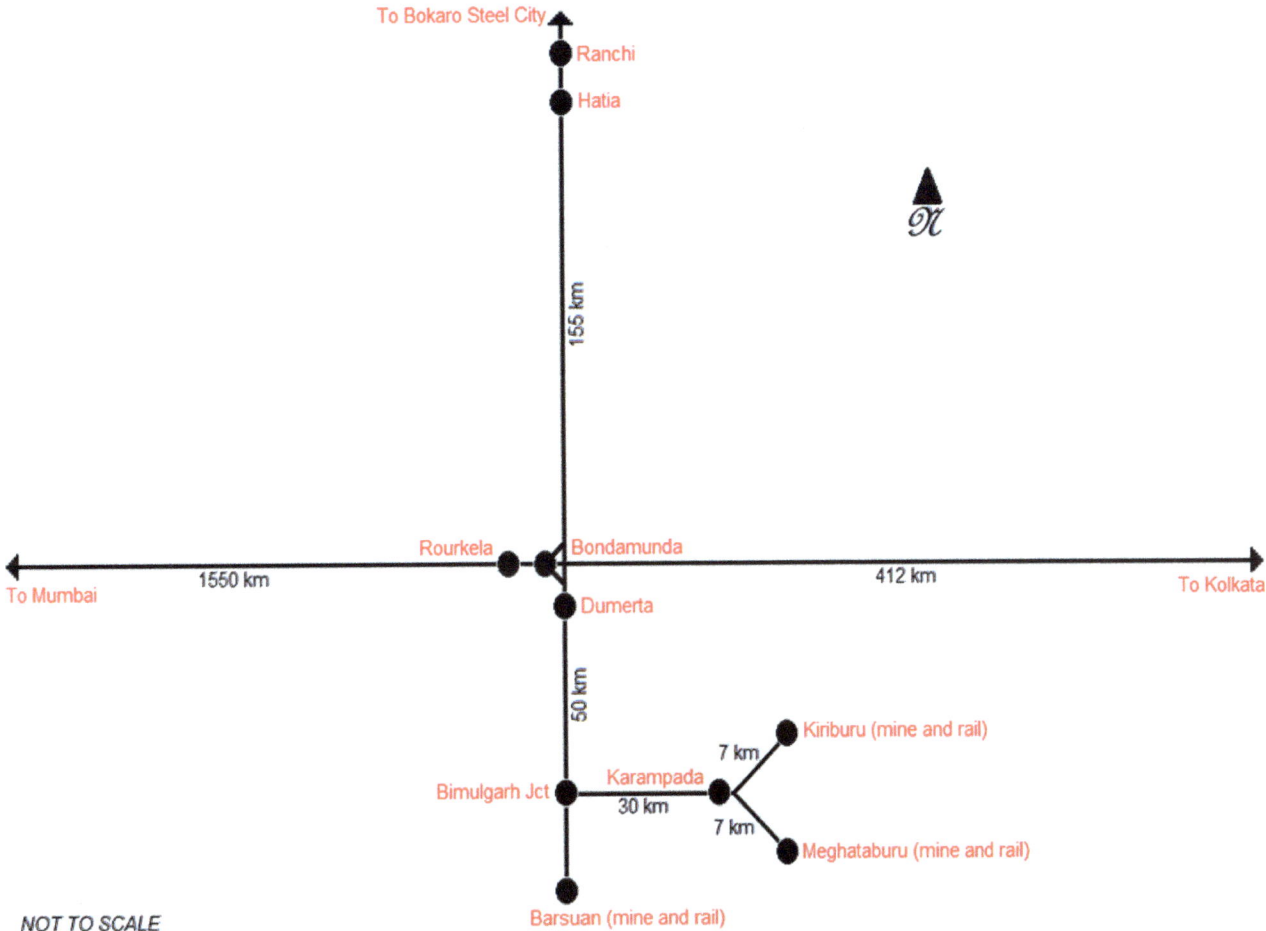

Figure 100: Diagram - Locotrol test train routes – South East Railway 1989.

in Guwahati. He's making steady progress fitting equipment brackets to the YDM-4[79] locomotives there. The YDM-4 is the 'junior' (Metre-Gauge) version of the WDM-2. He says the hotel is better than here in Rourkela!

I'm looking forward to meeting Wayne Barber when I get to Guwahati. Wayne is the Manager of Systems Engineering for New York Air Brake, who supply the air brake manifolds to Harris. Milt has brought him into the project up on the North East Frontier Railway because it will be the first time that LOCOTROL has been installed on vacuum-braked rollingstock. Wayne will oversee the installation of air brake equipment in Guwahati and troubleshoot any unique problems that may occur. Milt calls Wayne 'Mikey' (after a greedy kid on some US television commercial) because he will allegedly eat anything—and he has begun calling me that also ('Mikey II' actually). Can't imagine why! Truth is, Milt is jealous because I'm not so nervous about the food and can eat a more varied diet here in India than he can. He reckons Wayne and I will get on well together. I must say that when it comes to experimenting with food, I think Milt is (to use his own unique vernacular) a 'G*d*m flake'.

Sunday 20 August

Today is the BIG day. We run our first test trip. Excitement is high. We—for our part—want to bring it all together for these people. I think there are some here who are going to have to *see* a LOCOTROL train before

---

79 Indian Railways classification: '**Y**' = *metre-gauge/**D**iesel/**M**ixed traffic/**4**th version (model)*.

they really understand what it's all about. Most of the people we've been working with are breaking their necks to see it happen, but I suspect there are a few who would love to see us fall flat on our ass.

Well, the day started bad and got worse, although not for the system, which worked perfectly. They were supposed to be ready for us when we arrived at work; they weren't. Finally, we got the train together and were ready to go. But only as far as Dumerta—the end of the double track, just out of town. There was a train disabled in the section ahead and it took forever to deal with that. Eventually, we get 'Line Clear' from the stationmaster and are rolling at 1530—5 hours late. In addition to me, there is a two-man engine crew, a couple of locomotive inspectors, various engineering personnel, and a few technicians aboard the Lead units, and Gene and Lewis are aboard the Remotes, again accompanied by numerous tech people.

After an uneventful 3-hour run through superb jungle scenery and mountainous terrain, we arrive at our destination, Karampada. Here, the train will be split into two portions to be loaded: one at each of two nearby mines – Kiriburu and Meghataburu. We shut the system down and leave Karampada crews to split the train. We are taken by light engine up to the Kiriburu mine yard (where one half of the train will be loaded overnight), and from there by jeep up a steep mountain, past the actual Kiriburu mine, and on to the town which serves both mines. Unfortunately, it is now dark, and we can't see the spectacular jungle scenery.

We are at 3500 ft and spend the night in the SAIL (**S**teel **A**uthority of **I**ndia **L**td) guesthouse. It is just 5-years-old, but like everything else around here is already covered with mildew and the gardens overgrown – otherwise it is quite reasonable. We arrive close to midnight. A staff member has stayed up and cooks us fried chicken and rice. He brings it to our room with tea. Bless him; we are beat! Lewis goes off to claim one of the two rooms allocated to us. Gene and I eat and review the run before turning in. The place is air-conditioned, but we sleep under the mosquito nets just for the novelty of it.

Figure 101: Kiriburu rail yard and ore loadout facility.

Monday 21 August

We are up at 0500. This will develop into our worst day. By 0600 our transport hasn't arrived as planned. Staff are soon moving about the guesthouse. We breakfast in the dining room—just the three of us. At least we hadn't expected to be able to have breakfast: a small bonus. Later, we are guided out and through some woods to see—unfolded before us—the most magnificent view imaginable. From our elevated lookout, we feast on a vista of cloud-shrouded, forested hills, waves and waves of them rolling away before us. It is surely a let-off: we would otherwise have left here in the pre-dawn darkness unaware of our beautiful surroundings.

Our guide explains that there is elephant and tiger out there. Sated, we return to the guesthouse for coffee and I have a look in the office at the guest register. Two days ago, there was an Australian here as a guest of SAIL. His name was Freeman and he was registered as a 'consultant'. Could it have been one Gordon Freeman, who of course is now a freelancer following his unfortunate departure from the CEO's chair at Mt Newman Mining, and who knows more than just a little about iron ore production at one of India's export competitor's biggest mines?

Our jeep arrives at 0830 and we descend the mountain to Kiriburu yard where we board the consist at the head of our front portion (56 loads) and move off to dynamic-brake down to Karampada – about 7 km away. Our rear portion is supposed to be coming from the adjacent mine, Meghataburu. It isn't. We learn that they had a power failure during the night and couldn't load. In addition, it rained. Everything is wet up here. This means that the expected 2-hour delay extends to 10 hours.

We finally depart at 1830 and now I must take this 9000-tonne train down 20 km of twisting grade at a maximum allowable speed of 32 km/h, in the dark. Also, there is a tunnel that begins just outside yard limits and within which a 2% grade commences. While the train is in this tunnel, I am going to have to change from power into dynamic braking and make a minimum brake pipe reduction to settle the train. Having been on duty for 13 hours already, I could sure do with daylight just now. This would be helpful too, for the other drivers and inspectors in the cab who are supposed to carefully observe the required train handling techniques. Regarding this latter, Sen. Locomotive Inspector Ahmed and the engineers on board have left little to chance. They have delegated another of the locomotive inspectors the task of noting—on specially prepared sheets of paper where each line represents one kilometre of track distance—every throttle, dynamic, and air brake adjustment I make as we descend the mountain, and the relevant kilometre peg.

There are several 15 km/h slow orders on this section but everything goes well. At the foot of the 2% grade I vacate the engineer's seat and take up my position on a portable seat behind to guide him on the remainder of the trip. We have a trouble-free run and arrive at Dumerta—a small village just outside Bondamunda—at 2105. Lewis, Gene, and I have been going 16 hours. The system is shut down and the train is split to clear grade crossings in the yard.

The two locomotive consists will be manned overnight for security reasons, but we are driven back to our hotel. It is very late, and Gene and I sit in his room and eat a snack of leftovers from our lunch boxes before showering and turning in, exhausted. The reason we leave the train out at Dumerta is because there is no road long enough to accommodate it in the Bondamunda complex. Also, the Bondamunda railyard is a 'trap'; once you're in there you never know when you'll get out again. A section of double track begins at Dumerta and we use that to stable the train. In the morning, we'll continue from this location, bypassing the Bondamunda yard via a cut-off line and heading north to the city of Hatia.

Tuesday 22 August

I sleep through my alarm and the phone wakes me. I shave and dress in double-quick time and just catch a hurried breakfast with Gene and Lewis before grabbing some fruit, hard-boiled eggs, and curried parathas from the hotel kitchen, and climb into Malkit's car. We race off to Dumerta (reached by a very bumpy dirt road through rural paddy fields). Today turns out to be what Milt would call a 'piss-cutter'. For the uninitiated than means, 'pretty darn good!' There are people all over the place. My head-end crowd has swelled to eight, which is the maximum Milt has said I should accept in the cab at any one time. I should think so!

Here's a list of the head-end complement, spread over the three-unit consist (providing a new meaning to the term 'distributed power'):

- The engine crew (two senior locomotive drivers)
- The senior locomotive inspector and a colleague locomotive inspector
- A radio technician (who has brought on board a two-way radio about the size of the old STC units we used to have in the Railroad Ops vehicles at Mt Newman about a hundred years ago. It is powered by car batteries in a large wooden box with carry handles, and which communicates with the brake van via a huge HF antenna tied upright to the front handrail outside the cab)
- The senior divisional mechanical engineer
- An engineer from RDSO (the Indian Railways' **R**esearch, **D**esign & **S**tandards **O**rganisation, who have been responsible for ordering and specifying this LOCOTROL equipment)
- The South East Railways' senior electrical engineer (rollingstock), who is interested in putting LOCOTROL on their 25 kV WAG-5 (**W**ide gauge, **A**C electric, **G**oods, **5**th version) locomotives, of which they have a fleet numbering 350

I don't much care how many people there are, so long as they don't crowd me and the driver. The boys already have the train together, so we get the system up, do an air test, and set off after only a 1-hour delay. We haven't been going 30 minutes when we begin to get constant wheelslip from our Lead unit. This doesn't

Figure 102: Our hotel in Ranchi, overlooking the railyard. Get a front room!

bode well as we have *beaucoup* 1% grades coming up ahead. We have a sparky on the head-end consist and he comes up to the Lead unit and swaps the arc chutes on the transition and field shunt contactors. This doesn't improve things.

We are lugging along an extra unit on the head-end consist as insurance. The Senior DME doesn't believe me for sure when I say that we only require five units to take this 9000-tonne ore train from here to Hatia (they use three units for 4500 t). I ask them to put our Lead unit offline and to put the third unit 'on'. Some way up the grade we get three or four Ground Relays (which the Indians call 'power-grounds') on this third unit. They leave it online for a while and are rewarded with another GR. Now they decide to take it offline. We must put our slippery Lead unit back online as four units won't do it for us. To prevent it wheelslipping, we hold it down in Throttle 5 and 6. To compensate for this reduced head-end power, I use Independent Control to hold the Remotes in Throttle 8. This overloads them and we get Remote wheelslip. I jockey the Mode Selector in time with the WS light and we grind on up the mountain at 15 km/h. Then we get a Hot Engine on one of the Remote units. We struggle into Bano (halfway up the mountain range), pull to a stand, and throttle up the Remote consist to run the radiator fans faster—which cools the engines down—while we all detrain and wander across to the local market for a cup of chai.

During this stop, a foreman electrician comes up from the Remote consist and administers to the Lead unit. We have no more wheelslip problems, and no more hot engine problems either. The rest of the trip to Hatia is trouble-free. A passenger train is running late, and we are held outside one station to await its arrival. Other than that, we highball through every station, exchanging the token on the fly.

We arrive at Hatia in the late afternoon after a perfect trip (as far as LOCOTROL is concerned, that is). The engineers on board are so elated they want us to continue to Bokaro Steel City, where the train is to unload, but Gene and I have been going long enough. We decline the proposal.

The chief electrical engineer wishes to discuss putting LOCOTROL on their electric locomotives and has a 30-minute meeting with Gene[80] before we grab a taxi and hightail it to the Kwality Inns hotel in Ranchi (Pop. 900 000)—a kind of twin city to Hatia. I celebrate with a few Kalyani Black Label beers, Gene with Thumbs-Up cola, and we have a mighty meal in their excellent restaurant. We have completed installing the equipment, we have tested it statically in the shop, and now we've successfully commissioned it on the road through the toughest terrain the SER has to offer.

[Author's Note: The series of illustrations that follow are gradient/mileage charts Milt and I prepared from conducting preliminary runs on various freight and passenger trains. They depict the route from Bondamunda to Hatia, and contain some notes made later while running test trains. These charts and 'road notes' are utterly essential when one seeks to run a loaded 112-car ore train over seriously undulating and otherwise unfamiliar territory with unfamiliar locomotives having unfamiliar air brake valves. It was a big 'ask' (of ourselves) and we were keen to ensure we operated safely and effectively for the client. I hope they convey to some degree the sort of challenges that face the expatriate railwayman who ventures across the ocean to operate with minimal preparation and for a limited time on someone else's railroad. The charts are read from right-to-left for the loaded train.]

---

80 There was even a tentative enquiry about LOCOTROL on passenger trains! Milt and Gene managed to fend that away.

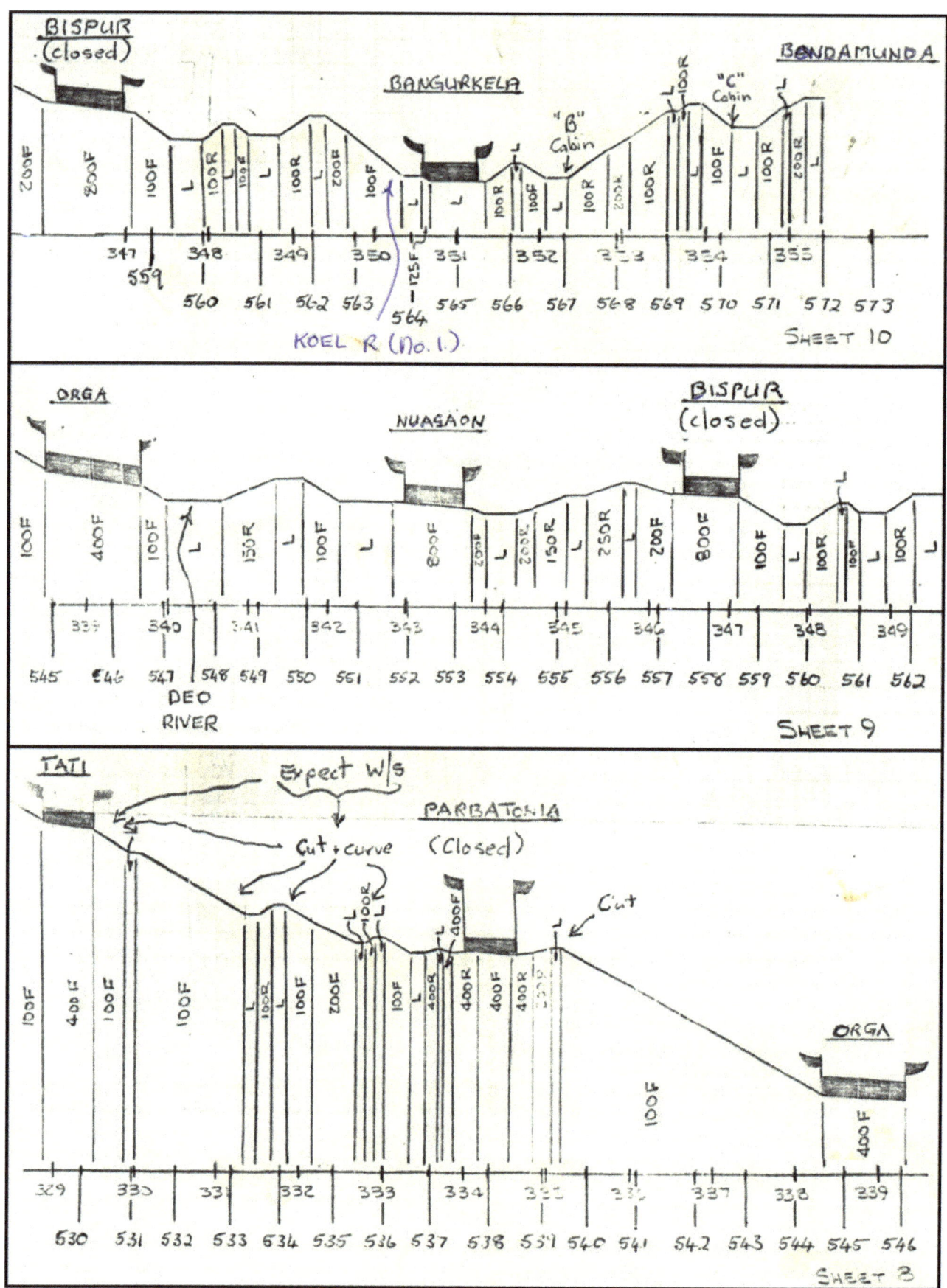

Figure 103: Gradient profile sketch 1.

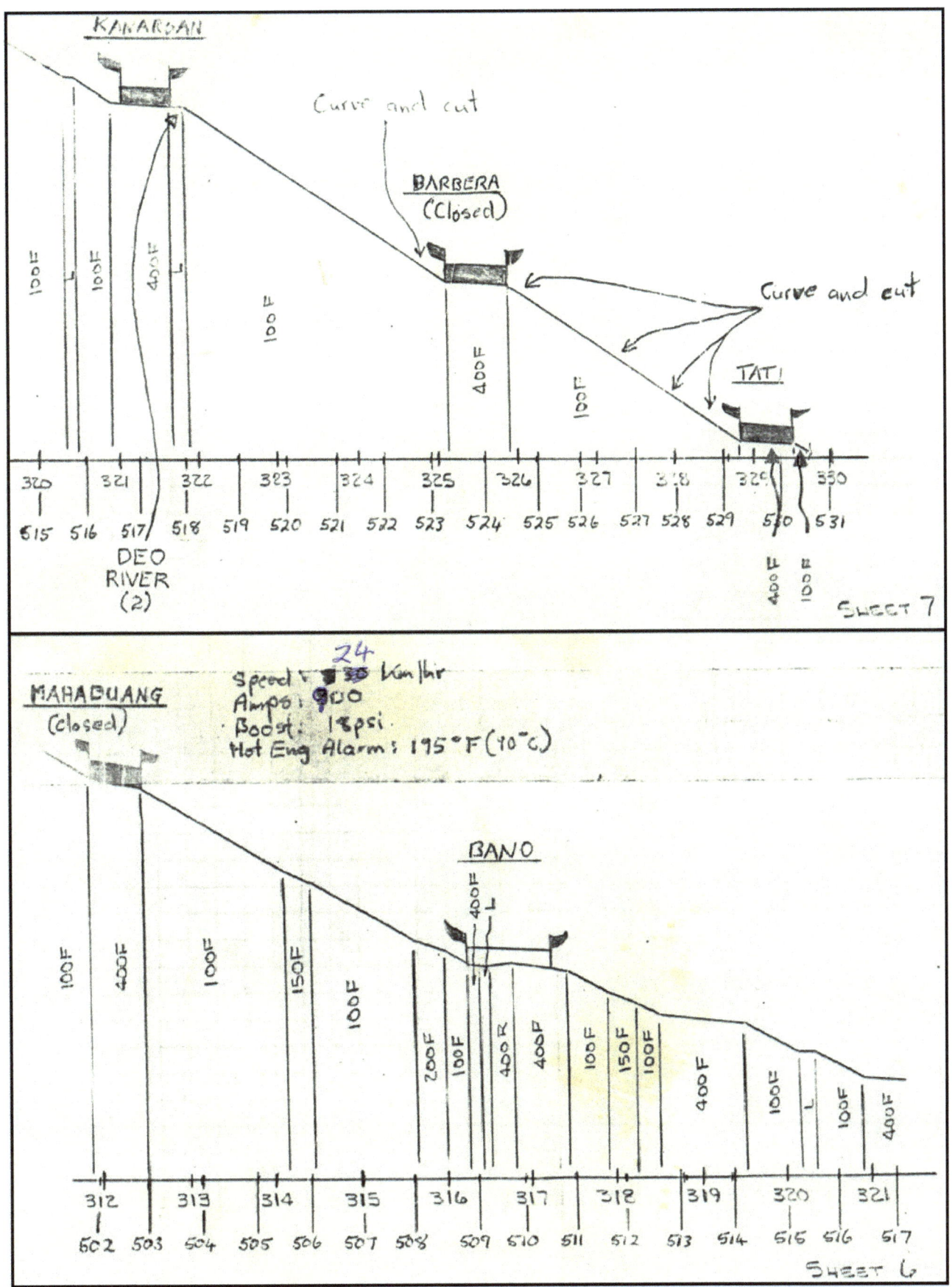

Figure 104: Gradient profile sketch 2.

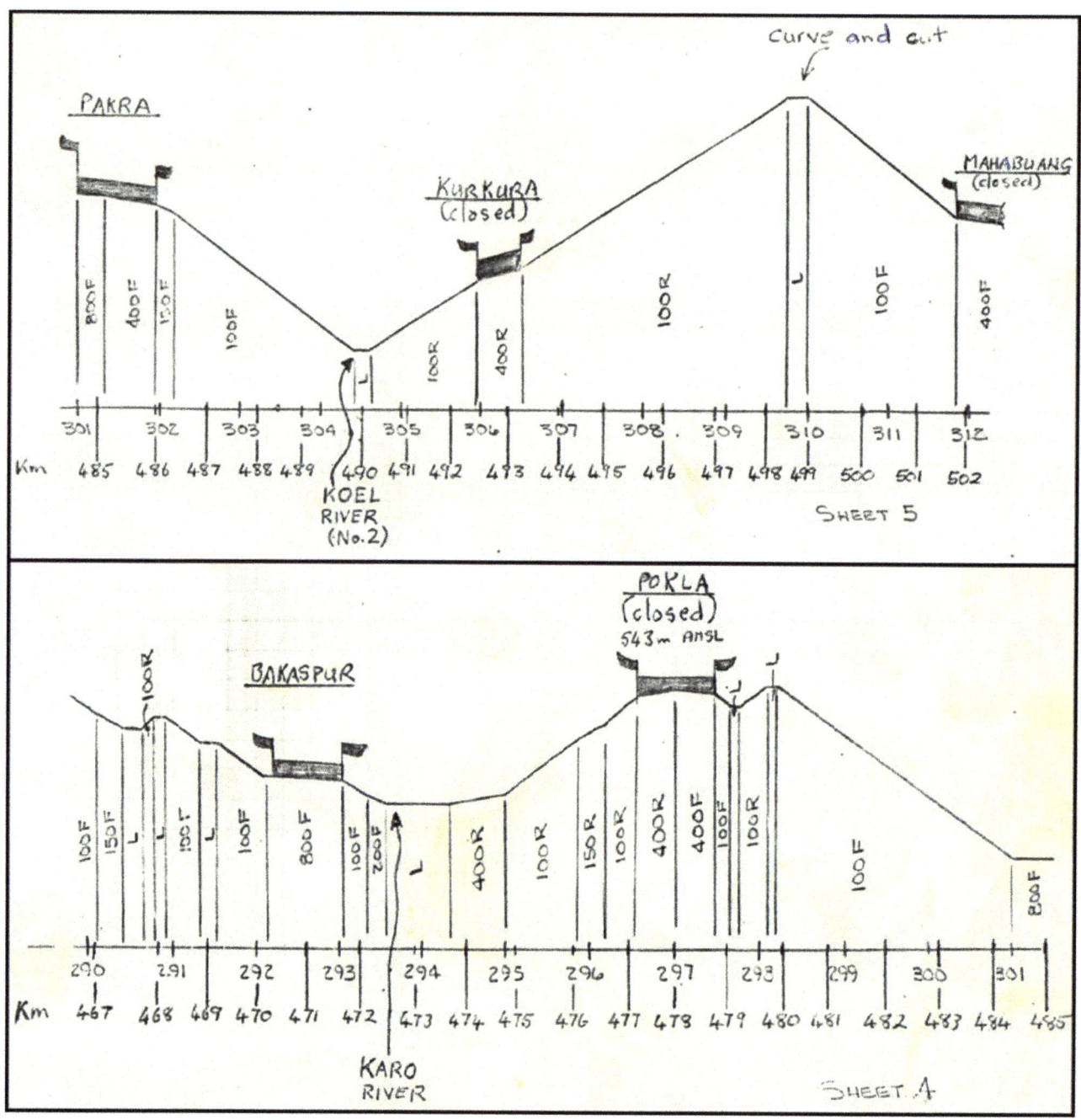

Figure 105: Gradient profile sketch 3.

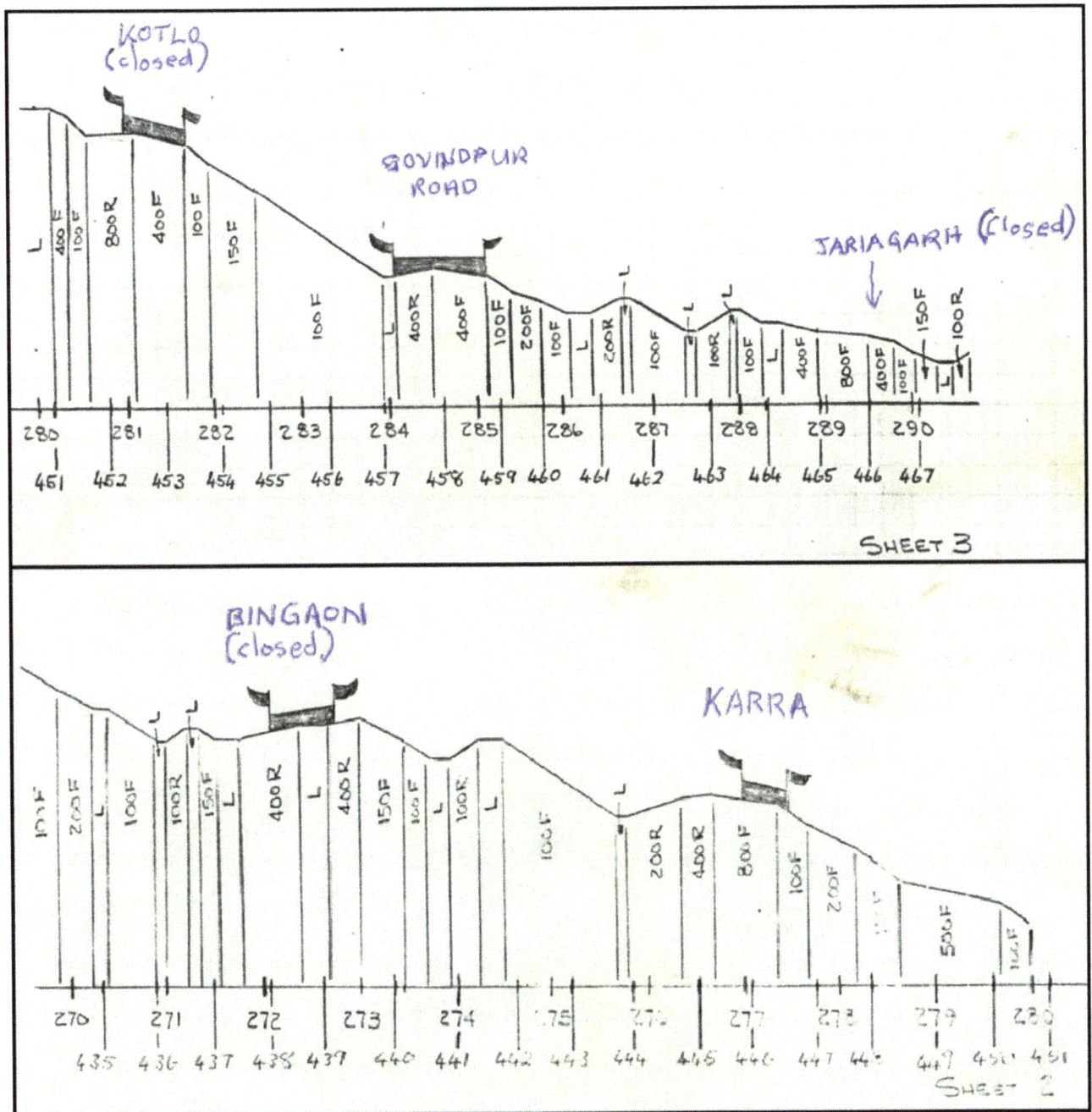

Figure 106: Gradient profile sketch 4.

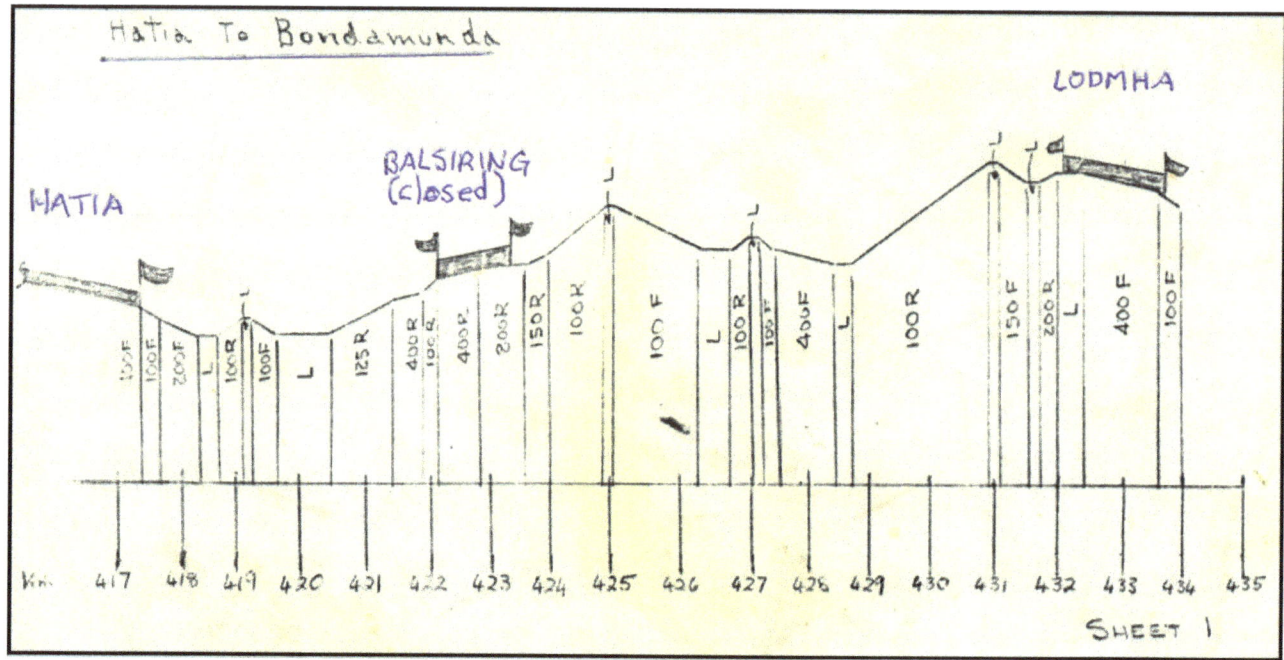

Figure 107: Gradient profile sketch 5.

Wednesday 23 August

This morning we sleep late (heck, we deserve it!) and depart Ranchi passenger station at 1315 on a passenger train, bound for Rourkela. It's just 7 km to Hatia and there we meet Mr Ahmed, our friendly Sen. Locomotive Inspector. He, Gene, and I board the locomotive for the ride home to Rourkela—5 hours away. Gene is enthralled with the scenery and takes many photos. He doesn't usually get to ride the head-end and I can see that he is enjoying this excursion. The ore train was split at Hatia last night and continued to Bokaro in two sections. Our Kwality Inns Hotel was hard beside the yard at Ranchi and we saw the two trains—one section a few minutes behind the other—roar through town shortly after we had checked in. The 112-car rake will return to Bondamunda as two trains over the next day, following which we shall schedule another empty trip to the mine(s).

Figure 108: Lewis Cox and I pose for Sen. Insp Ahmad at Dumerta prior to departure for Karampada

Wednesday 30 August

We have completed another successful 3-day road trip. Once again, the system worked perfectly, and I think the folks here are convinced. In fact, discussions are being held about a new contract to fit the equipment to electric locomotives. Gene Smith pulled out for Kolkata on Saturday where he is to meet up with Milt and hold talks with SER officials regarding the electric locomotives. Following that, Gene returns to the USA. He has included the locomotive maintainers—who will be working on LOCOTROL in Guwahati—in his school down here in Rourkela, so he is not required for this second phase of the project in Guwahati.

Lewis Cox and I are left to do the final road trips. On Saturday night, he and I host the SER Chief Mechanical Engineer, the two Senior Divisional Mechanical Engineers (Diesel and Operational), and some of the RDSO engineers to a very enjoyable dinner at our hotel. Later that evening, the two of us join the CME to travel overnight with him to Karampada in his private car. It is a most magnificent old railway office car, with a wood-panelled lounge complete with bar and sofas as well as private sleeping berths with full bathroom facilities. We are not offered the use of the bar, as liquor is not permitted by the old gentleman's religious beliefs.

Figure 109: Returning on a passenger train from Ranchi, Lewis enjoys his first locomotive cab-ride. RDSO's young engineer (and one of Gene's star technical pupils), Sarkar, nearest the camera, is along for the ride.

The car is stabled in the dock platform at Rourkela railway station. We board and are served tea by the steward. After a chat, the old gentleman excuses himself and turns in. Lewis and I do likewise and soon I feel a gentle bump as the power couples on. We run light consist (3 × units) plus office car to Karampada, arriving about 0530. Although the car is not air-conditioned, there are plenty of electric fans to keep the air circulating and we both sleep well. The ride is very smooth and quiet—almost as if the crew have special instructions not to blow the horn excessively. We are appreciative.

The following day—once again marred by operational delays (including a broken drawgear on the rear portion of the train which had been pushing back prior to coupling on to the front portion)—we get a wheel in the late afternoon and run the loaded back to Bondamunda where once again we stable at Dumerta. Again, I ran the train down the 2% grade from Karampada yard to fine-tune the technique to be used. Although I had studied and was referring to gradient profile charts for the route, I was feeling my way down this grade on these trips.

On the Monday morning Lewis and I are interviewed on television out at Dumerta before departure. We have the usual retinue of officials and the usual crowd of drivers and inspectors (some sitting outside the cab on the walkway to eat breakfast) so we do our tests and set off—one big happy family. We have a good run

Figure 110: The lead portion of 56 loaded cars awaits the arrival of the trail portion at Karampada. The Kiriburu and Meghataburu mines are in the distant hills seen above the train. Railway offices are seen in the yard to the right, but the village of Karampada—and the railway barracks—are up the hillside to the lef

through the mountains to Hatia where once again the train is split to continue the 125 km to the steel mill at Bokaro.

Lewis and I again spend the evening at the Kwality Inns Hotel in Ranchi. For me, my third-floor window is the perfect vantage point to catch the steam action around Ranchi Junction. Passenger trains coast into the station, with its four island platforms, and blast out at full throttle. The station also has a Narrow-Gauge (600 mm) dock platform, and yard activity occurs immediately across the road from the hotel. The following evening, we catch the express back to Rourkela, travelling in an air-conditioned First Class saloon reserved for railway officials on duty. We arrive on time at 0200, and repair wearily to the Radhika Hotel.

Thursday 31 August

Milt has flown back into Kolkata, and calls me from there today. He wants me to complete another road trip (this should be by Tuesday next week) and then pull stakes and go to Kolkata for a couple of days before hightailing up to Guwahati to begin work there on the next portion of the project. I haven't seen Milt since he left here to fly to hospital in New Delhi about 6 weeks ago. I'm looking forward to a change of venue, as Rourkela and the hotel here are getting a bit old. I'm also looking forward to a big steak and a green salad at the Oberoi Grand. My wife has been holidaying with her parents in New Zealand and is flying home to Perth, Western Australia on Friday. Looks like an airline strike in Australia means she may have to bus it from Perth to Port Hedland: a 15-hr ride.[81]

Monday 4 September

This phase of the project here in Rourkela is ending. When his EPROM[82] programmer arrives from Customs in Kolkata, Lewis will make some software changes to the system. Meanwhile, we've run another road trip from the mine to Hatia. This time—in a vain effort not to become involved in the interminable delays that seem to bedevil the loading of these ore trains—we have arranged to travel by road to Karampada when they have two rakes loaded and ready to couple together. On a warm summer morning Lewis and I set off with Malkit and a buddy of his to drive to Karampada. As we get close, we find our way (a bush road-cum-track) blocked by a flooded ford. There is nothing for it but to return to the nearest siding on the nearby railway and call for assistance. We obtain help at Rakshi, a siding where iron ore cars are hand-loaded by local villagers. The stationmaster gets on the blower, and while we relax in the shade and eat lunch, they send a light engine down from KPD to pick us up.

Figure 111: Local villagers unload ore from mine trucks at Rakshi as we await the arrival of our transport to Karampada.

---

81 The 1989 Australian airline pilots dispute was one of the most expensive and dramatic industrial disputes in Australia's history.
82 See Glossary.

The locomotive drivers are still a bit hesitant about bringing the train down the 2% grade from the mine. I've run the train down myself to work out what to do and to show them, as it's quicker (and safer) than explaining while they're running it themselves. The communication process sometimes leaves a bit to be desired and I feel that I should work the train handling process out myself before I start delivering instruction. Several 15 km/h slow orders on the grade make the process quite interesting.

I should point out here that these speed restrictions are usually the result of track damage due to derailments. One can always see the aftermath of these mishaps since the wrecked cars are not picked up but are lifted clear and dumped in the jungle alongside the track or down a bank. Local villagers then purchase them on contract and move in to cut them up for scrap. Some of the wrecked cars have been there for some time and have trees growing through them!

On Wednesday, Lewis and I went up to the mine to bring another train down to Bondamunda. Lewis's role is the same as was Gene's: to ride the Remotes and provide technical support should the equipment malfunction. This will be our last road trip. On Thursday, we continue to Hatia with it, then return at night again via passenger train № 89 Down, arriving at Rourkela at 0200 on Friday.

A sleep-in is the order of the day because at 1425 we're out of here via train № 12 Down, to Howrah station in Kolkata. After 9 weeks in Rourkela I'm looking forward to four indulgent nights in the Oberoi Grand Hotel.

ROURKELA (facts applying at the time of writing)

Until 1960, Rourkela was an insignificant village along the South Eastern Railway main line from Calcutta to Bombay. Today it is a flourishing industrial city where one of India's major steel plants is located (slap in the centre of town). It was the first among several large such plants to be built with foreign expertise to tap abundant local mineral resources (coal, iron ore, limestone). Nowadays all steel plants are run by SAIL, the **Steel Authority of India Ltd.** The area of the city is about 90 sq km., the population is 483 000, and the altitude is 718 ft (220 m) asl. Temperatures in winter range from 7 to 33°C (44.5 to 91.5°F), and in summer from 13 to 45°C (55.5 to 113°F). Rainfall is 59 in (1500 mm), being monsoonal from July to September; and we've seen plenty of that. The city has a modest paved airport and is linked by turbo-prop services with Ranchi and Kolkata several times per week. It is 412 km by rail to Howrah (Kolkata), and 1550 km to Victoria Terminus[83] (Mumbai). Highway connections are not what we're used to!

The Rourkela steel plant was set up in collaboration with four German firms and an Austrian firm and was inaugurated in February 1959. It is one of India's major industrial undertakings and produces large steel pipes for conveying crude petroleum, steel plate, and steel sheet. Initial capacity was 1 mtpa; it is now 4.5 mtpa. The plant gets almost all the raw materials it requires from the surrounding area. Iron ore comes from three mines within 80 km, limestone from a quarry 32 km distant, coal from somewhere south of here, and most of the power from a hydro-electric scheme at Lake Hirakud, 214 km to the west. This lake was created by one of the world's largest earthen dams; I would have liked to have seen it. Other electricity requirements are fulfilled by the plant's own, huge coal-fired power station which also supplements the city's consumption. All resources arrive by rail in unit trains, both diesel-electric- and electric-hauled. The steel plant consists of coke ovens, a by-product plant, a blast furnace and a gas cleaning plant (this complex spews out the most bilious, noxious-looking stack emissions you can imagine, so I'm very doubtful that anything gets 'cleaned'). To train the large number of skilled workers and technicians required for the operation of this plant, a technical institute was established in 1959 and a regional engineering college in 1961.

---

83 Now called Chhatrapati Shivaji Maharaj Terminus.

INDIAN RAILWAYS (facts applying at the time of writing)

Indian Railways—wholly owned by the government of India—covers the entire length and breadth of the country with one of the world's better transport networks. It has an approximate route length of 67 400 km—the world's second-largest railway route network—and employs 1.3 million people. It is the largest employer in the country and one of the biggest employers in the world. It has USD26 billion in annual revenue and around USD2.8 billion of debt. About 20 000 trains are run every day serving 7350 stations, moving 8.26 billion people and more than 1.16 billion tonnes of freight. The system operates on three gauges—Broad ('Wide', as they call it), Metre and Narrow, the last being of negligible extent and minor importance. All the trunk routes are served by Broad-Gauge while most feeder lines are Metre-Gauge. Before 1950 several private companies owned the railways in different parts of the country. The English influence is still very evident both in architecture and operating philosophy. In 1950, these companies were nationalised to form Indian Railways. Nowadays, the organisation is controlled by the Railway Board under the overall supervision of the Minister of Railways. After its incorporation, Indian Railways was divided into nine Zonal Railways and 52 Divisions. Each Zonal Railway is headed by a General Manager and each Division by a Divisional Railway Manager. The Zonal Railways and their headquarters are: CENTRAL RAILWAY (Mumbai) 6486 km; EASTERN RAILWAY (Kolkata) 4281 km; NORTHERN RAILWAY (New Delhi) 10 971 km; NORTH EASTERN RAILWAY (Gorakhpur) 5163 km; NORTH EAST FRONTIER RAILWAY (Guwahati) 3763 km; SOUTHERN RAILWAY (Madras) 6729 km; SOUTH CENTRAL RAILWAY (Secunderabad) 7138 km; SOUTH EASTERN RAILWAY (Kolkata) 7075 km; and the WESTERN RAILWAY (Mumbai) 10 224 km.

There are five major production units: INTEGRAL COACH FACTORY (Madras), CHITTARANJAN LOCOMOTIVE WORKS (Chittaranjan), the WHEEL & AXLE PLANT (Bangalore), the DIESEL COMPONENTS WORKS (Patiala), and THE DIESEL LOCOMOTIVE WORKS (Varanasi). Apart from its commercially viable routes, IR operates several strategic routes and unremunerative lines as a part of national defence policy or to fulfil social obligations. Passenger trains are classified into Local, Suburban, Ordinary, Fast, Mail/Express and Super-fast. They build their own diesel-electric (Alco) and electric (ASEA, Siemens) locomotives and still use steam in certain areas, particularly on local passenger trains. For all its faults (and looking at them from a 20th century, heavy-haul viewpoint, there are many of these) Indian Railways really is a most magnificent railway system.

South Eastern Railway

With this LOCOTROL project at Rourkela, we deal with the South Eastern Railway (written without a hyphen) rather than Indian Railways as such. In Guwahati we will be working with the North East Frontier Railway. In 1988–1989 the SER carried 99 million tonnes (over 27 000 t every day). The target for next year is 105 million tonnes. Every day the SER loads over 12 500 wagons and moves over a million people. It serves two refineries, two aluminium plants, three ports, six steel plants, 12 cement plants, several iron ore mines, hundreds of collieries and, of course, many more key industries, as well as a huge number of passenger stations over its extensive 7000 km network. Rail travel is relatively cheap: our 414-km journey to and from Kolkata in an air-conditioned sleeper costs just Rs241 (about AUD19.00).

Friday 8 September

Today I leave Rourkela and I'm quite happy about that. Lewis has some software changes to complete and will follow when that is done. I'm travelling on № 12 Down, the *Ispat Express* ('Ispat' being the Hindi word for steel). This train runs every weekday and makes a daytime trip from Howrah to Rourkela and return—11 Up/12 Down. I am riding the cab of this 25 kV electric locomotive. Sen. Insp. Ahmed will accompany me to his division change point at Chakradupur.

The crew have padded flip-down squabs for seats, but they don't make much use of them. The driver's assistant stands over by his cab door—head out the window and hand on the horn valves (they have two —'loud' and 'not-quite-so-loud'). Many people are about the right-of-way and he sounds the horn more or less constantly while calling signals to the driver. Driver Prasad also eschews his seat. He stands easily at his control desk and runs the train at 120 km/h—constantly rotating the 32-notch controller. The road is double-track main line with three-position colour-light signals. It is jointed rail and, although 5 ft 6 in gauge, the locomotive jumps around much as a Dx class GE U26C does in New Zealand when running above 90 km/h. At Chakradupur, I bid goodbye to my good friend Insp. Ahmed and return to my a/c sleeper where other sitting passengers have been guarding my luggage. We slide through a darkened car yard and into Howrah at 2200.

I spend five nights in the Oberoi Grand and eat delicious sirloin steaks every night for dinner. Their salads are good, although there is no lettuce. The marinated mushrooms are delectable, as are the tiny, whole-pickled aubergines. It's also a treat to have some wine. There are four restaurants here, one excellent lounge bar, and a bar outside at the pool. The Ming Restaurant (genuine antique vases on display) is Szechuan Chinese, The Rotisserie is European a-la carté, The Mughal Room does Indian cuisine (it's decorated like a Maharajah's palace), and the Garden Cafe is fusion. There is also a very good Tearoom upstairs that is open during the day.

I go shopping near the hotel, but the (very old) New Market behind the hotel is brilliant. Lewis and I make a foray into this huge, sprawling rabbit warren. It goes forever, and foreigners such as ourselves are well advised to latch onto an official market guide, who, for Rs20, will take us to all the shops we wish to go to (especially the ones run by family members) and will generally ensure we can find our way out of the place again. It is indescribably filthy and either stinks disgustingly of animals and their mess or smells deliciously of aromatic scents and spices depending upon which corner you turn. The part of the market that is devoted to livestock (sheep, goats, pigs, chickens) is—even to this farmer's son—not a pretty sight. Our guide takes us to all the silverware shops and I eventually find what I want for a friend's 25th wedding anniversary gift. Lewis is looking for a new briefcase but doesn't see what he wants.

Tuesday 12 September

Harris' local agent, Mr Kundu, takes me to Assam House to steer me through the rigmarole of getting my Restricted Area permit (required by all visitors to the northern state of Assam), and to get our visas extended to the end of November. This literally takes hours. I feel sorry for the foreign backpackers I see there (including one from New Zealand to whom I chat) who are all struggling through this procedure without a 'local' to assist as I have. Once again, the Indian administrative bureaucracy is the obstacle—mind-numbingly slow.

Wednesday 13 September

Lewis and I fly up to Guwahati: a 1-hour hop on an Air India Airbus A310 widebody. There, our taxi sustains a flat tyre coming in from the airport in the rain. I meet up with Milt again—also Wayne Barber, who is an extremely personable chap. Milt looks old. The leg problem has knocked him about and he uses a walking stick now. He doesn't look to have recuperated completely. We have a long talk about my future. He would like me to go to Florida to talk to his boss, John Boucher.

Thursday 14 September

Out to the diesel shop at New Guwahati. Power fluctuations and outages are part of life here. Lewis has trouble getting the simulator up and running. I take a school for drivers and inspectors. There are many problems with electrical grounds on the YDM-4 locomotives. The track is Metre-Gauge (3 ft 3 in). The LOCOTROL trains will consist of one Lead unit and 32 'crossing total'[84] (about 16 bogie box wagons), then one Remote unit and 18 crossing total (nine bogie boxes, I think). Milt reckons we won't need our portable radios—he says we'll be able to yell at each other from Lead to Remote! I believe that one reason for testing LOCOTROL on these little trains is for the North East Frontier Railway to prepare themselves for running larger oil tank trains in the future. This is an oil-rich state.

The trains are vacuum-braked, and I don't think LOCOTROL has been integrated with vacuum braking anywhere either before or since. Wayne submits me to an impromptu air brake general knowledge quiz and appears satisfied with the results. I can breathe again. He is having fun on the air brake side of things – especially with the flow sensor and feed valve. We hope to run a test trip from Tuesday for about 4 days. We will take our accommodation along with us as part of the train. The run will take us from Guwahati, north-east to Lumding Jct from where we'll make a couple of return trips south to Badarpur over the 1-in-37 (2.7%) grades. Our hotel in Guwahati is the Brahmaputra Ashok, located right on the banks of the awesome Brahmaputra River, and the view from my room is magnificent. The river is about a mile (1.6 km) wide here and flows at around 2 or 3 knots by the look of it. This is about the narrowest part of the river and there is also one of the very few bridge crossings near here. It's a double-decker of many spans carrying both national rail and road links.

Tuesday 26 September

We're heading out on our commissioning trip today. We start from the New Guwahati diesel shop at 1030 bound for Lumding. Unfortunately, no load could be attached due to a strike by yard staff, and a vigorous debate ensues. Result, we depart sans freight but with a LOCOTROL train consisting of Lead unit 6190 followed by our accommodation (the general manager's inspection car) and the crew accommodation (ordinary sleeping car), then Remote unit 6177, and the GM's private car bringing up the rear. At least the crew will be getting some running experience. Milt is doing the honours on the head-end and shares the cramped cab-space with the locomotive crew and various brass: Milt on a stool sandwiched between the driver and the rear cab door. I get to relax in our inspection car and view the wonderful scenery.

For the record, our on-board complement consists of: K L Nath, Driver; C L Gogi, Diesel Assistant; A N Sen, Deputy CME North East Frontier Rly; Y N Dhawan, Deputy Director RDSO; B Paul, Chief Motive Power Controller (promoted from Loco Insp.); Sohan Singh, Locomotive Inspector; P K Kundu and R K Das, Electrical Mechanics Grade 1; H Chakraborty and S Singh, Mechanics Grade 1; Badal Barua, steward and cook (our car); and another 'Barua', officially designated as Helper. Some of the brass are accommodated in the GM's car attached to the Remote unit.

We make rapid progress to Lumding, 180 km distant, over easy grades across paddy fields at track speed of around 80 km/h. Along the way I see evidence of sidings being extended to accommodate LOCOTROL trains. As I've mentioned, we have the exclusive use of the General Manager's inspection car. It has two beds,

---

84 For the purposes of defining the operating lengths of passing tracks ('crossing loops'), Indian Railways refer to overall train lengths (or portions of train lengths) using their basic four-wheel wagon length as a unit. So, a train or rake comprising 16 bogie boxcars—of a length equating to approximately 32 standard four-wheel wagons—is considered as 32 *crossing total*.

two bunks, a fold-out sofa, and a fold-down bed. Also, two bathrooms with shower and western-style toilet, a living-cum-dining room with fridge, a small observation lounge with instrument bench and fold-down bed (my bedroom), and a kitchen with coal stove and quarters for the steward. We will spend 5 days and nights on this car. It goes everywhere we go, including the road trips.

Milt and I will alternate on the Lead unit and Wayne, Lewis and Sarkar (our Indian electronics engineer who is being trained on the system) alternate on the Remote. Coupled behind our car at the head of the train is a Second Class sleeper with facilities for the engine crews, mechanics, electricians and inspectors who are riding with us. Attached behind the Remote unit is the GM's private car, which accommodates the top brass when they're not riding on the Lead unit. Upon arrival at Lumding, our car is shunted to a back track in the yard for the night. Badal, our steward, cooks us supper[85] and we partake of a few 'quiets' before sacking in. We go to sleep with the sounds of the yard engine (steam) thumping up and down beside us and sleep fitfully as yard shunting activity continues throughout the night. Milt has decided to take the first shift on the head end tomorrow.

Wednesday 27 September

I sleep in and am only dimly aware of the activity around me as—in the wet pre-dawn darkness—our car is shunted onto the head of our freight train, the power couples on, and with Milt in charge the LOCOTROL system is cranked up and the air test done. No need for unnecessary haste, so I am still dozing when we pull out of town at 0650 in pouring rain.

We are soon pounding along in fine style and a few glimpses under my blinds confirm that the weather is clearing and the scenery, once again, is captivating (I could almost feel guilty about being paid for this!) so I arise and have some breakfast. Wayne Barber and Lewis Cox were both up at the same time as Milt and are riding the Remote unit. Our crew today: Driver S P Baidya, Diesel Assistant S B Chakraborty, Loco Inspector Sohan Singh, and Y N Dhawan from RDSO. We have the Lead unit, 13 bogies, the Remote unit and nine bogies: 44 crossing total for 1038 tonnes.

Lumding (LMD) to Badarpur (BPB) is 185 km. There are 37 tunnels and 18 km of 2.7% down-grade. The line is historic: being not only constructed as a metre-gauge track in the last part of the 19th century to move tea, coal and timber from upper Assam to Chittagong port (now in Bangladesh), but also as an engineering marvel because of its traverse of difficult terrain across the Barail Mountains. The line runs through thickly-forested country almost all the way, with spectacular views from steep mountain-sides and numerous viaducts.[86] The tribal locals are engaged in subsistence rice farming and tea growing. There are frequent, small crossing stations—all staffed—and many trains to cross, including crowded express passenger services. Safeworking is by electric token and lower-quadrant semaphore signals. We can expect the trip to take between 10 and 13 hours.

At Maibong, 65 km into the journey, Milt comes back to the car and hands over to me. I grab my bag, walk up to the Lead unit, settle myself onto the padded stool behind the driver and we set off again. This will be my

---

85 Having arrived at Lumding, we were wondering what we'd be having for dinner. The mystery was solved when we saw Badal returning across the railyard with a live chicken from a nearby market. This he proceeded to slaughter on the concrete floor of the galley that, thankfully sloped in a concave manner, to a drain hole. The grate was lifted, and all entrails and fluids went down onto the track. He prepared a delicious chicken curry with roti and condiments, which we relished with a beer, while trying not to think too much about it all.

86 The entire NEFR in Assam has been converted to broad-gauge (the term 'wide-gauge' is no longer used). The steep and sinuous section of metre-gauge line that we traversed—between Phiding and Ditokcherra—has been abandoned, having—in 2014/15—been bypassed by the new BG line that itself includes many tunnels and bridges. To this day, I feel fortunate to have seen and operated over the old MG route.

lot until we reach BPB later tonight. My stint includes the 2.7% descent of the North Cachar Hills: however, Milt has done his homework and his calculations—as ever—prove accurate.

At Haflong Hill (HFG) we have a brief respite from the climb as we await a double meet. This lovely place, standing at 3000 ft asl is the only 'hill station'[87] in Assam. It has a large community and I notice many old British-style buildings – no doubt being relics of the Raj. Three kilometres further on we reach the summit at Jatinga (JTG). This picturesque place is famous for the Blue Vanda, apparently the only blue orchid in the world. Also, we are told of the world-famous bird mystery that is a notable annual feature during September to October (that is to say, now). Apparently, many birds of colourful feathers rush to the place only to commit suicide by starving themselves to death. I never did find out if anyone knew why.

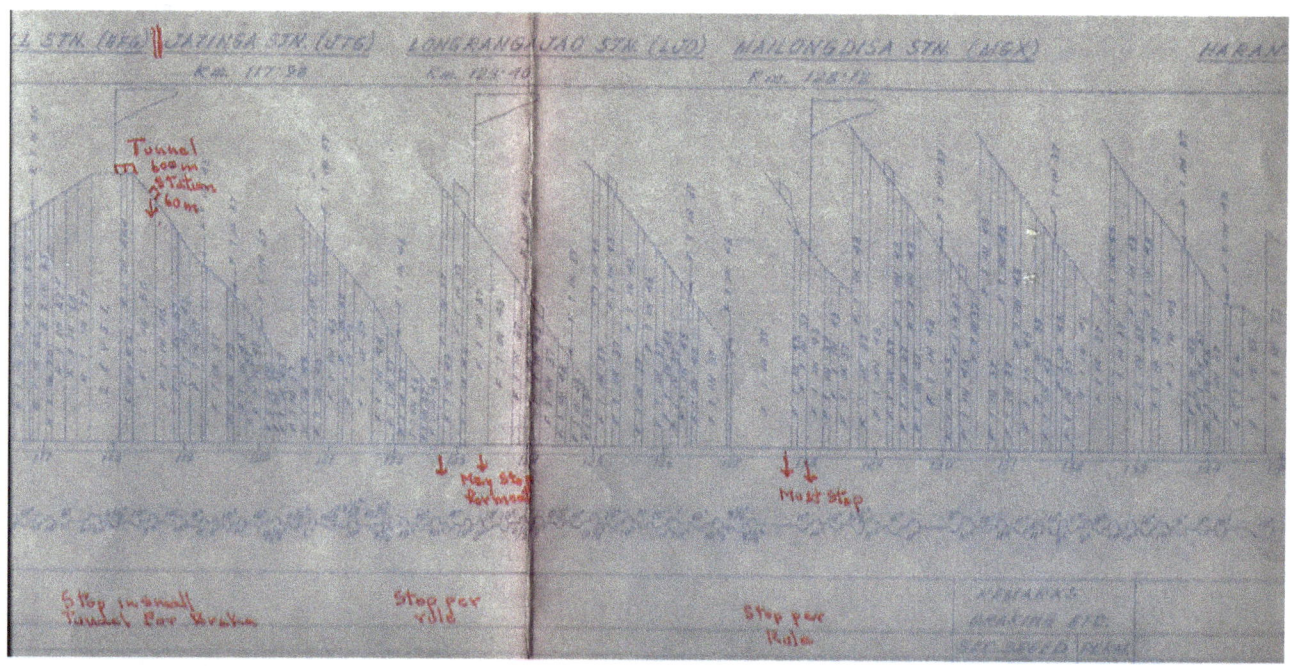

Figure 112: Indian Railways NE Frontier Railway track profile (curve and gradient) diagram used by Milt and I when running between LMD and BRP. The flags represent station locations.

Operating rules specify that all trains must stop at JTG before descending the grade to Harangajao (HJO), to have all handbrakes loosely pinned down. Milt comes forward from the car and remonstrates with the rollingstock engineers. He tells them that with LOCOTROL, air brake control is improved to the extent that this requirement will be unnecessary. The brass are dubious but eventually agree to allow it on a trial basis (that's why we're here, after all). We set off into the 600-metre tunnel situated right at the crest of the descent. As soon as the entire train is on the down-grade I am required to make a stop, then release and continue. This is a demonstration requirement of Indian Railways. I do this and duly receive a radio call from Milt to *'Get us out of this G*d*m tunnel before we choke to death!'*

The descent at the maximum allowable speed of 18 km/h is accomplished without drama. I take the train down the grade myself, as it sometimes takes too long to communicate to the driver what our train handling requirements are. By doing this, I can test out Milt's calculations and develop a train handling strategy for them to follow. Milt has calculated such things as the relative friction of the brake shoes on the wagons and the dynamic braking ability of the locomotives versus gradient and curvature and has estimated that I'll need a minimum brake pipe reduction only (as read on the locomotive; the train itself gets a vacuum brake

---

87 Mountain villages where the British rulers would go to escape the oppressive summer heat.

application) and about position 4 or 5 on the Selector (dynamic brake) handle. He's correct, and I have no trouble settling the train down and manipulating speed by modulating the D/B. With the minimum reduction set, I need full D/B to slow to 10 km/h for the bridges, then by easing D/B gently to prevent excessive draft slack action, I find the train will accelerate back to 18 km/h quite quickly. All this while 'maintaining braking'[88]. This is one steep little mountain!

We continue, with many crossing delays, and finally arrive in Badarpur at 2000. Once again, we spend the night in the yard (not surprising really, when your 'hotel' is a railroad car), this time close to the station. As with LMD, there is considerable steam activity that ensures we don't get too much sleep.

Thursday 28 September

We depart BPB at 0500 and run back over the 'hill' to Lumding, arriving at 1900, a trouble-free run in glorious weather. The 2.7% climb—which is the reason LOCOTROL is warranted on this section—presents no problems. The brass are starting to relax. Lewis and Wayne catch the overnight passenger train back to Guwahati to get some work done on the second set.

Friday 29 September

Milt and I make another run over the hill section to BPB with 1000 tonnes. This time I have the chief motive power controller, Mr Paul, take the train down the 1:37 section. He will be responsible for a lot of training when we leave.

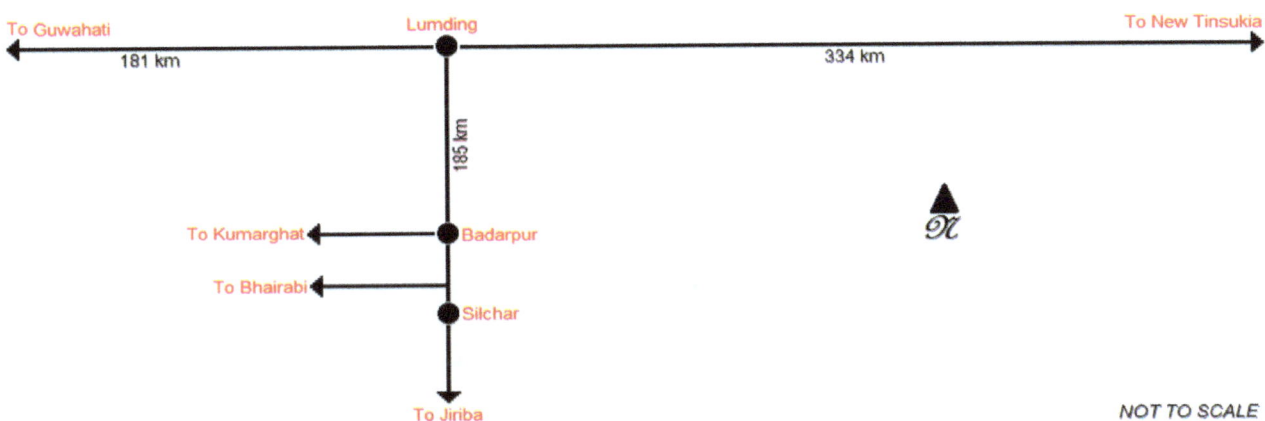

Figure 113: Locotrol test train routes – NE Frontier Railway 1989.

Saturday 30 September

We get out of BPB again at 0500 (although this time we stay in bed while the Indians fire up the system under the supervision of Inspector Singh, and brake-test it themselves). We are at Damchara (DCA), 30-odd clicks out for our first meet, before I wander up to the cab. They have everything well in hand. We get into LMD around 1700 and are shoved around the yard for some time by an asthmatic old steam switch engine before leaving at 2200. We're a 'Special'—just two LOCOTROL-equipped locomotives running 'conventional', and two business cars. We go to bed and I find it difficult to get to sleep as we make an extremely fast run to Guwahati, although waiting for meets means we don't arrive until 0600. We were probably only doing 80 km/h, but I think that much of the sensation of speed comes from the old-style jointed rail that I'm not used to these days… and the curves. It felt just like being in a sleeper on № 227[89] and flying down through Horopito or

---

88   This is an air braking term for maintaining a continuous air brake application over an extended distance on a descent.
89   In New Zealand, the (then) overnight Limited Express between Auckland and Wellington, the *Northerner*. New Zealand Railways was first corporatised (in 1982) then privatised and—eventually—became the state-owned enterprise, KiwiRail. There is no longer an overnight passenger train on the North Island Main Trunk.

Tangiwai: GREAT! Later, Milt mentions too, that it felt as if the wheels were barely touching the rails.

Being Sunday morning, we go straight to the hotel and rest up. Tomorrow, it's back to the diesel shop to complete the second set and do back-to-back testing.

GUWAHATI (facts applying at the time of writing)

With the ancient name of Pragjyotishpura, meaning 'The light of the east', Guwahati is the gateway to and the nerve-centre of North East India, a region consisting of seven states—including Assam—situated on the banks of the mighty Brahmaputra – 'the son of Lord Brahma'. Surrounded on either side by a ring of green hills, the city is a veritable darling of nature. Guwahati has an air about it – a unique, magical charm that locals suggest is perhaps the magic of the demon king, Narakasura, who built the ancient city. A guidebook I have purchased mentions a similar atmosphere, stating that, '… maybe it emanates from the Navagraha temple (the temple of nine planets), the seat of astronomical studies in ancient India—where great astronomers perform their miracles even today', or you may discover it at sunset from atop the Bhubneswari Hills (as I did—fantastic!). The guidebook continues '… as you look down upon the crimson ripples of the Brahmaputra caressing the pale pink city, suddenly its lights flicker to life and your heart lights up with divine ecstasy…' Who would I be to take issue with that?

And what of this mighty river itself? It flows across the state of Assam from the north-east to the west dividing it equally into two parts. With its many tributaries—numbering about 120—the river provides a water system throughout the plain areas, is a major source of irrigation and waterpower and when in spate is, unfortunately, a cause of tremendous conflict between human ingenuity and the forces of nature. The total length of the river is estimated at a little less than 3000 km and its drainage area greater than 580 000 sq km. It rises behind the Himalayas in Tibet (as the Tsangpo) and flows in an easterly direction for about 800 km before turning south into India. At this point it is joined by several tributaries and widens dramatically to flow down the length of Assam before turning south into Bangladesh for a further 200 km to its confluence with the Ganges. The river's effect on Assam is to produce a profuse wealth of everlasting greenness that I am told exists throughout the year. But back to Guwahati…

Situated 55 m asl, the climate of the city varies from 34°C (93°F) in summer to 10°C (50°F) in winter with an average annual rainfall of 8 in (203 mm). The city is well-connected by air, road, and rail to all parts of the country, and is described as a fast-growing city of commercial and industrial importance. Its municipal area is 215.73 sq km (83.3 sq mi) with a population of about 200 500. The city has been an important trade centre from ancient times. It was originally fortified and is naturally protected by the river and the natural amphitheatre of the surrounding hills.

A Dual-Gauge (Broad- and Metre-) railway runs in from the west, providing the connection with the rest of India. The B-G track runs through the city and ends at the New Guwahati diesel shop on the eastern outskirts (refer also to Footnote 86) The railway bisects the city and is very busy, as suburban passenger services—often steam-hauled by ancient American-built Alco locomotives—share the route with regular freight and shunt movements. Up to nine B-G passenger trains arrive and depart daily. The remainder of the state's railway is M-G. The western suburb of Maligaon houses the headquarters of the North East Frontier Railway. I was intrigued to notice a spur leading off the main line and into the headquarters complex. One of my engineer-students informed me that it served the general manager's office via a private platform. I like their style!

Our digs are the Hotel Brahmaputra Ashok. It is a unit of the Assam Ashok Hotel Corporation (a joint enterprise of the India Tourism Development Corporation and the Government of Assam). It is licensed to exchange foreign currency and includes a restaurant/coffee shop, a reasonable bar, a 'party' room and a banquet hall. It is centrally air-conditioned, has three-channel closed circuit TV and a four-channel music system – neither of which can be guaranteed to work all of the time. There's a 24-hour switchboard, a doctor on call, a travel counter, a shopping arcade and even a British Airways office. Tariff for our double (twin) rooms is Rs375 per night (AUD30.50 in 1989).

Tuesday 3 October

The interminable delays continue. Milt has found that the units have an incorrect dynamic brake circuit – something to do with a local modification made at some time in the workshops here and never documented. The result is that they don't have a drawing of it and the LOCOTROL system isn't compatible. Lewis makes a small error of judgement and burns out the last spare transistor that is suitable. He and I take off up-town and spend half a day shopping for some new ones. Lewis is then required to modify the LOCOTROL D/B board to make it work. Then it rains (units have to be moved inside), then they stop for tea, then there's a power failure, the compressor on the 6191 needs new valves, the units have many electrical grounds (how about zero resistance?), there is a day off for a public holiday, and now—just to really get us smiling—there's a possibility that we'll have to return to Rourkela! Tonight, the prevailing mood as we have after-dinner drinks in Milt's room is of a generally darkish hue. Milt and Lewis have a robust discussion, and we all polish off Milt's litre of JW Black Label (normally reserved for completion of the project) plus quite a few beers.

Friday 6 October

'Anutha day, anutha dollah.' It's the culmination of a portion of the project and Milt and I are in high spirits. At 1400 we will leave Guwahati and fly to Kolkata. Unfortunately, we will then have to continue to Rourkela as they have been in touch and think they have a problem. Milt has seen this all before (new owners of LOCOTROL being unsure/afraid of it, forgetting their training, and not reading the tech documentation) and isn't too perturbed regarding the likelihood of a 'problem'. He says that it doesn't matter how much information you give a new customer, there is always something they don't understand and eventually a simple glitch (like not following the set-up procedures properly) frustrates them and they say, 'To hell with it!' So, he and I will go back although we already have an idea what the problem may be.

It eventually transpires that this particular 'simple glitch' is that my instruction about making effective[90] air brake applications has been misinterpreted in the translation. In discussing air braking with locomotive personnel, I have stressed that making an air brake application of no more than a Minimum (brake pipe) Reduction is not sufficient, and that at least one further reduction needs to be made and allowed to equalise before initiating a release. So, these guys have been making TWO healthy reductions following the initial Minimum (in other words, diligently making three consecutive reductions of brake pipe pressure), and the amount they've drawn off has been bringing BP pressure below the threshold for activation of the LOCOTROL 'Low BP' safety feature. The result for them has been a penalty Emergency brake application[91] each time, which they have interpreted as a fault. We would soon clear up the confusion.

Milt isn't thrilled with extending the stay though. He'd visualised leaving Guwahati and India. Lewis stayed behind in Guwahati with Sarkar to complete some software modifications. He won't come to Rourkela unless he's required. The plane is very late arriving from Kolkata, with the result that we don't depart Guwahati until 1700.

It's dark when we land at Kolkata. The airport here is called Dum Dum, after the area in which it is situated.

---

90 An 'effective air brake application is one where the total brake pipe pressure reduction is of sufficient intensity to properly activate the car operating valves (variously called triple valves, control valves, or distributing valves) and the exhaust allowed to completely cease—allowing relevant pressures to equalise—prior to making a release. An effective application is the only proper application.

91 An 'Emergency' brake application is one initiated by an immediate exhausting of all BP pressure – not by a controlled, partial reduction. It is designed to occur more rapidly and to be of greater intensity than a Service application. A 'penalty' brake application is a Service or Emergency application initiated automatically because of a pre-set limit having been infringed (in this case, Low BP).

This area has lent its name to the infamous 'dum dum' bullet. The small-arms factory in which it was invented in 1898 by Captain Bertie Clay is located nearby.[92] The dum dum bullet was designed to punch a fist-sized hole in Afridi tribesmen who were apparently undeterred by conventional ammunition. Apparently, nobody had ever explained to them that they were supposed to fall down when shot, so Captain Clay set about inventing something that would knock them down.

We arrive amidst the Durga Puja[93] festival. The situation is chaotic, with huge crowds and bamboo shrines illuminated with fluorescent lights everywhere. I am told they are celebrating something to do with the goddess Kali who is the most respected and feared of all Hindu gods and after whom one of the three original villages that were the genesis of the settlement takes its name. She is known as Kali the Terrible and appears with devilish eyes or a tongue dripping blood with snakes entwined around her neck or with a garland of skulls. The mythology has it that Kali was the wife of Siva. Siva the Destroyer, Brahma the Creator, and Vishnu the Preserver constitute a divine group at the head of the Hindu religion. When Kali died, the grief-stricken Siva placed her corpse on his shoulders and went off stamping around the world, working himself into a frenzy. The other gods decided he had to be stopped before his rage destroyed the whole world. So, Vishnu flung a knife at the corpse cutting it into 52 pieces which were scattered across the earth. One of her little toes landed beside a great river in Bengal and a temple was built there along with a village. The river was the Hooghly—at one time a natural branch of the Ganga (Ganges)—and the village was called Kalikata.

And so, we are back in Kolkata. It takes the taxi almost an hour to get from the airport to the Oberoi Grand Hotel, such is the crush of humanity in the street. But what of this magnificent and frightening city? It is a problem city of the world. It has insoluble challenges that grow more pressing every day. The population is well over 12 million (only Tokyo, London and New York contain more people). Thousands sleep on the streets at night. The poverty is dreadful. There is violence, anarchy and raging Maoist Communism. I have read that Kolkata, while not the largest city in India, is commercially, industrially and intellectually the most important. It is the richest city in India, is set in one of the most ancient cultures known to man and is one of the youngest cities in the world (New York was founded 81 years, and Montréal 50 years before Kolkata).

Apparently, Kolkata is tolerable in winter – on that I can't comment. It's pretty crook in summer, I know that. Between March and the start of the monsoon the temperature can get to 49°C (120°F). It will stay above 37°C (98.5°F) for days on end and the humidity will be 100%. The monsoon arrives in June and lasts for 4 months, giving the city most of its annual quota of 64 in (1625 mm) of rain! I've seen it described as the '… filthiest climate on Earth'. Sitting sodden, astride the Tropic of Cancer, Kolkata lies within that wide, flat piece of delta country that contains the massive outflows of both the Ganga and the Brahmaputra Rivers. The Hooghly River—astride which the city sprawls—is a diversion of the Ganga (*GUNG*-ga: only ignorant foreigners know it as the *GAN*-gees). This whole area of thousands of square kilometres is basically untamed tropical marsh. The only firm ground is by the riverbanks. When the rains come, the area becomes a huge inland sea and this, of course, is the tragedy of Bangladesh – which cops the lot. It is a perfect breeding ground for malaria, and it is on this bog that the British created their capital of India! They say that nothing but commercial greed could possibly have led to such an idiotic decision.

---

92 The name derives from the Persian word 'damdama', which refers to a raised mound or a battery. It is part of the Kolkata urban agglomeration. During the 19th century, the area was home to the Dum Dum Arsenal, a British Royal Artillery facility.

93 An important festival in the Shaktism tradition of Hinduism that reveres and pays homage to the Hindu goddess, Durga and other major deities. Nowadays, the importance of Durga Puja is as much as a social and cultural festival as a religious one.

Sunday 8 October

Our local agent, Mr Mukherjee, manages to get us a two-berth air-conditioned cabin on train № 5 Up, the *Bombay Express*, departing Howrah Station at 2200. We have had two exceptional days in the Oberoi Grand—plenty of beef and just a little alcohol to accompany it. Decent beer too. Milt is adamant that he and I will not spend more than 2 days in Rourkela. He believes they will want us to ride some trains and he is equally adamant that we will not do that. He is quite certain that they don't have any major problems requiring our presence and that they are quite capable of running their own LOCOTROL trains.

Monday 9 October

We pull into Rourkela on time at 0630 and are met by our young driver and friend, Malkit Singh, who is almost beside himself with excitement. We check into the Radhika before having breakfast and driving out to Bondamunda. A 112-car LOCOTROL train has been assembled out at Dumerta, so after a meeting with Senior DME Dev Rudola we proceed out there. We spend 5 hours going through a full test regime – Milt on the Lead and me on the Remotes. Nothing amiss! By this time the engine crew has their time in and has been relieved. Milt tells the officials that the train can go, and we wait long enough to see it pull out, bound for Bokaro Steel City.

The next morning, we spend 3 hours with Mr Rudola at the Bondamunda shops, discussing the project and their new LOCOTROL equipment. I have written a manual for their drivers and Mr Rudola is very pleased with it. Milt reminds them that they will have to keep the equalising reservoir circuit on these locomotives tight. We hear that the loaded train has arrived safely at the steel mill. This is what we want to hear! It is proof for them that they can run LOCOTROL trains without us being around. However, Lewis is going to have to come down and clear up a few points of confusion regarding the software system with their engineers. Milt and I depart Rourkela on Tuesday afternoon on № 12 Down, the *Ispat Express*, and arrive back in Kolkata at 2200.

Wednesday 11 October

I book a flight to Hong Kong via Bangkok on Saturday 14th to meet up with my wife, but we are frustrated again! Mr Mukherjee, Harris Corporation's agent here in Kolkata has let us down – in fact it's getting to be common these days. Weeks ago, he was supposed to get some details for me of what I had to do to obtain an Income Tax Clearance. This is documentation that one must have in order to be allowed to leave the country after being in it for more than 30 days. Essentially, it proves that you've been cashing your own funds while in the country and not earning rupees.

Naturally, after four months, I have reams of traveller's cheque encashment certificates from banks and hotels, etc. to prove that this has indeed been the case (you get one whenever you cash a Thomas Cook). Now, with only days to go before I'm due to leave the country, Mukherjee arrives at the hotel with a Mr Roy—a lawyer—who has provided a written opinion that under Indian law I will be liable to pay income tax on my earnings while in the country (my consultancy fee has been paid monthly into my Australian bank account by Harris Corporation). Mr Roy's professional opinion is definitely *not* what our traveller's information—that we have from several sources including the Canadian embassy—tells us. Mukherjee, Roy, and I go to see an Income Tax Officer (whose name, it transpires, is also Mukherjee!). Long story short… my contract provides for me to have a fortnight's R&R in Hong Kong where I am scheduled to meet my wife (who is already en route) and it is becoming apparent that the Indian administrative bureaucracy is working at full steam to invent hoops for me to jump though.

I spend a further week in Kolkata trying to get out of the place. The details of that story are not for recounting here, but include a midnight taxi-ride to the Bangladesh border (actually two: on the first one I arrived at the jungle border crossing during a transport strike and had to return to Kolkata for 2 days), a rickshaw ride to Benepol, a bus ride to Jessore, and a domestic flight to Dhaka. Suffice to say that my wife has been to Hong Kong and has endured our holiday there alone. I finally catch up with her in Singapore and we have 2 days before returning to Australia.

Milt was booked to leave the same day as me, but via New Delhi, Frankfurt, London and Manchester – there to visit Davies & Metcalfe before returning to Florida. He has asked me to join him on the next major project—spending a year in Iran—but this project is fated to become a victim of the General Electric takeover of Harris Corporation, and I shall be left high and dry in Australia just short of signing a contract. But that's another story… for now I'm just happy to be going home.

Postscript

Years later, I noticed Devendra Rudola's name on Facebook and renewed contact with him. He gave me a brief rundown on Indian Railways' LOCOTROL and his career after our project was completed. Dev wrote…

From Bondamunda, I went to the Diesel Locomotive Works at Varanasi in 1990. It was a sort of reward posting, since I was stated to have done some good work and because of which I had received the Railway Minister's Award. This was quite a surprise since normally, few [at the top] *cared for people working in such remote and godforsaken places. Anyway, after manufacturing diesel engines at DLW for close to 5 years, I was drafted onto the Ministry at the Railways Board in New Delhi. I was there for another 5 years after which I was elevated to the Chief's rank. During my tenure as Assistant Divisional Railway Manager, Delhi Division, RITES*[94] *won an international bid for concession of the central portion of Mozambique Railways in 2004 and I was hand-picked for the role of Director – Rollingstock & Operations there. I worked on this project for about 4 years after which the Essar Group offered me the role as their CEO in Mozambique. So, it's now 7 years that I have been in Moz.*

Insofar as the South Eastern Railway is concerned, we eventually ran 11 LOCOTROL trains consisting of 116 wagons and six locos. But it pains me to inform you that after I left Bondamunda, the project was shelved and not one more train was run. It was all due to filthy politics between the Mechanical and Operating Departments. The Operating Department felt peeved that the Mechanical Department piloted the project. They thought LOCOTROL operations should really have been their *baby. As you perhaps know, inter-departmental bickering is the biggest bane of Indian Railways, and that is so unfortunate!*

As to the North East Frontier Railway, I was told they didn't run a single LOCOTROL train after your team left, though their people had come to Bondamunda for instruction. Someone told me the reason, which is eluding me at this point in time.

So, all you can do on an expatriate project is fulfil your corporate contract. After that, the politics of your client's organisation and government can change anything. C'est la vie!

---

[94] RITES Ltd (earlier known as **R**ail **I**ndia **T**echnical and **E**conomic **S**ervices) is an engineering consultancy, specialising in the field of transport infrastructure. Established in 1974 by the Government of India to provide consultancy services in rail transport management to operators in India and abroad, the organisation has since diversified into planning and consulting services for other infrastructure, including airports, ports, highways, and urban planning. RITES claim to have executed projects in over 62 countries on every major continent.

From Deno's many accounts of the trials and tribulations of foreign installation projects (and not a few domestic ones as well), it was apparent to me that our experiences in India were pretty much par for the course. For that reason, I thought some more images from my photo album (although I don't pretend to be much of a photographer) might serve to provide context and nuance to the preceding story. A LOCOTROL installation wasn't just about screwdrivers, computers, and driving trains. Every day bought some new scene or experience to record and cherish (or not). These kinds of experiences occurred—in one way or another—on pretty much all of Deno's overseas projects.

Figure 114 (above): The copper tubing that connects the pneumatic functions of the dual control stands in the WDM-2 cab is evident beneath the floor panels. Pneumatic tubing connecting locomotive systems with the Locotrol Control Console can be seen in the back of the short-hood control stand.

Figure 115 (above): The Air Brake Manifold is installed on a vertical surface in the short-hood of the WDM-2. The cylindrical black object below it is the Equalising Reservoir.

Figure 116: The Lead unit Logic and Relay Interface cabinets are mounted side-by-side above the Main Generator.

Figure 117: The Lead unit Logic and Relay Interface cabinets are mounted side-by-side above the Main Generator.

Figure 118: Engineer's long-hood control stand on the WDM-2. Shows the Air Brake Console installed into the control stand and the Control Console on top. Both control stands were equipped with an Air Brake Console. The Control Console could be moved and connected atop either control stand. The image also shows the Equalising Reservoir gauge (calibrated in both psi and Kg/cm2, that Deno insisted upon) installed atop the control stand.

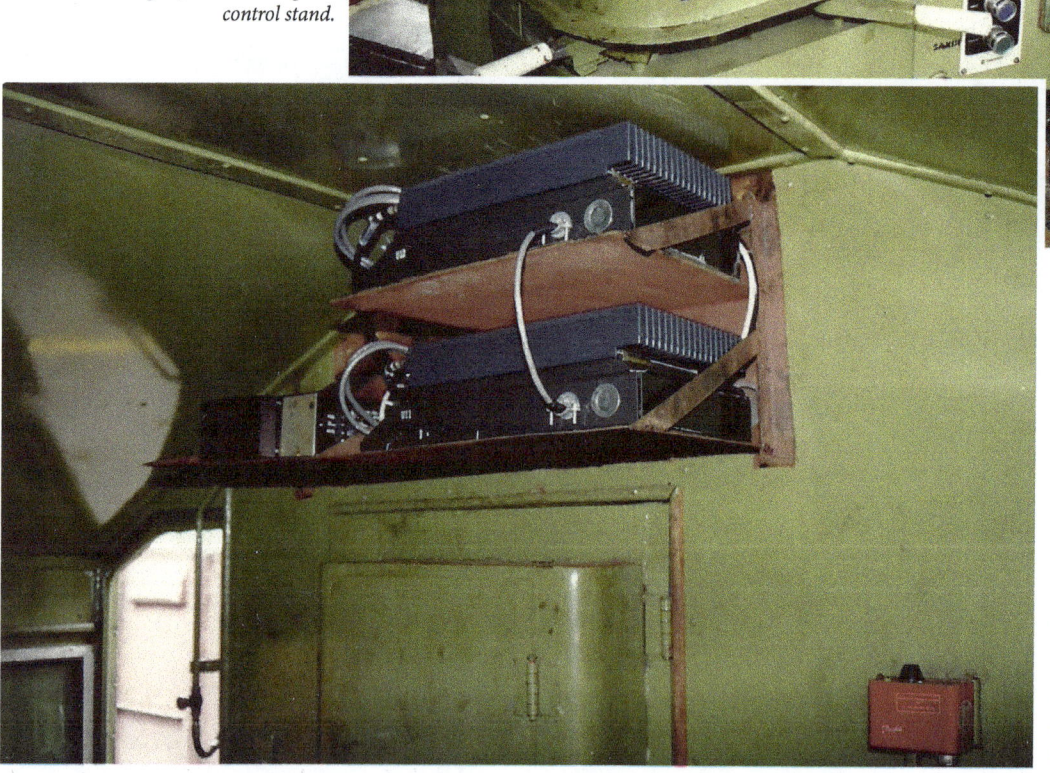

Figure 119: The SER's interpretation of the installation specification for the UHF radios (FM half-duplex, 3 kHz bandwidth) wasn't pretty… but it sufficed.

Figure 120: Senior Locomotive Inspector Anwar Ahmed conducts 'safeworking' duties with the station master in a station office. The Neale's Ball Token instruments are seen behind.

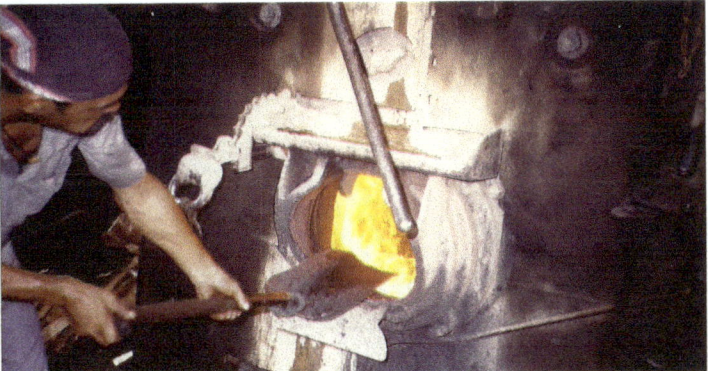

Figure 121: Fireman aboard SER WG 2-8-2 locomotive shovels some of the 19 tonnes of coal in the tender into the 46 sq ft firebox. The amount of 'shoveling-forward' required in the tender meant that two firemen were rostered on these locomotives. The driver's regulator gland leaks muddy water down the backhead of the boiler. The author was 'learning the road' on this passenger train, courtesy of Senior Locomotive Inspector Ahmed..

Figure 122: 'Downtown' Rourkela in 1989. Railway station in the centre, Radhika Hotel is the tallest building, slightly to the left, and the SAIL Rourkela Steel Plant beyond.

Figure 123: Part of the workers colony built at the time the Rourkela steel works was constructed. This is 1989… it is much improved in 2020. The Ispat Hospital in the distant centre, behind trees.

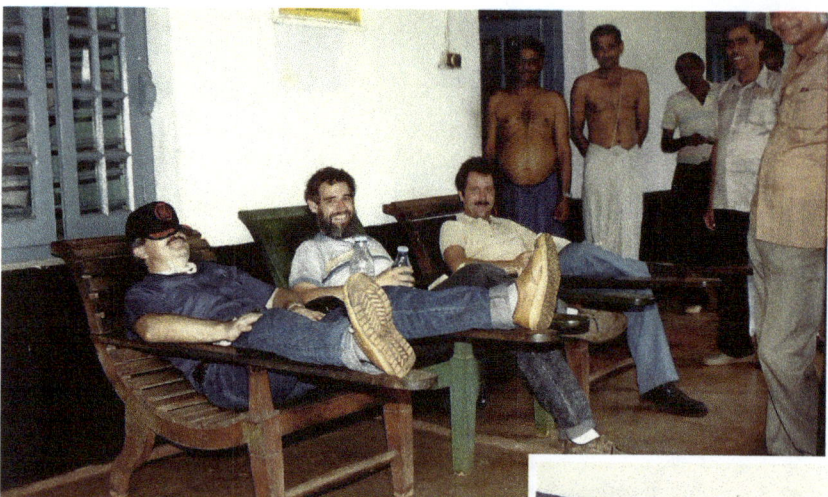

Figure 124: Me, Gene, and Lewis ham it up at the Karampada train crew barracks, where both locomotive and traffic staff stayed. The big chairs are made for relaxation and the extended armrests are indeed for your legs! Off-duty railwaymen like to change into their casual 'lungi'. Sen. Inspector Ahmed at far right.

Figure 125: The Karampada train crew barracks, where we waited out some of the ore-loading delays. I think, from memory, we did spend a night here on one occasion. I seem to remember a big dormitory and lots of snoring. I remember they had a cook employed there, and the food was simple but delicious!

Figure 126: View from the crew barracks across part of Karampada village, down to the railyard. Some of the Meghataburu mine infrastructure—7 km distant by rail—is visible near the skyline. The ore pit runs along the ridge in this location. The Kiriburu mine, further to the left, is on the same orebody.

Figure 127: Our 112-car test train descends across paddy fields. Look directly up from the two cows to see the Remote units..

Figure 128: Rail action and a local market from my front-facing room at the Kwality Inns Hotel in Ranchi. A wide-gauge passenger train pulls in behind a WG-class steam locomotive. The interlocked narrow-gauge trackage seen on both sides of the wide-gauge main line was steam-operated in 1989. Look closely at the locomotive tender and notice the remnants of its 19-tonne coal load near the angle-sheet. The requirement for constant shoveling-forward was the reason these locomotives carried two firemen.

Figure 129: The Kiriburu ore-train loading sidings. Four 56-car rakes can be accommodated and loaded from the central travelling loader. The pit and town are over the ridge at left.

Figure 130: SER support staff wait while loading is completed. The 56-car rake will be taken down the 7-km 2% grade to Karampada and coupled to a similar rake from the Meghataburu mine to comprise our LOCOTROL test train.

Figure 131: The view from the ore-train loader at Kiriburu. The limited capacity of the BOXN wagons when loaded with wet iron ore is evident. When conveying coal or limestone, these wagons are fully-loaded.

Figure 132: Everything is wet at Kiriburu. This territory is a part of the extensive Saranda Singhbhum Range and there's a lot of iron ore around here. Workplace Health & Safety did not attract the same focus in 1989 as in Australia; the state of housekeeping on the train loader left a lot to be desired.

Figure 133: Railway housing in Bondamunda precinct in 1989.

Figure 134: Descending the mountain range from the mines toward Bimalagarh Junction on mostly 2% grades, we encounter the remains of numerous runaways. Our challenge is that the SER people are looking to see how LOCOTROL will make this safer!

Figure 135: Another runaway! The derailed rollingstock is left in-situ for local villagers to recover and sell for scrap.

Figure 136: Another runaway and derailment!

Figure 137: Part of the ore-handling infrastructure between the Kiriburu mine (over the ridge) and the crusher and train-loading facilities.

Figure 138: We're 'in the hole' for a mixed train at Renjeda. On this service, you can sit in a carriage and pay a fare, or you can…

Figure 139: Our empty test train, bound for the mines, is overtaken by the daily local passenger to Barsuan.

Figure 141 (below): We have arrived at Hatia on our first loaded test train from the mines. We shall detrain here and go to the Kwality Inns Hotel in Ranchi to await a return passenger service to Rourkela. The train has been split and the two portions will continue separately to the mill at Bokaro Steel City. To demonstrate the capabilities of LOCOTROL, Deno had them make-the-break ahead of the Remote consist, then using remote control from the Lead unit, he reversed the rear portion back to a switch, then ran it ahead to come up beside the Lead portion. He then warned the SER people not to try this themselves! Pictured here is the complete contingent of SER trades and operations staff who travelled on the 6 units. I am congratulating the engineer, Sen. Inspector Ahmed is to my left with neckerchief, and Milt can be seen below the handrail of the third unit..

Figure 140 (above): The author and our complement of electricians pose for the camera. Note the battery in the carry-box. This is for voice radio communications between SER staff on Lead and Remote units..

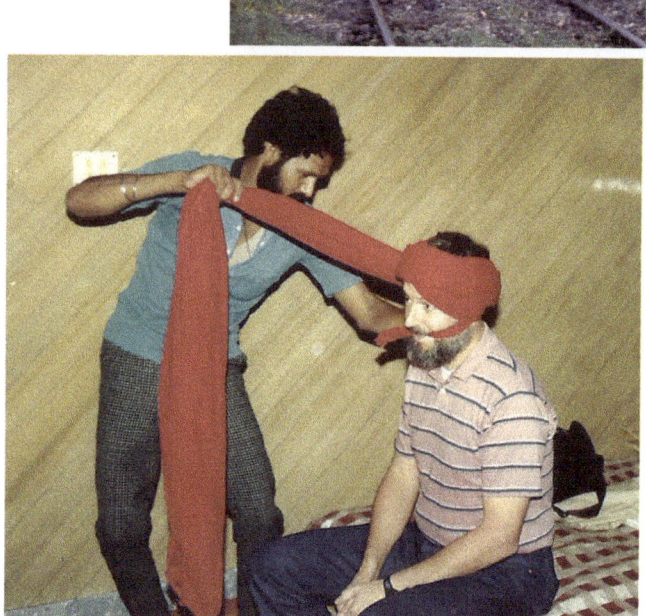

Figure 142: Malkit Singh, our industrious and much-appreciated driver for the two months in Rourkela, demonstrates on Gene Smith how to make a turban.

Figure 143: Milt got one too, and poses with Malkit.

Figure 144: Due to delays to the project at Rourkela, we were unable to pack our equipment in time for air or rail transfer to Assam. We eventually got it all onto a truck. Photo shows the trucker about to depart from outside our offices at the Bondamunda Diesel Shop on the 1200-odd km, two-day journey to Guwahati. Malkit, despondent at our imminent departure, walks toward me.

Figure 145: SER train № 12 UP, the Ispat Express about to depart Rourkela for Kolkata. The author rode the locomotive.

Figure 146: LOCOTROL radios, Air Brake and Control Consoles, and ER gauge installed on the metre-gauge YDM-4 locomotive. Not pretty, but that was up to the NFR.

Figure 147: View of the back of the engineer's control stand, YDM-4 locomotive.

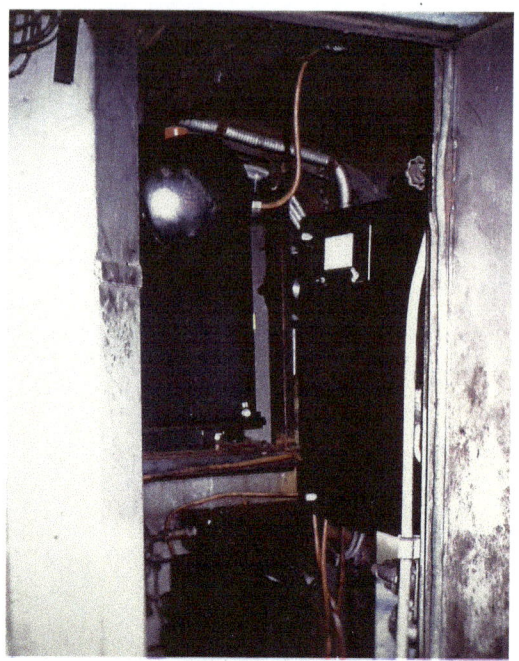

Figure 148 (above): LOCOTROL Logic and Relay Interface boxes, and below them, the Air Brake Manifold installed in the short hood of a YDM-4 locomotive..

Figure 149 (above): metre-gauge Alco steam action in the yard near the New Guwahati Diesel Shop. Viewed from the signalbox.

Figure 150 (right): Morning view across the Brahmaputra River from my window in the Hotel Brahmaputra Ashok in Guwahati.

Figure 151 (above): View across downtown Guwahati from the Ghandi Mendap, atop Sarania Hill. Rail yards and station, and Nehru Stadium to the left.

Figure 152 - NFR HQ in the western Guwahati suburb of Maligaon West in 1989, from the Kamakhya Hill Top.

Figure 153: Some locals at an undetermined location gaze with bemusement at the foreign locomotive engineer while others are equally bemused at the foreign photographer. Milt wears his lucky Budweiser hat that is mandatory apparel on every project.

Figure 154: Milt and NFR engineering personnel confer while we are stationary for a meet at Haflong Hill. Their procedure is to drop handbrake levers here for the coming descent. Milt proposes that with the air brake control afforded by LOCOTROL, they won't have to. Milt was right.

Figure 155: We make a stop on the Dayang River bridge at the bottom of a grade, so a group photo can be taken. This excludes those riding on the Remote unit and adjacent staff car, which can be seen on the bridge behind Milt's head. To Milt's right is Y.N. Dhawan, Deputy Director RDSO, and on his left is Mr Paul, the Chief Motive Power Controller. Locomotive Inspector Sohan Singh is in the turban with his arm on the shoulder of Barua, our cook and attendant. Our locomotive engineer, K.L. Nath, is crouching with the red pen in his breast pocket. Milt may not look as tense as he feels in this image. He has a morbid fear of snakes and had to be coaxed down off the locomotive for this photo. When he did finally alight, he carefully walked exactly in my footsteps..

Figure 156: A view from one rail level to another in superb mountainous, jungle territory.

Figure 157: This view shows one of the run-off sidings on the 2.7% grade between Haflong Hill and Damchara. As you approach a Distant signal, you have to sound the horn to tell the station master you have your train under control, upon which he will set the runoff points for the main line. The photo is unable to convey just how incredibly steep the runoff is. I could not imagine how they laid the track up those runoff sidings..

Figure 158: The 'Observation end' of the NFR GM's Inspection car. Apart from some equipment, this was my bedroom, and, when the bed was folded away it was where Milt or I rode when one or other of us was not on the Lead locomotive. It was a wonderful place to ride!

Figure 159: One of the two other bedrooms on the car. Part of the lounge/dining room can be seen through the door. Milt slept in this one.

Figure 160: This image is out of sequence. It's another view of part of the Ranchi railyard across from our Kwality Inns Hotel. This image shows yard steam action on the 2 ft 6 in network then extant. The trestle in the background facilitated the (bottom-discharge) transshipment of bulk minerals—I forget which—from 2 ft 6 in wagons directly into 5 ft 6 in wagons below... an entire rake at a time!

Figure 161: A simplified map of Guwahati shows the layout of some of the places I photographed. I'd say about 10% of the city is depicted, and about 5% of the roads and streets therein. The New Guwahati Diesel Shop is off the right-hand side of the map.

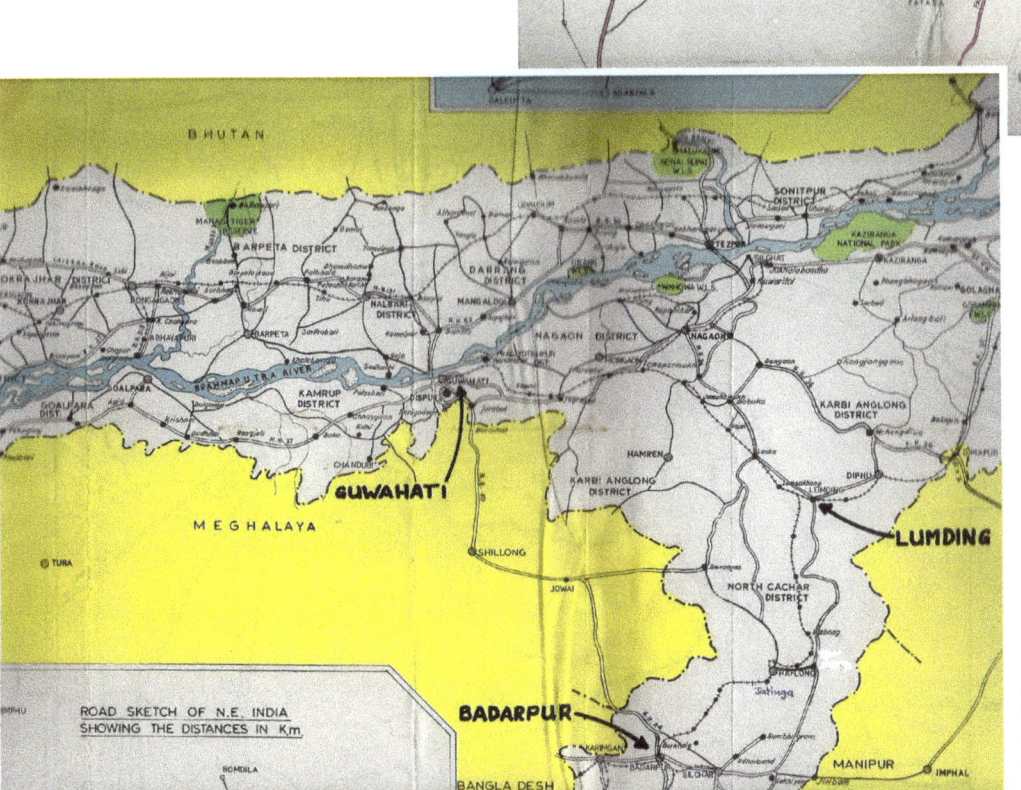

Figure 162: A very badly-presented map of that part of Assam relevant to our metre-gauge project. But it was the best I could find to show the right places. Rail distances are given in Figure 113.

Address: Hotel Radhika, Bisra Rd., Rourkela
Dated 5 August 1989

TRAVELLING ON LOCOMOTIVE FOOTPLATE OF LOCOTROL HEAVY HAUL TRIAL SPECIAL GOODS TRAIN — INDEMNITY BOND AS PER RULE 136, GOODS TARIFF No. 33 Part I

In consideration of the undersigned, namely F.E. Moffat, representative of M/s Harris Corp., USA, being granted permission to travel in Bondamunda – Bokaro steel city and Bondamunda – Karampada sections on locomotive footplate in connection with heavy haul trials with 9000 tonne goods trains for field testing of Locotrol equipment, I do hereby undertake and agree that the President of India, his agents and servants shall be free from all responsibilities for any injury or loss to me, or to any property of whatsoever kind accompanying me, however occasioned during the journies for which the permission is granted, while I or the aforesaid property is within the limits of any of the Railways concerned. i do hereby further undertake to indemnify the President of India, his agents and servants for any loss or destruction of, or damage to any property of any of the Railways concerned, that may be caused through any act, or ommision on my part while so travelling on locomotive footplate in connection with heavy haul trials.

I also hereby undertake that I shall not interfere with or obstruct the Drivers in question in performance of his duties and shall also obey all reasonable directions which the driver in question may give me, subject to the bye-laws and other General rules and reulations for the time being in force on th Railways concerned or any of them.

Witnesses

1. (D.K. Rudola) 7/8/89
   Sr DME(D)/BNDM; S.E. Rly.
2. (R.P. Gupta) 7/8/89, AME(D)/BNDM, S.E. Rly.

Name: F.E. Moffat
Designation: Ant. Road Foreman of Engines.

Office of Sr. DME (Dsl), SE Rly Dsl Shed, Bondamunda, Orissa

To,
    The GM (Optg),

Figure 163: This was one document I had to sign to carry out my role. I have misplaced the other – an authority to drive IR trains as required by the project.

5

HANDLING A Locotrol TRAIN

[Author's Note: I think a description of how a LOCOTROL train is handled is an essential subject for inclusion in a historical record such as this. Herewith, then, a second chapter from my career.]

Introduction

As tonnages on the single-tracked Mount Newman Mining Co (now BHP Iron Ore) railroad in the Pilbara region of northern Western Australia increased, operations management began to consider possible solutions to the need for extra capacity. Electrification and double-tracking were investigated, as was the use of remote-controlled distributed power. Since the magnitude, at the time, of the capital expenditure for either overhead wiring or double-tracking made these options look less likely in the foreseeable future, it was decided to investigate the use of LOCOTROL.

In 1975 and 1976, MNM inspected LOCOTROL operations on Queensland Rail's coal network and Charlie Parker was invited by MNM to visit Perth and Port Hedland, WA, and to discuss Canadian Pacific's experience with the technology. As a result, MNM decided to purchase two sets of LOCOTROL 105SS equipment from Harris Controls' then Australian agent, Evans Deakin Industries of Brisbane, Queensland. Two ore cars were modified as remote-control cars and the equipment duly installed therein and on four Alco C636 locomotives, as well as the maintenance equipment at the Locomotive Service Shop. Instruction was provided, and a number of trains operated to test the usefulness of the technology. A large-scale acquisition was not proceeded with at the time and on-track use of the two systems for a number of years was sporadic. Today—having operated LOCOTROL III and now LOCOTROL XA—the BHPIO Railroad continues as one of two distributed power operators in Australia's iron-ore producing Pilbara region.[95]

During the mid-1980s, iron ore sales increased, and it was decided to equip for a full seven-trains-per-day (each way) schedule of DP operations. The opportunity was taken to purchase the latest version, LOCOTROL II, and seven sets of this equipment were installed on seven locomotives and five remote-control cars (two of which were recycled from the old LOCOTROL 105 operation), and the required training and instruction provided. By 1988 the company was operating a full LOCOTROL schedule, with the train configuration originally set as 3 + 270 + 3 + 270 (2940 metres/9646 feet freestanding; 2981 metres/9780 feet stretched). Operations with this configuration were successful; however, two issues surfaced. Some passing tracks were not long enough for 270 cars and the majority of others were barely long enough. It was found that if a train arriving for a meet was not taken right up to the Roll-By marker in as bunched a state as possible and held there with an air brake application, the last car was liable—after the train came to a stand—to move backwards

---

95 The Roy Hill iron ore mining railroad—having commenced operations in 2015 and also exporting through Port Hedland—operates LOCOTROL XA, and can be expected to eventually integrate it into a fully-automated train operation.

under the influence of drawgear rebound and roll its last axle across the insulated rail joint applicable to the track circuit for the Departure signal behind it.

Although there was never any chance of the last car being able to physically foul the main line, such an incursion would put the main line Departure signal back to Stop and if this occurred as the loaded train was running through, it was guaranteed to excite its crew! Drivers of empty 270-car trains entering passing tracks soon learned to do so using Remote throttle against head-end dynamic braking or Idle power, and to hold their train bunched on a brake application. The second issue with 270-car operations was germane to dividing the train in the port yard for dumping. It was found that rake size with the train split in the yard was not optimum for the dumping of the different ore-types often transported in the one train. For these reasons, it was found convenient to reduce the LOCOTROL trains to 3 + 144 + 2 + 96, and this proved to be a realistic solution, becoming the standard train size for many years.

The following account has been adapted from one written by the author and originally published in the Australian journal *Railway Digest* in April 2005. It chronicles a singular moment in time: a loaded ore train run from the mining town of Newman to the Indian Ocean terminal of Port Hedland in 1989. All empty trains in that era ran directly from Hedland to Newman, were divided and loaded—sometimes with a variety of ore-types—and these portions remarshalled into full-sized trains for return to Port Hedland. Single-driver operation was in the future, as were fly-in/fly-out employee arrangements, rostered 12-hr shifts, major new satellite mining operations, 6000 hp AC traction, and LOCOTROL III. This, then, is how it was during the early years of full-schedule LOCOTROL operations on the MNM Railroad. Join us in the cab for a ride down the *Tdjilla Trail*.[96]

౿🙰

Newman to Port Hedland

It is 5:30 am in the company town of Newman (pop. 6500) in the remote northern Pilbara region of Western Australia. Today I am operating a loaded LOCOTROL train to Port Hedland on the return leg of a two-day trip running iron ore unit trains on the Mt Newman Mining Railroad.

I have allowed myself an hour and 45 minutes to shower, dress, breakfast and be out of town on our 7:15 am schedule. The soft, blue night light in the hallway of the rail-crew quarters doesn't prepare me for the fluorescent glare in the lounge room as I enter and mumble greetings to other crewmen ranged around the room in easychairs. My colleague, Jim Ward is already up—pulling on socks and steel-capped shoes and drinking tea. Together we walk to the nearby company mess hall. It is a typical high-summer morning in Newman. Crows screech their unlovely cacophony from the gum trees and at this hour the air is still soft and cool. The bush flies are as active as usual but to the uninitiated there is no hint of the 40+°C (104°F) temperatures to come.

Our meal is eaten in a mess hall empty at this time except for catering staff, and once finished, we collect an *Esky* full of food and the crew vehicle, and drive ourselves the 5 km to the mine-site to board our train, with engine 5630 (a GE Dash 8) leading. The train has arrived overnight as we slept and been loaded by a Newman

---

[96] 'Tdjilla Trail' is the title of a documentary filmed for BP Australia in 1969, to celebrate the construction of the 430-km MNM iron ore railroad. Tdjilla means 'snake' in the local aboriginal dialect and the film included a song and original lyrics sung by Jimmy Little, an indigenous Yorta Yorta man from New South Wales and an accomplished musician, actor, and teacher. See https://www.bpvideolibrary.com/record/150 (retrieved Sept 2020).

yard crew, and it looks like we'll be away on time. Jim and I will share the running of the train by dividing the trip into quarters. It is normal for the loaded run from Newman to Port Hedland—426 km to the north—to take about 8 hrs. The run is frequently made non-stop for a running time of 7 hrs or less—normal loaded train speed being 65 km/h (40 mph) when powering, with a maximum permissible track speed of 75 km/h (47 mph) where appropriate when coasting or dynamic braking over undulating territory.

The preferred train handling method is to utilise freewheeling where possible, consistent with controlling slack action and this works well on a railroad that is essentially downhill. In theory it should be possible—assuming a non-stop run, no speed restrictions or delays, and good dynamic braking on the two consists—to make the trip with no more than five air brake applications although a theoretically perfect trip is, of course, seldom achievable. With our journey taking us from an elevation of around 600 m (1969 ft) to sea level this is hardly mountain railroading, but with a train of this extreme mass (approximately 30 000 tonnes or 33 070 US short tons) the gradient profile—although favouring extended use of dynamic braking—by no means permits a driver to 'set-and-forget' the controls. Both the track profile and daily on-track occurrences vary sufficiently and combine such that locomotive drivers get plenty of opportunity to actively 'handle' the train. Perhaps one difference to LOCOTROL operations between this railroad and one operating over more varied topography is that most of the subtle control variations effected to regulate train slack are made by modulating dynamic braking effort rather than a power throttle setting. Nonetheless, locomotive engineers obtain plenty of opportunity to practise both Independent Control and Independent Motoring[97].

Since Jim will run the first sector today, he stows his gear and slides into the right-hand seat while I carry out the important task of transferring some of the contents of our food cooler to the cab fridge. As the train has been stabled for some time since being loaded, Jim carries out a brake pipe continuity test. He makes a 70 kPa (10 psi) equalising reservoir reduction, waits until the brake pipe exhaust has ceased and pressure has settled, then turns the LOCOTROL Mode Selector to ISOLATE position. This action cuts out the Remote feed valve and Jim waits a few seconds until that fact is reported to him via a Control Console indication, whereupon he resets the Mode Selector to MU, (**M**ultiple **U**nit [or synchronous] operation) arms the Remote feed valve by pressing the FV switch and pulls the Automatic brake button to initiate a Release. When the feed valve IN light illuminates on the console and the End-of-Train monitor evidences a brake pipe pressure increase we are satisfied, and Jim radios the dispatcher[98] at Port Hedland.

'Hedland, three-zero.'

'Fifty-six thirty, go ahead.'

'Good morning, Pop. Twenty-one eighty-seven on two-road at the mine. Two-forty high-grade. Good to go.' [train serial number – for the dispatcher's requirements, and 240 cars]

'Okay, morning Jim. You're right to depart. First meet Kalgan.'

'Roger-roger.'

The exchange is necessarily brief. Dispatcher Pop Mitchell has the morning 'rush-hour' to contend with: the crews of track maintenance Hi-Rail[99] vehicles are calling for clearance to either 'on-track' or 'off-track'; a ballast train crew is getting cranked up at the Redmont line camp hard up against the Chichester Range; an earthworks gang building a train-loading hardstand at Gidgy siding wants to foul the track; and there are other trains competing for attention. Jim sets about getting us under way while I thumb through the

---

97 Remember the difference! See page Footnote 13
98 During this era, the dispatcher's formal title at BHPIO is 'Traffic Controller'.
99 An abbreviation of **H**ighway-**Rail**way, for a road vehicle that would normally operate on a road (e.g. truck, bus, tractor, digger or 4WD utility) but is modified to also operate on rail track.

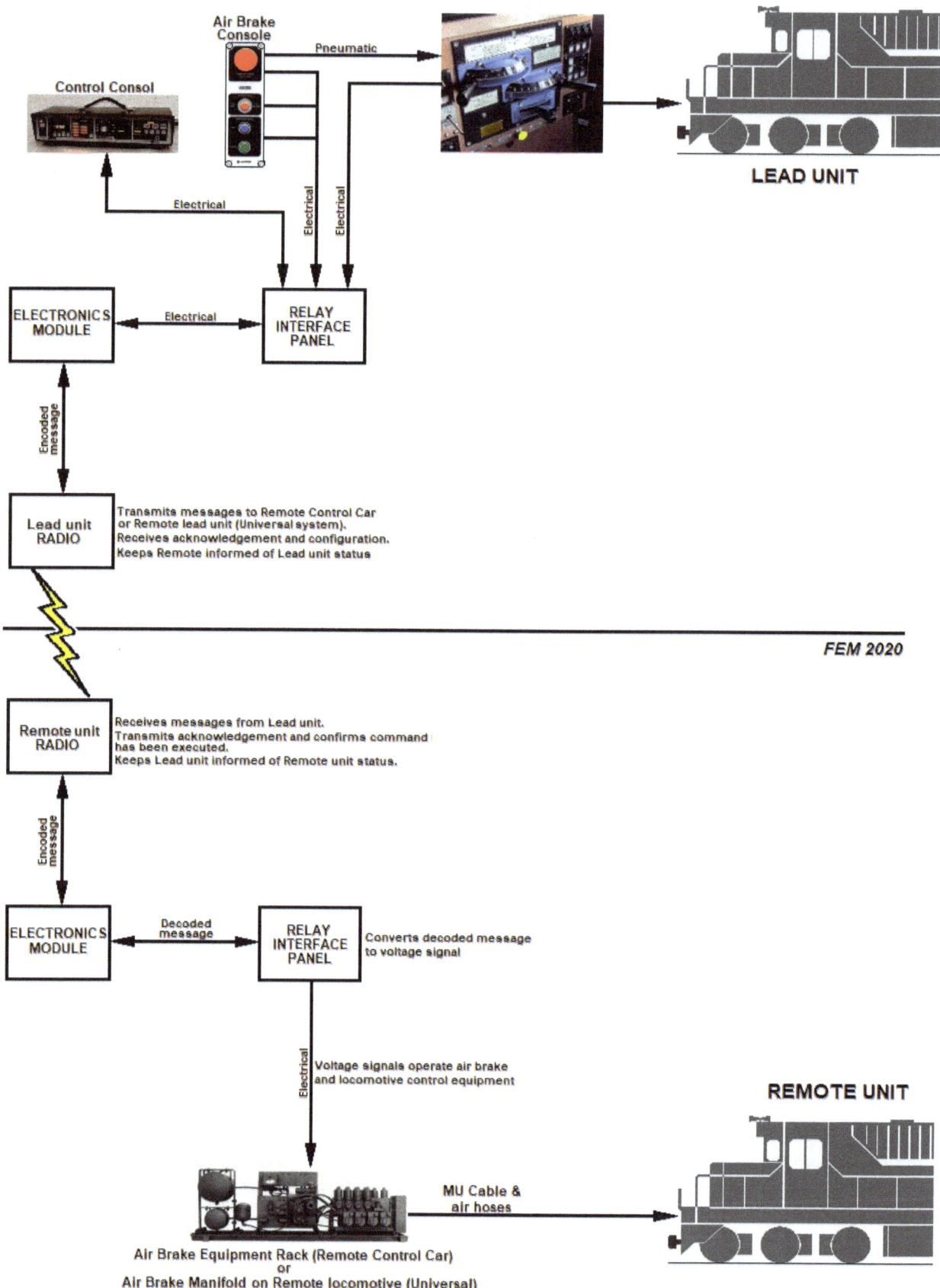

Figure 164: Block diagram – LOCOTROL II. The author.

locomotive repair log to see what's been ailing this unit recently. This is a LOCOTROL II train and we have 240 cars (2643 m or 8672 ft stretched) with each car containing between 100 and 120 tonnes of ore. We have an all-GE consist on the head-end consisting of a GE CM39-8, trailed by two C36-7Ms and an all-Alco Remote consist comprising two MLW 636 units spliced by the LRC (**L**ocomotive **R**emote-**C**ontrol car).

The train has been sitting on the № 2 load-out loop, anchored by five sets of Independent brakes and all indications show a fully-charged brake pipe. Displayed within the cab are brake pipe pressure readings from three points in the train: the Lead unit (via the brake pipe pressure gauge on our locomotive control stand), the

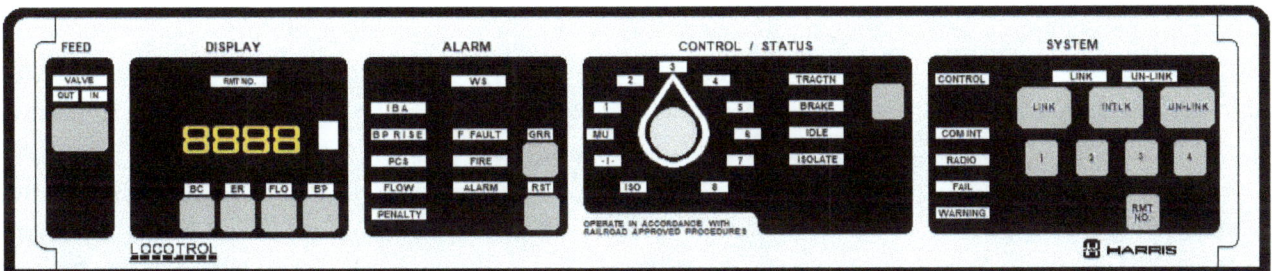

Figure 165: Locotrol II Control Console. The author.

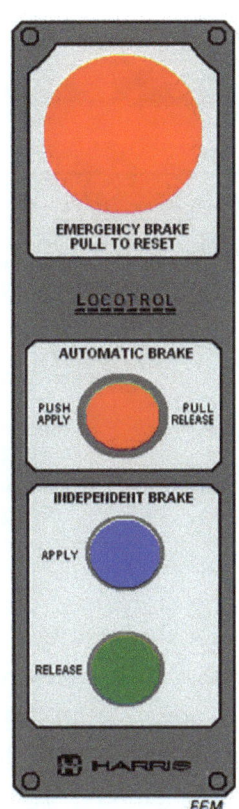

Figure 166: Locotrol II Air Brake Console. The author.

Remote units (via the LRC and presented on the driver's LOCOTROL Control Console), and from the last car (via the in-cab *Digitair* display for the end-of-train monitor). Our pressure looks good at 630 kPa (90 psi) from end-to-end and so with repeated stabs at the Independent brake RELEASE button, Jim gently bleeds off the locomotive brakes on the Remote consist while holding the head-end applied with the Independent brake valve handle.

Five sets of Independent brakes could hold the train on this particular departure road, but three cannot. With velvet smoothness, our 30 000 tonnes of ore train begin to settle in against our Lead consist, and with a protesting groan from brake shoes hard against locomotive wheels, we begin to creep forward. Jim places his selector handle into Set-Up to create the electrical control circuits for dynamic braking and slowly eases the Independent brake handle to Release. With the whole train thus on the move we will not receive a damaging run-out of drawgear and coupler slack. Over its entire length, this train has a total of some 40 metres (132 feet) of free play in the connections between the cars. It is towards controlling this slack action that the major part of our effort in running the train will be concentrated on our trip home.

Parked at the highway grade crossing, the Newman-based load-out crew 'eyeball' our train as we roll out of the yard, checking for stuck or dragging brakes. We snake out onto the main line, with light dynamic braking holding speed to the 20 km/h (12.5 mph) yard limit. Once we are clear on the 'main', Jim slips the units out of dynamics and opens the throttle. The load-out crew have radioed that the train looks good and that all Remote units were revved-up[100] in dynamic braking as they traversed the grade

---

100 Unlike today's microprocessor-controlled locomotives, these earlier Alco units utilised a mechanical fan, driven from a right-angled gearbox on the engine driveshaft, to cool the dynamic brake resistor grids during braking.

crossing. At this stage—as Jim begins to advance the throttle—I exit the cab and move back along our trailing unit walkways to check on how they're loading (a check of the in-cab turbo-boost gauge on each unit) and to examine their repair books. Behind us in the distance, the exhaust haze above the Remote units indicates that they are also leaning into the train.

With the train stretched, Jim is quickly into 8th notch[101] as I return to the cab and serve a cup of coffee. On the favourable grade, we accelerate to track speed and Jim carries out a running brake test to validate our train air brake operation and to get a 'feel' for the train. Holding Throttle 8, he pushes the Automatic brake button once, and checks his air brake readings for the expected Minimum brake pipe pressure reduction of 63 kPa (9 psi). He watches the equalising reservoir needle drop to around 570 kPa and sees the Remote consist brake pipe pressure do likewise. The cab display unit for the EOT is mounted at 'eyebrow' height above the windscreen and we both watch the digital display as it unwinds, finally settling at 567 kPa – spot on! As he sets his Automatic brake application, Jim holds the Independent RELEASE button depressed with his thumb. This action ensures that the Automatic brake application does not occur on either the Lead or Remote locomotive consists, which are still powering. The brake pipe reduction is increased to about 110 kPa (16 psi) and Jim waits until we can feel it taking effect before reducing the throttle to Notch 6. This action is important with a train of this length and weight, and ensures it remains stretched since he will be powering up to full throttle again after releasing.

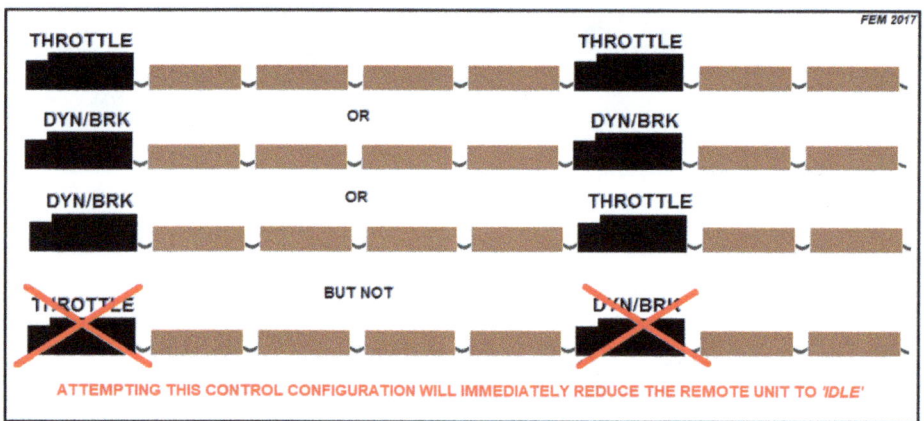

Figure 167: Graphic – Locotrol permitted synchronous and non-synchronous operation. The author.

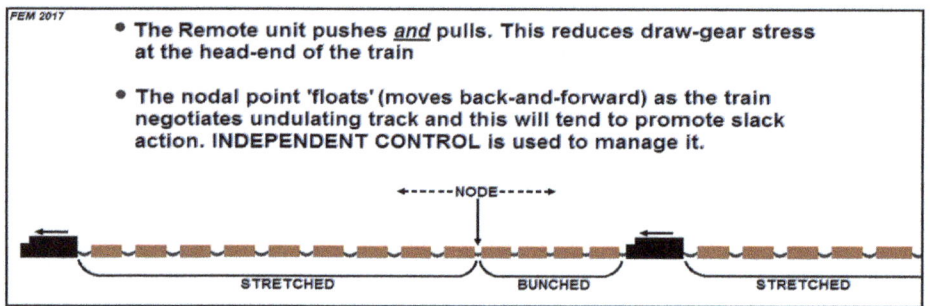

Figure 168: Graphic – The Locotrol 'node'. The author.

In fact, though, the entire train is *not* stretched. Between a quarter and a third of that portion of the train ahead of the Remote consist is being 'pushed' by these units, so between 30 and 40 cars are bunched rather than stretched. There is a point in the train at which this 'buff' or pushing state between the cars becomes a 'draft' or pulling state. The forces at this point are theoretically neutral and the location is referred to as the 'node'[102]. The node tends to float up and down the train ahead of the Remote consist depending on the locomotive engineer's control actions and undulations in the track profile. Control of the node (including ensuring it doesn't move *behind* the Remote consist) is also part of the engineer's task in preventing excessive slack action.

---

101 Notch 8 position is full throttle.
102 See Chapter 3.

With distributed power operations, all throttle and brake control actions made by the driver are transmitted by radio link to the LRC and replicated on the Remote units. For this reason, train air brake manipulation is via a push-button Air Brake Console rather than the normal drivers brake valve, although the locomotive brakes on the Lead consist can still be operated using the Independent brake valve handle. Thus, brake applications and releases are initiated at the

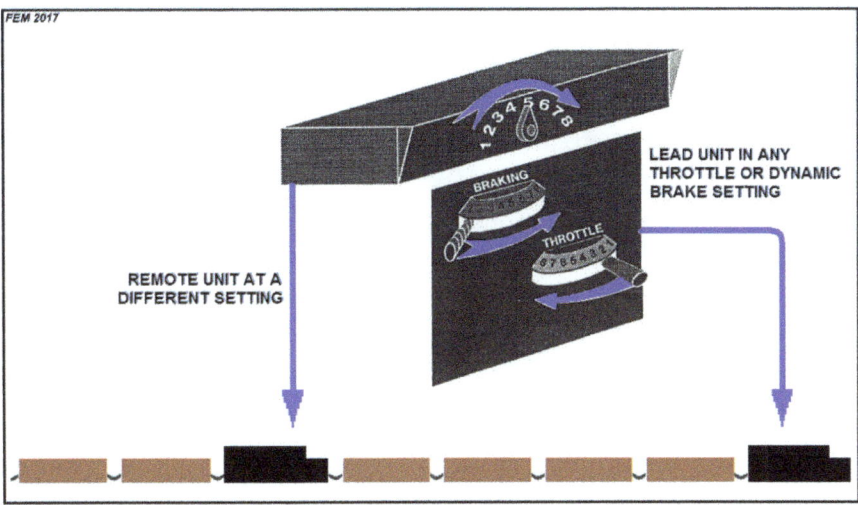

Figure 169: Graphic - Locotrol asynchronous operation: 'Independent Motoring' and 'Independent Control'. The author.

Remote units (the LRC) as well as at the Lead unit. Being depleted at two points in the train, the brake pipe reduction occurs swiftly, and the brake application is rapid, reducing speed relatively quickly. At around 45 km/h (28 mph)—with the reduction settled and effective—Jim pulls on the Automatic brake button to initiate the Release. This command is also transmitted to the LRC, and the brake pipe is recharged from two points.

Figure 170: With two Dash 8s leading and another two as Remotes (just visible among the trees at the extreme left of view) a 240-car 'loaded' rolls out of track 2 at the Newman yard. Photo Tom Winterbourne, MNM Public Affairs. The author's collection.

In concert with the ABDW air brake control valves on the ore cars, this provides a fast and positive brake release along the train and within half-a-minute speed begins to increase. As we approach our maximum permitted track speed, Jim notches back on the throttle and eventually shuts off power. After waiting a few seconds for the slack to settle along the train, he moves the selector handle to again set up dynamic braking. This control action bunches the slack in the train except that—contrary to running under throttle—the 30 or so cars ahead of the Remote consist are now 'hanging away' from it and are thus stretched. Again, the 'node' will float up and down from the theoretical point of neutral force by several car-lengths in either direction as Jim adjusts Lead and Remote dynamic effort independently as the train negotiates undulations in the track profile. Barring any unforeseen delays, we will now coast or run in dynamics for the next 70 km (44 miles), with Jim varying dynamic effort as required—much as you would modulate the throttle when running in power—to maintain track speed.

In fact, today we have been given no speed restrictions for this distance and so we shall do just that, provided our 'meet' is in the clear at Kalgan. While generally downhill, this section is definitely not level!

Figure 171: A 240-car loaded ore train with 3+2 all-Alco motive power departs the Mt Whaleback mine near Newman, WA in 1987.
R Hepburn, the author's collection.

Meanwhile, at 75 km/h, we bear down upon Kalgan and spot our meet tucked safely into the 2886 m (9470 ft) passing track. This empty train was the midnight departure from the port and the Newman-based crew aboard are almost home as they detrain to roll us by. Forty minutes later, as we roll around the curve past the Quarry 6 ballast siding and the site of the old Sandhill railroad line camp, Jim eases out of dynamics, takes

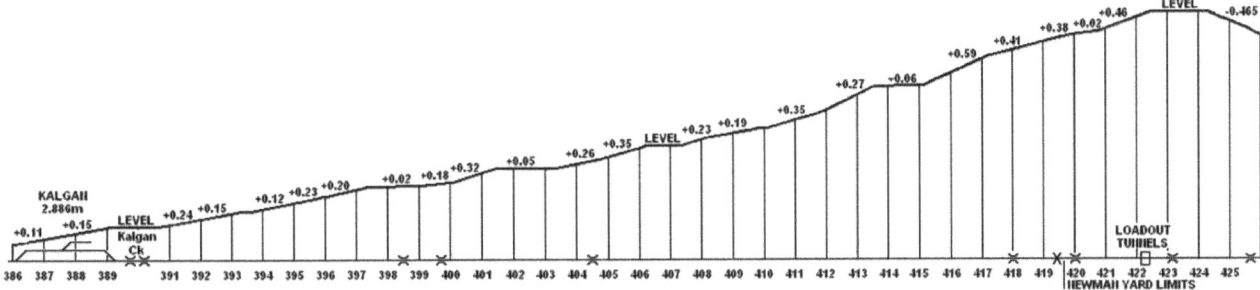

Figure 172: MNM railroad track profile from Newman Yard (426 km) to Kalgan passing track (388 km). The depiction of gradient is exaggerated to better show the undulating nature of the track that has such an effect on trains of this size. The author.

power, and begins to widen the throttle. Ahead is a dead-straight climb at an average 0.09% (1-in-1000) up an alluvial river fan to the Weeli siding location, and in order to maintain track speed on this grade Jim will have to get our combined 18 300 hp (13 650 kW) into full power as quickly as train handling considerations permit.

A blast from the locomotive horn restores reality as Jim warns some slow-moving cattle to clear the way. Our 'office' for today, one of the first four GE Dash 8 locomotives to be built and operated outside of the USA is—at 196 tonnes (432 000 lb) and 3900 gross hp—at this time the largest, heaviest and most powerful diesel-electric locomotive in Australia. Now, as we roar at full throttle through Weeli, Jim is rising to hand the running to me. Taking the hot seat, I glance quickly at air brake gauges and the LOCOTROL Control Console to familiarise myself with the 'state of things'.

The engineer's work-station on these locomotives has been designed with a balanced 'wrap-around' control-stand arrangement. A flat desk arrangement as had recently become fashionable in the US was not our preference, thus to the left is the familiar vertical control stand, somewhat abbreviated due to the absence of a vertically-mounted air brake valve, but incorporating the engineer's LOCOTROL Control Console with its various operating switches and displays. Arranged on a 'dash panel' ahead of the driver are the load meter (presenting an indication of the current draw of the traction motors), speedometer and air brake gauges, and to the right—against the cab side wall—is a sloping console on which are mounted the air brake controls (refer to Figures 54 and 55 on page 101).

These early Dash 8s are equipped with the 26-L air brake schedule, but the driver's brake valves are the W30D/G (Automatic brake; **D**irect or **G**raduable release) and W30IND (locomotive **IND**ependent brake). Designed and manufactured by Westinghouse Australia they are a significant modification of the contemporary 30-CDW valve designed for desk mounting – the train (Automatic) and locomotive (Independent) portions being physically separate and able to be mounted as required for a flexible cab layout. These MNM locomotives are Australia's first installation of this equipment and full advantage has been taken of the design intent, with a sensible arrangement of equipment.

The Automatic brake valve handle is not utilised in distributed power operations and is latched in the RELEASE position. To accomplish remote control of air brakes at the LRC, a set of air brake push-buttons—on the Air Brake Console—is used and this assembly is logically integrated into the forward Control Console alongside the aforementioned W30 brake valves. These push-buttons pilot the operation of solenoid-equipped air brake manifolds fitted to both the Lead unit and (via the secure radio link) the LRC.

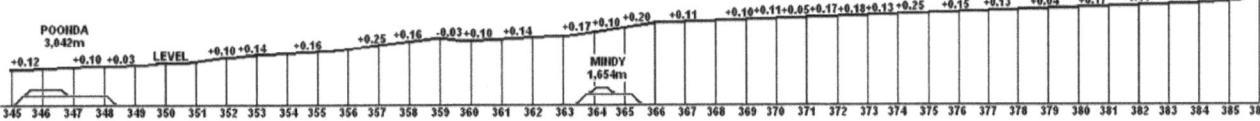

Figure 173: MNM railroad track profile, 386 km to 345 km. The author.

We are approaching the crest of the grade up and over the east branch of Weeli Wolli Creek and thereafter some undulating country. I reach to the LOCOTROL Control Console and spin the rotary Mode Selector switch from 'MU' to the '8' position. The Remote units continue to power in the 8th (full) throttle position, but I now have *Independent Control* over them, and I will begin to use that feature soon. It is axiomatic that a heavy train requires power to ascend a grade but not necessarily to descend it and that with a very long train it is quite possible to have a portion of the train going uphill while the rest of the train is still on a down-grade or vice-versa (there are also some locations on the railroad where a train can be draped over four separate grades (up-down-up-down). This is where the function of *Independent Control* is used. As the head-end of my train crests the grade and runs through Weeli, I begin to reduce the throttle: however, the major part of the train is still on the upgrade and continues to require power if we are not to slow down excessively. Thus, using the Mode Selector switch, I hold the Remote consist in full throttle until I judge they too will be approaching the crest (now behind me), at which point I commence to throttle them back as well, one-notch-at-a-time.

In this way, varying the Lead and Remote power levels as required, I traverse the undulating track over the various branches of the creek with as much control over in-train forces as possible. When the track profile permits, I will place the Mode Selector back to its 'MU' position in which the Remote consist synchronously follows all of my throttle or dynamic brake inputs.

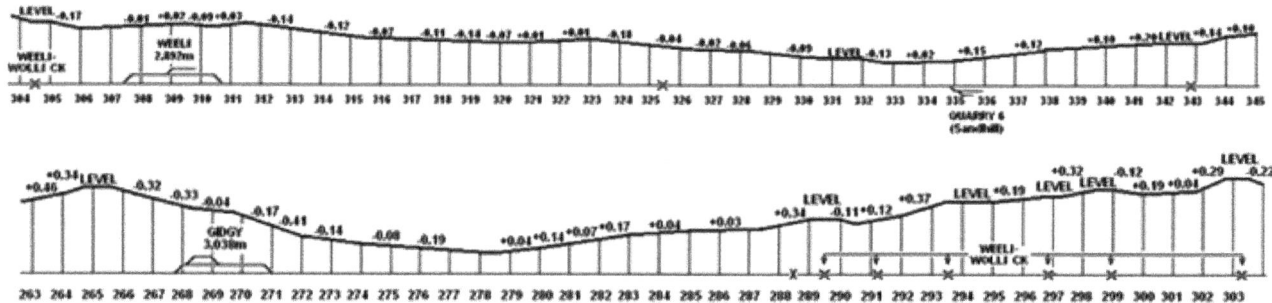

Figure 174: MNM railroad track profile, 345 km to 263 km. The author.

But now the dispatcher is calling us on the radio, and it is not good news. The 0300 departure (empty ore train) from Port Hedland is running late and production requirements mean that its passage to the mine is to be expedited. For Jim and I this means that we will not have a non-stop run home today: indeed, we will be required to stop on the main line at Gidgy—the next siding location—to allow the empty train to 'run through' via the passing track. Apart from the inconvenience of having to bring this monster to a stand, the consequent restart means a slow uphill drag. For the company, there will be both a delay in our schedule and a considerable fuel penalty incurred. However, that's railroading!

We approach Gidgy on a 0.41% (1-in-240) upgrade. The track levels considerably to 0.04% for two-thirds of the length of the main line and then rises again at 0.33% (1-in-300) through the north switch and for another 3 km before cresting. We come pounding up the grade and by the time we pass the Arrival signal, I have reduced to Throttle 6 and am holding about 600 amps. Using the kilometre pegs as markers, I run up the main line for 1000 m before making a minimum air brake application. Using *Independent Control*, I can hold Throttle 6 on the head-end and reduce the Remote consist to Throttle 4. As the light initial brake application begins to take effect, I judge my distance-to-run to the Roll-By marker—which is 100 m before the Departure signal and is my stopping point—and reduce power one notch at a time. While doing this, I use the Mode Selector switch to independently reduce power on the Remote consist, keeping it two notches lower than the head-end as I do so. This is in deference to the fact that the lead portion of the train is now on the upgrade approaching the

Departure signal, whereas the Remote group and rear portion of the train are still on the more level section.

By the time we draw to a stand, I have the Remote group in 'Idle' but on this occasion have misjudged on the head-end and am still holding Notch 2 power. To close the throttle at this point would cause a vicious and damaging electrical arc to occur in the high voltage control compartments on each head-end unit as the power contactors are forced open by control air pressure, so I rotate the hump-control (slow speed) knob to its 'Minimum' position to allow the amperage I am holding to decay. Now I can safely close the throttle (the contactors open with a gentle 'click') and reset the hump-control to 'Off'. The hump-controller takes its name from its original design function, which was to provide precise control of very slow speeds during hump shunting operations. Nowadays we use it for loading trains and at any other time a controlled very slow speed is required, such as while dropping track ballast. Now I increase the brake pipe pressure reduction to 100 kPa (14 psi). The train will sit here on this application until we depart – the Independent brake alone being insufficient to hold us on this gradient.

Jim and I alight to stretch our legs and also to give the units a walk-around from ground level. Soon, the empty train approaches and trundles through the turnout at 35 km/h (22 mph). Jim and I take up station—one either side—to roll it by. *Tang-tang-tang-tang-tang...* the crew ring a friendly greeting on the locomotive bell. Monotonously, the 240 cars rumble past (note: empty trains tend to 'rumble', whereas loaded trains glide, almost silently) and with a quick glance to check that the EOT monitor and tail-lights are in place, Jim reports '... *Roll-By's good...*' on the portable radio and we climb the ladders and prepare to move out.

I reduce the Independent brake to about 200 kPa (30 psi) brake cylinder pressure and release the Automatic brake. The basics of train air brake operation are that car brakes apply when the brake pipe pressure is reduced and they release when that pressure is restored, thus the beauty of air brake control on these 'radio' trains is in the two points of brake pipe exhaust and recharge. This makes a 240-car train easier to handle in this regard than one of our conventional 192-car trains with head-end power only whereby all brake pipe depletion and supply occurred exclusively at the lead locomotive. The ABDW control valves on each car have a feature known as Accelerated Release, which propagates a rapid serial response action back along the train (and in the case of the Remote consist, forward of it as well).

From these handy attributes, we note within 30 seconds the Digitair cab display indicating the restoration of brake pipe pressure on the last car. Now I can fully release the Independent brake and open the throttle. I have the Mode Selector in 'MU' and the Control Console displays the throttle setting on the Remote units as I notch out. Using the appropriate membrane switch[103], I select the brake cylinder display to check that the Remote Independent brake has released. With the Remote display,

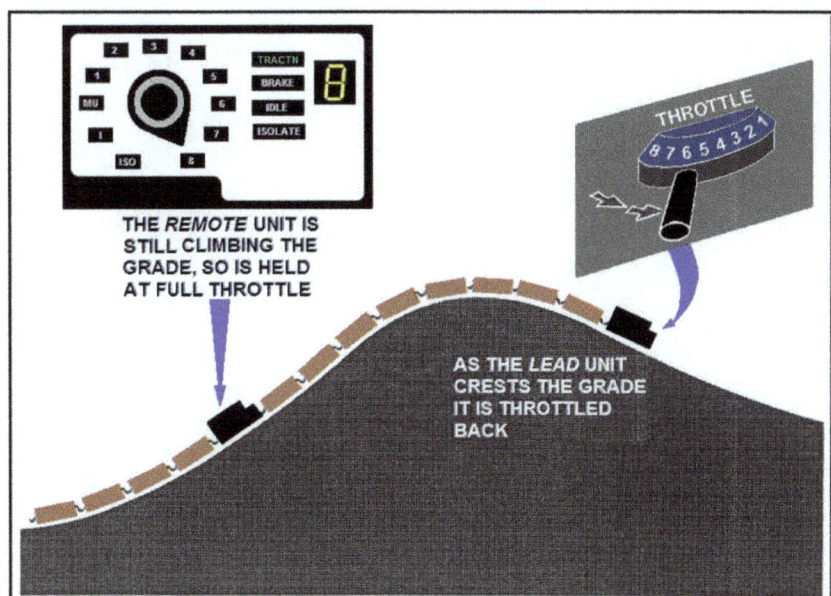

Figure 175: Graphic - Locotrol asynchronous operation: Cresting a grade. The author.

---

103   See Glossary.

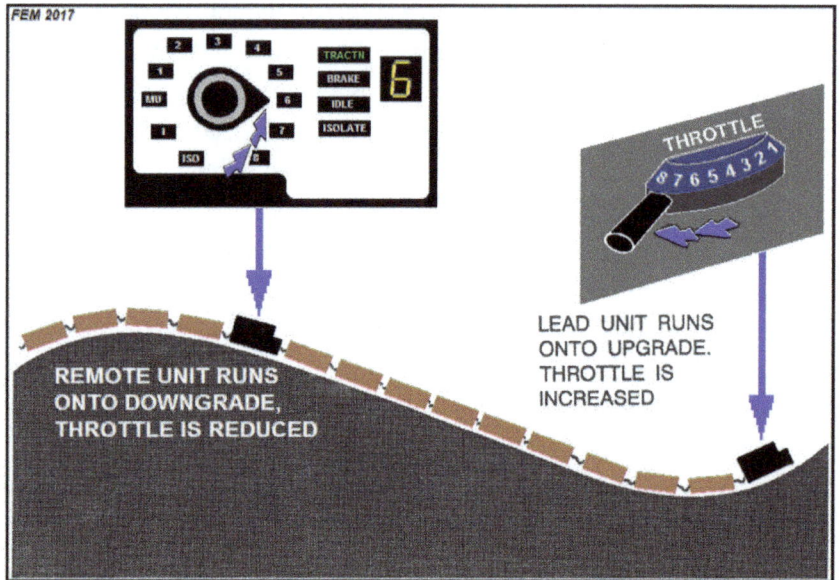

Figure 176: Graphic - Locotrol asynchronous operation: Negotiating a sag. The author.

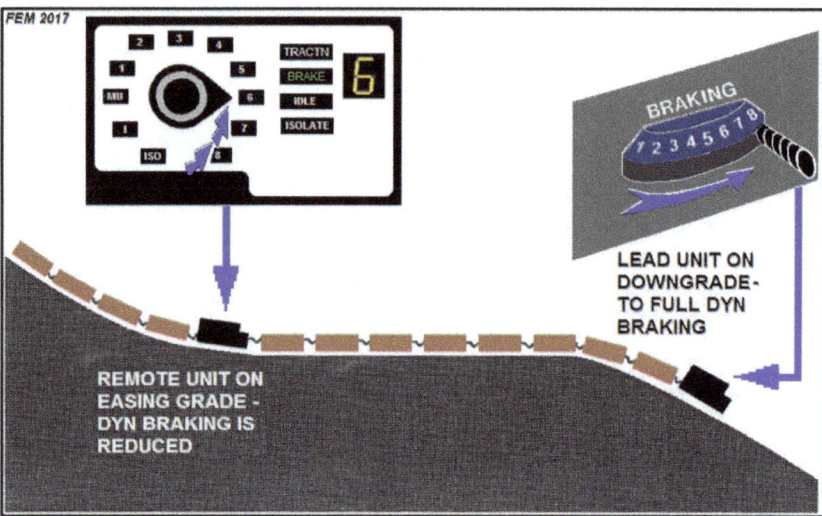

Figure 177: Graphic - Locotrol asynchronous operation: Negotiating an undulating descent. The author.

I can also select equalising reservoir and brake pipe pressures, as well as obtain an indication of the brake pipe charging flow rate at the LRC. We pull high amperage as we get a roll on this uphill start and I am alert for any wheelslip indication from the Remote consist. Such an indication—should it persist—would necessitate some Independent Control of throttle to temporarily reduce power back there.

Some 3 kilometres out of Gidgy we crest the grade, having managed to accelerate to 43 km/h (27 mph). I immediately begin reducing power on the Lead units while holding Throttle 8 on the Remotes. Gradually reducing the Lead units to Throttle 1, I begin to back off on the Remote consist power also as I judge it to be approaching the crest behind me. Now I can close the throttle and slip the Lead units into dynamic braking, holding a low initial effort. I continue to throttle back gradually on the Remote units and soon I have them in Idle and coasting. By gently increasing the head-end dynamic braking effort, I lightly bunch the front portion of the train and as we roll onto the down-grade, I increase

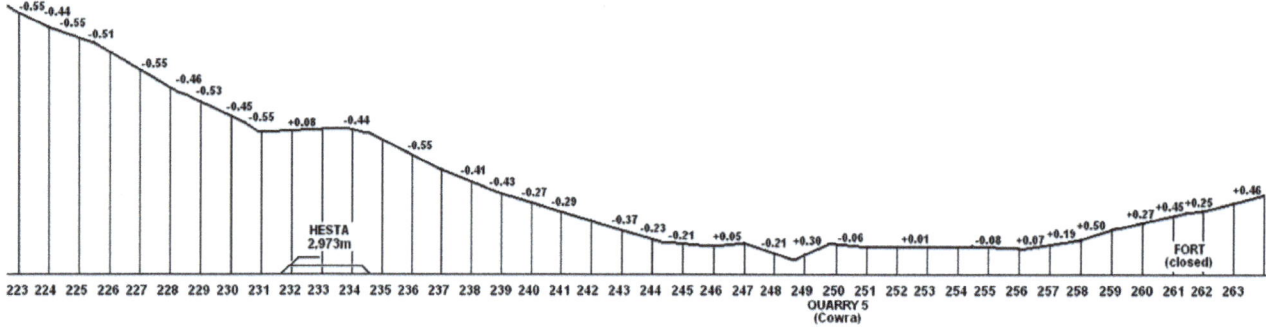

Figure 178: MNM railroad track profile, 264 km to 223 km. The author.

dynamic effort separately – using the selector handle to modulate Lead unit braking and the Mode Selector for the Remotes. Once I have full dynamic braking on both consists, I move the Mode Selector from Braking '8' position back to 'MU' for synchronous operation.

Now we descend onto the Fortescue River flood plain and—traversing a wide 2620 m (80 chain) radius curve that provides an excellent rearward view of the train from my side—we cruise past Quarry 5 and the derelict site of the one-time Cowra line camp. I get into full power as we head for the long embankment leading up to Hesta siding in the foothills of the Chichester Range, at an average gradient of 0.36% (1-in-280). On a comparative basis, a gradient of this magnitude may not appear as much of an obstacle: however, by the time we stamp across the south switch at Hesta we are down to 25 km/h (15.5 mph). A rearward glance as we lean into the reverse curve through Hesta reveals the mighty spectacle of five high-horsepower locomotives hurling exhaust gases skyward as their prime movers pound at full throttle under full load to shove and drag this 30 000-tonne monster over the range. To me, the sight is stirring and dramatic – a worthy modern-day display to rival those I recall from the bygone days of steam.

The gradient through Hesta eases significantly and provides a brief respite from the long drag. Barely the length of the train, this short let-up enables us to accelerate to around 40 km/h (25 mph) and is usually sufficient for the Alcos and Dash 7s in our consist to make forward transition – but not for long. As we leave Hesta behind, we begin to climb the range proper and the train settles down to a 23 km/h slog. With the grade averaging 0.50% (1-in-200) against us, our Lead locomotive is some 14 m (47 ft) higher in altitude than our last car. This part of the trip takes about 30 min, with manifold pressure (turbo-boost) maxxing out at around 32 kPa, and as we grind slowly around the final 600 m (17 chain) radius reverse curves with flanges squealing, the load meter—pegged on 1200 amps—has been in the orange band for longer than it should have.

However, we have conquered the Chichesters yet again and now the track levels briefly across the top of the range through Shaw siding. As we crest the grade, I am mindful of the horsepower concentrated at our head-end: three modern GE units as against the two older Alcos located mid-train. As the front of the train comes onto the easing grade through Shaw, it will tend to accelerate while the Remote units, 144 cars back, are still working hard at their balancing speed. Now, remember the 'node'? As a distributed power train crests a heavy grade at a high level of tractive effort and begins to accelerate, this nodal point begins to move back towards the Remote units (that is, the front portion of the train begins to stretch out). Since the Remote units—still

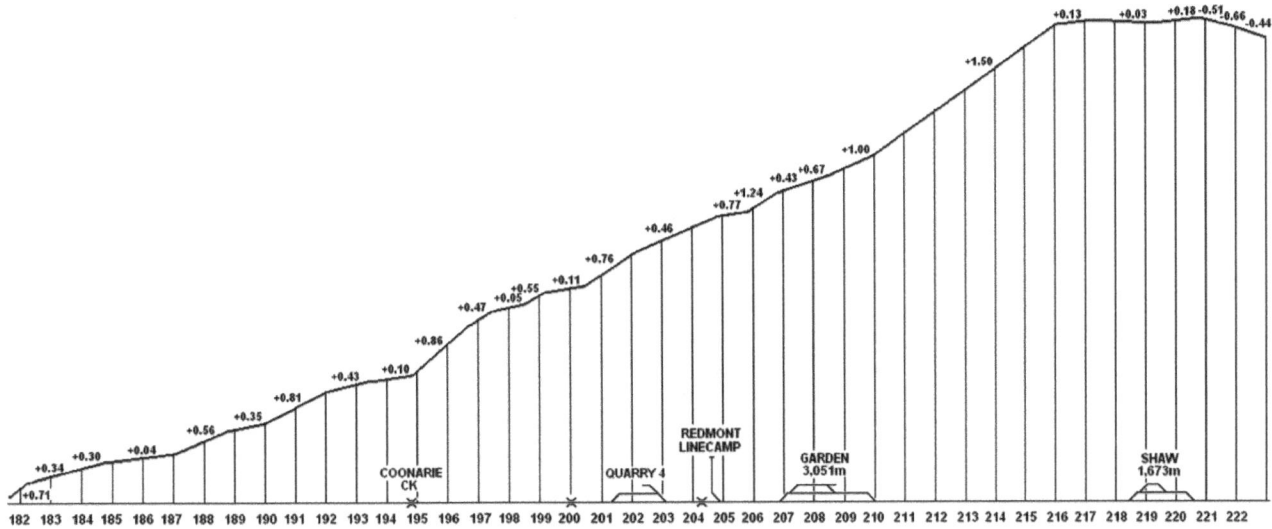

Figure 179: MNM railroad track profile, 223 km to 182 km. The author.

leaning into the grade—cannot attain the same rate of acceleration at this time, it is quite possible that the node might move behind them. This would be undesirable since the drawbar pull of the Remote consist is then added to that of the Lead consist with the result that an excessive draft force will be imposed upon those cars immediately trailing the Remote consist. If such a force should prove unsustainable, then a drawgear yoke or a coupler knuckle will break, and the train will part and be disabled. For this reason, as we come through Shaw, I have reduced power on the Lead units to Throttle 6 while using the Mode Selector to maintain full throttle at the Remotes.

Our journey is now half-completed, and I hand the train back to Jim. He has made tea and occupies the engineer's seat, having just completed his meal. He now has the responsibility of running the train over what I consider to be the most interesting and challenging part of the journey – the 100 km (60 miles) from this summit down and across undulating country to the 120 km peg where I shall again take the reins. Immediately ahead is a short, sharp descent at 1.5% (1 in 75) for about 6 km. Thereafter, the descent eases but overall, we will lose some 300 m in altitude over this 100 km sector and will run in dynamic brake for 90% of it.

But now Jim has our Lead units in dynamic braking as we drop 'over the top' and he goes to *Independent Motoring* on the Remotes, holding Throttle 6 for a short while before notching off and selecting dynamic braking there as well. The operation of the Remote consist is accomplished by using all of the skills any train driver must apply – that is, by exercising judgement in throttle and brake operation, by detailed knowledge of the physical characteristics of the track ahead (and behind!) at all times—collectively known as 'road knowledge'—and by understanding train dynamics and the tractive effort characteristics of the motive power. Handling a distributed power train smoothly means knowing how far back your Remote units are and where they are at all times relative to the Lead units and with respect to the track profile at any given location.

Jim now has full dynamics on both Lead and Remote consists but as more of the train comes onto the downgrade, speed continues to increase and he is obliged to make a complementary air brake application. The descent from the summit is sinuous and located in cutting for much of the distance. Under these circumstances, the Lead and Remote consists can lose radio continuity – a condition known as *Communication Interrupt*. With this in mind, Jim has gone to Independent Control and has advanced the Mode Selector to Braking 8 position to lock in full dynamic braking at the Remotes. He is glancing frequently at the Control Console, alert for a *Comm Int* indication. Should this occur, Jim knows the Remote units will remain in dynamic braking until the system can re-establish continuity. With a total brake pipe reduction of around 100 kPa and about three-quarters dynamic brake effort, our huge train slides down the northern face of the Chichester Range and through Garden siding with the dynamic cooling fans at full song. A work *Extra* with empty ballast cars is in

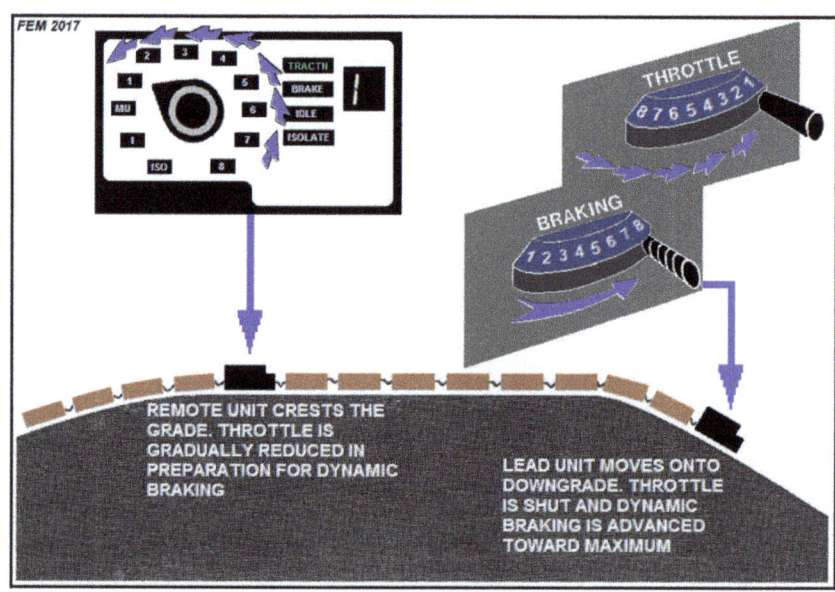

Figure 180: Graphic - Locotrol asynchronous operation: Transition from heavy upgrade to severe downgrade. The author.

the passing track and its crew—standing well clear—give us the 'once-over' and report that we '... *look good, Jim, and two good dynamics on the Remotes.*'

Apart from visual inspections by train crews waiting 'in the hole', all track maintenance and other personnel who are near a passing train will observe it and communicate with the locomotive crew or the dispatcher if necessary. In addition, no effort has been spared in providing the technology to monitor train status en route. Every siding has a dragging equipment detector outside each mainline facing switch and there are four Hot Bearing/Hot Wheel detectors ranged along the length of the route. These devices make an infra-red scan of each passing train for overheated axle roller bearings or hot wheels (signifying dragging brake shoes) and transmit both an alert to the dispatcher and a radio tone to the train if anything is detected.

In this case the engineer will already be in the process of stopping the train by the time the dispatcher calls to provide the car number, axle number, and side (east or west rail). The non-driving crew member can then proceed to the exact car and carry out an inspection. At other times, a train can suffer a burst or parted brake pipe hosebag or any one of a number of alarm conditions on the Remote units. All such occurrences cause considerable delay since a crew member must walk back to look for and inspect the problem. Such an expedition is no fun when it's 46°C (115°F) in the shade, you're carrying a long, heavy 32 mm spare brake pipe hose in one hand and a monkey wrench and portable radio in the other, and there's 47 bush flies walking over your face. Then again, there is often a nearby track maintenance Hi-Rail vehicle travelling on the access road and they will always come to the scene and assist.

The advice from the crew in the passing track is gratefully-received since we will have no direct indication if one of our Remote units decides to go back to Idle at any time. To properly handle his dynamic and air braking, Jim needs to have a feel for whether he is getting all or part of the dynamic braking he is calling-for at the Remote units.

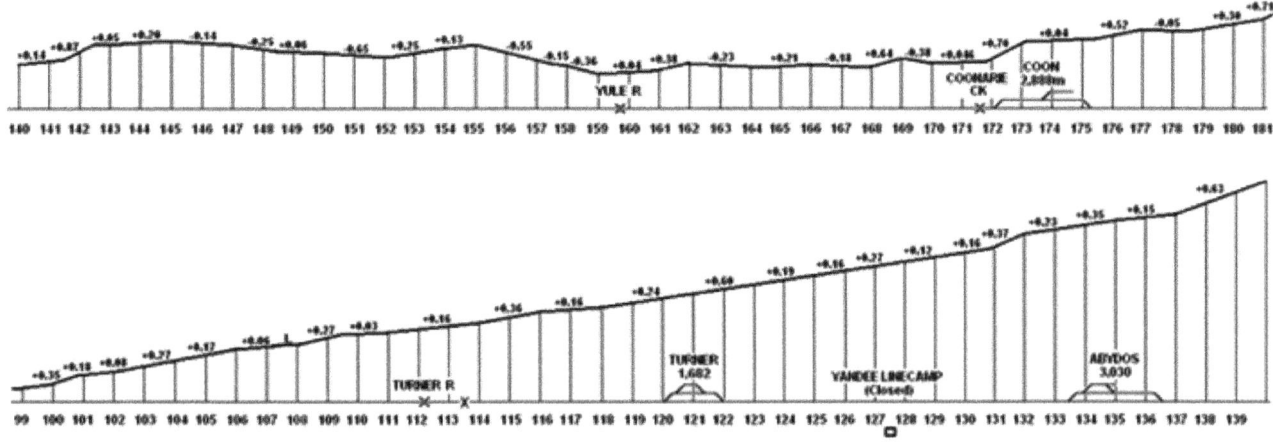

Figure 181: MNM railroad track profile, 181 km to 99 km. The author.

At the 197 km area, and with speed reduced to around 55 km/h (34 mph), Jim releases his air brake application. We roll down over Coonarrie Creek and another brake application is required at the 191 km peg to check our progress. As we round the right-hand curve at the 177 km and approach the Coon siding there is a good view back along the train—showing it draped over three separate gradients—that serves to illustrate the challenge of running these trains. We roll across undulating country to the Yule River bridge then power up for the climb up the 155 km bank.

At the 145 km peg, Jim is again in dynamic braking as we drop down towards Abydos siding. Presently, we pass the site of the old Yandee line camp—closed after being accidentally burnt to the ground following a kitchen fire—and are approaching Turner siding. This location was once a passing track but is nowadays a single-ended, switch-locked spur used by track maintenance machines. It's on a steep little pinch of track, 0.60% (1-in-166) when the average around here is nearer 0.2% (1-in-500), but with five good dynamics Jim has things well in hand when he hands back to me for the final 120 km to the port.

Ahead is a 25 km/h speed restriction due to re-railing on the Turner River bridge. Three kilometres before the commencement I have the train in full dynamic braking at 75 km/h. With 2000 m (6562 ft) to run I make a minimum brake pipe reduction and watch the Digitair display drop by 7 kPa (1 psi) increments to 557 kPa (81 psi). The brake pipe quickly equalises, and I follow with a further small reduction. The Digitair display drops again and stabilises at 500 kPa (73 psi) – a total 120 kPa (17 psi) reduction. Slowly, as energy is dissipated, our speed drops until, at 40 km/h, I make the Release. As this occurs, our speed continues to reduce and the extended range dynamic braking on our Dash 8 lead unit assists considerably in controlling the train as we run over the slow order.

Our journey continues without incident over gently undulating country as we descend across the coastal plain towards the port. The gradient averages 0.25% (1-in-400), but several short, sharp descents of 0.47% (1-in-210), 0.53% (1-in-190), 0.80% (1-in-125), and 0.56% (1-in-180) ensure I remain attentive. The general train handling strategy is to coast where possible over this territory: however, our 2600 m length requires—more often than not—a touch of power here or an easing of dynamic braking there on either the Lead or Remote consist to ensure there are no excessive accelerations within the train since constant gradient variations affect our progress. The smallest such variations are amplified in their effect by the sheer mass and kinetic energy of the train.

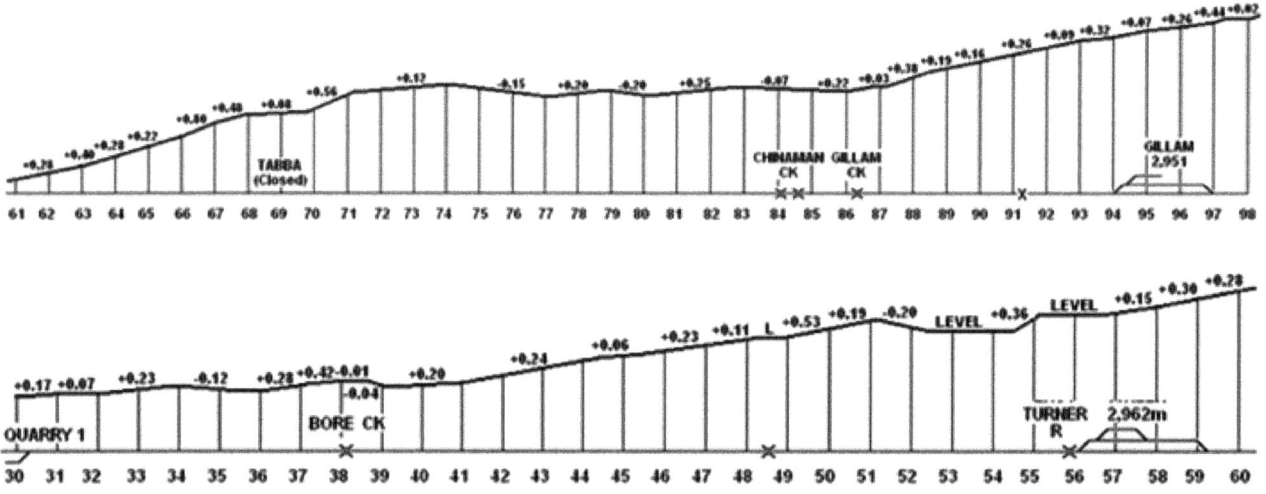

Figure 182: MNM railroad track profile, 98 km to 30 km. The author.

At night-time, the lights of Port Hedland and the Nelson Point railyard are visible from 70 km (43.5 miles) away, but as we near our destination in the early afternoon we are 30 km out before we are able to identify such community landmarks as the nearby solar salt stockpile—shining brilliantly against the blue sky—and our company's towering crusher buildings. On a high tide, one can clearly discern the white superstructure of a giant ore carrier alongside the wharf – looking as if it is floating on land.

At track speed, we curve through Bing—the last siding—on the very outskirts of company satellite town, South Hedland, still with dynamic brake fans howling. Even full dynamics will not be sufficient to slow us down for the Goldsworthy Mining Ltd railroad diamond, so I make a minimum reduction at the 16 km peg and settle back to see how she runs. Despite the fact that ore trains are usually of common length, they do not always weigh the same. Add to this the possibility of an idling or dead unit (or units), the pronounced effect of cross-winds, and the presence of temporary speed restrictions at various locations and you have enough variables to require some modifications to one's train handling actions on a daily basis.

There is an intuitive intellectual process of discernment called 'judgement' that is an indispensable part of any good locomotive drivers' repertoire of skills. When I talk to trainees about this subject, I put it like this: *'Judgement in train handling is a process by which we calculate distance-to-run versus present velocity and the rate-of-change of that velocity, to initiate throttle and brake control movements sufficiently in advance such as to have the desired effect at the target location.'* This may be somewhat verbose—and it invariably results in enthusiastic laughter—but my point is usually made.

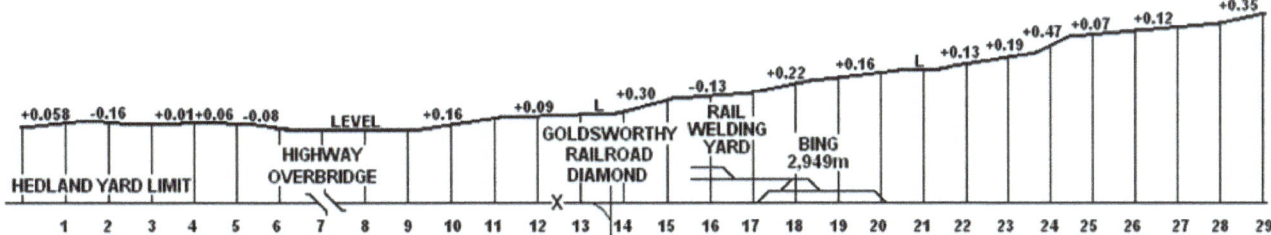

Figure 183: MNM railroad track profile, 30 km to 00 km. The author.

I am now striving to display that judgement as I make the second part of the split reduction and attempt to bring our speed down to 30 km/h at a point 200 m from the diamond before releasing. We clunk over the diamond at the regulation 20 km/h (12.5 mph) and I switch to Channel 3 and call the Yard:

Figure 184: GE Dash 7 and Alco/MLW power plus LRC cars hold the ready tracks at the Locomotive Service Shop at Port Hedland. The author

'Hedland Tower, good afternoon. Three-zero over the diamond. Two hundred and forty high-grade. Serial number two-one-eight-seven.' Back comes the response; 'Three zero, g'day. Clear onto 53 road. Make the break. Engines to shops.'

From this exchange we can expect the Yardman to meet us and split the train immediately ahead of the Remote consist, following which we shall cut off our Lead consist and run the units onto the servicing pad outside the Locomotive Service Shops.

Journey's end. We roll sedately into the port yard reception area, acknowledging the wave from the car examiners stationed either side of the train as they 'roll us in'. I pull down to the port end of our allotted road and with dynamic braking on the head-end and Throttle 1 power on the Remotes, come to a stand on a light brake application. This action bunches the train ahead of the Remote consist and permits the Yardman to lift the uncoupling lever. He is presently calling us to 'ease away' one car-length and Jim proceeds back to our trailing unit. The tower asks us to '... *leave the air in it...*' and I can see a shunt locomotive waiting for us to clear so it can couple on and move these 144 cars to the dumper.

I make a 120 kPa brake pipe reduction, move the Mode Selector to 'Isolate' position and wait for the Remote feed valve to drop out. When the brake pipe exhaust has ceased, I signal Jim and he moves in and closes the brake pipe angle cocks between our units and the first car. The Yardman has done likewise at the front of the Remotes and I can now *Unlink* the LOCOTROL system without the air dumping on the lead portion of the train. The Lead consist though—as per the software program—goes into Emergency when I do this, and I am required to reset and reclaim my air before we can proceed. Finally, Jim can uncouple us, and we move off to the Locomotive Service Shop where we select a vacant road and spot ourselves beneath the sand towers. We clean up the cab and climb down, laden with food cooler and gear. The locomotives have spent approximately 60% of this trip in dynamic brake and will have each consumed about 6500 litres (1717 gal) of fuel for the round trip. They will be checked, serviced and made ready to do it all over again later this evening. Jimmy and I walk to the Yard Office to sign off. We have safely delivered another 26 000 tonnes of ore for export, using LOCOTROL II distributed power and some $30 million worth of locomotives and rollingstock. It'd be fair to say, we're satisfied.

6

DEMONSTRATION TRAINS

As LOCOTROL projects have been completed for various customers over the years, those railroads have often requested (or the project team have proposed) that an outsize train be run as a one-off demonstration to all who might be interested – but particularly for media and publicity purposes. These trains are inevitably a challenge for locomotive engineers since there is generally no prior training for handling trains of these proportions. Harris Corporation personnel would be present in the cab to provide guidance but, inevitably, there exists an obvious disconnect between the instantaneous mental processes of both the Harris instructor and the locomotive engineer assigned to run the train, and the natural challenge presented to the instructor of the powers-of-coordination of a separate person. For these reasons, such trips did not always proceed as planned and a train parting was always a potential outcome, and sometimes occurred. Deno's personal credo was that such a trip made without mishap of any sort was to be considered a 'success'. If there was any other outcome, regardless of the train eventually reaching its destination, the trip was *not* to be considered a success, since the abilities of LOCOTROL had not been effectively demonstrated as intended. Some of the more outstanding examples of outsized demonstration trains are presented in this chapter.

Norfolk & Western Railroad (USA)

I acknowledge Bob Loehne of the Norfolk & Western Historical Society and Clyde Taylor, then N&W General Foreman Mechanical at Williamson, for being able to produce this account.

During the early fall of 1967, the Norfolk & Western railroad had been experimenting with longer and longer coal trains, including the use of mid-train, radio-controlled locomotives (RCS). N&W had previously run several 250-car trains carrying 24 000 tons of coal and they quickly upped the ante: 280, 300, 420, 425 (third week), then 450 cars. As the test trains had grown successively longer, so too had the distance between the Lead and Remote locomotives. Eventually, on a frigid 15 November 1967 morning at Iaeger (*YAY*-ger), West Virginia, strings of 150, 150, and 200 loaded coal cars were patched together and Pocahontas Division Road Foreman of Engines, Q A Goff Jnr, gently notched out to get train Extra 1757 underway for the 256-km (159-mile) run to Portsmouth, Ohio, to set a world railroad record that would last for almost 22 years. This run was going to be for more than just bragging rights. In addition to wanting to see how many cars one train could handle, the railroad was also looking at how far the remote-control radio signals could travel, how reliable they would be if they made it—especially considering the tunnels—and whether the radio-control system in total would be as reliable a part of the train as the locomotives and cars had already proven to be.

The 500-car train ran in the fourth and final week of testing and met with mixed results. For sure, the tests had proved that six EMD SD45 locomotives (21 600 total hp) could handle the load on that forgiving territory and the expectation of reasonably compliant behaviour from 500 cars proved surprisingly accurate, although two cars developed stuck brakes, two cars sent false alarms that activated Hot Box Detectors, and a coupler broke on the restart at Williamson, WV.

The support systems, though, proved frail. Towards the end of the day, intermittent interruptions to radio communications had an effect upon brake pipe pressure that was believed to be the cause of the two stuck brake problems. But crews were aware of the situation and nursed the train the remaining short distance to Portsmouth. This train got through without major incident, but had it been forced to stop in a track section with road crossings, it would have tied up all grade crossing traffic for more than 4 miles. The broken coupler occurred in the Williamson yard and, thus, did not block any streets; Clyde Taylor oversaw a hasty coupler change-out. This run would not have satisfied Milt Deno's definition of a successful test-run.

One interesting aspect of the 500-car train was one of N&W's contingency measures. Taylor and a few other railroad officials drew up plans to cope with a variety of possible emergencies, of which the response scenario to a drastically-failed hopper car was particularly swift and dramatic. On command, a carman was to bang open the hopper doors and a front-end loader would then lift the car out of the way as it emptied onto the track. A bulldozer was to push the coal off the track, and the train would back up, re-couple, and be on its way as soon as possible. Loehne figured that N&W felt the black diamonds were cheap, whereas a blocked main line and N&W pride were both expensive. The recovery manoeuvre was expected to take less than 2 minutes once equipment arrived at the scene. Taylor thought they could do it in 1 minute if there was enough space to easily manoeuvre the front-end loader and bulldozer.

Built to a 3 + 300 +3 + 200 + caboose configuration, the train measured 6.5 km (4 miles/21 425 ft) and conveyed 44 000 tonnes (48 584 US short tons) of coal.

Sishen–Saldanha railway (South Africa)

On the 26–27 August 1989, Spoornet (now called Transnet Freight Rail) operated an iron ore train comprising 660 cars plus a tank car and caboose over the Sishen-Saldanha line. Measuring in at 7.3 km (4.5 miles) hauling a 69 393-tonne payload, and grossing 78 967 tonnes (77 720 tons), the train utilised 16 locomotives: nine 383-kN Class-9E electric locomotives and seven 2750 hp Class-34 (GE U26C) diesel-electric units in two discrete power groups. It should be noted though, that this lash-up represented some built-in redundancy—since not all of the units were under power at any one time—and neither was it a distributed power train, both power groups being manned. It was though, a test of the ultimate possible capabilities of distributed power, as Transnet would eventually go on to install LOCOTROL III systems on a fleet of the Class-9E electric locomotives.

A 7.3 km-long train has a lengthy brake pipe, and running this train would have required some well-developed train handling skills, not to mention an advanced degree of coordination between the drivers of the two power groups. However, one aspect would have been in the operator's favour since these ore wagons were operated in 'married pairs': 660 ore cars really meant 'only' 330 sets of air volumes that had to be charged and then recharged after an air brake release, greatly simplifying the air braking task.

Ultimately, Transnet Freight Rail did operate a LOCOTROL test train. In August 2004, a 3.9 km-long, 34 300-tonne ore train was operated with LOCOTROL III distributed power. The normal revenue ore trains on the Sishen–Saldanha line are unique for the manner in which locomotive consists are multi-coupled as an electric/diesel-electric combination. Class-9E locomotives—being the ones fitted with LOCOTROL equipment—serve as the master unit of each mixed consist. The trains comprised a Lead consist with three Class-34 diesels trailing the master unit and two separated Remote consists (each with two diesels) also trailing a Class-9E master unit by means of a *slimkabel* (smart cable). Each train was therefore essentially comprised of three separate 114-car trains coupled together.

Figure 185: A Transnet Locotrol train covers easy territory about 100 km (62 miles) from the coast at Lamberts Bay. Capturing even the whole length of this scheduled Locotrol train is difficult to accommodate in one image. Wikipedia

Operating very large trains is an extremely serious business and when things go wrong—be it a failure of rollingstock, track structure, or train handling process—the vast kinetic energy involved inevitably results in a spectacular and expensive calamity. On 22 July 2010, a serious derailment caused a 5-day traffic disruption. At kilometre 203 between Loop 4 (Knersvlak, near Vredendal) and Loop 5 (Saggiesberg), 107 loaded wagons were derailed, some falling from a bridge over a dry watercourse. Two Class-9E electric locomotives and one Class-34 diesel-electric, almost buried in the spilled iron ore, were reportedly damaged beyond repair. The train was of the usual configuration of three portions of more than 100 wagons each and four sets of locomotives, including units at the rear.

Fortunately, there were no injuries, the locomotives involved being unmanned Remotes. The author's information is that these locomotives comprised the second Remote consist, the derailed cars being in the third portion. The incident was thought to have resulted from a train operating error that caused the second Remote group to not receive a radio command signal from the Lead unit. As a result, this Remote consist continued in an inappropriate power-setting when operating speed was reduced at the head-end, telescoping the train.

Figure 186: Sishen-Saldanha railroad. The derailed train mentioned in this narrative. Transnet

Robe River Iron Associates (Australia)

The Robe River project to apply LOCOTROL III to Robe River Iron Associates' GE Dash 8 locomotives was conducted in the field by Milt Deno and Gene Smith. The intended use of LOCOTROL was to unman the Pusher consist that was used to assist all loaded iron ore trains from the mine and terminal (at this time being Mesa J, south of the company town of Pannawonica) over the 100 km (62 miles) to Siding Two at the railroad summit atop the Chichester Range. There was no intention to use the technology to increase the size of RRIA trains.

The project was notable for four reasons. Firstly, the Pushers ('Bankers' in Australian parlance) were *automatically detached* on the run to come to a stand on the main line at Siding Two while the train continued unassisted, mostly downhill, for the 100 km to the port at Cape Lambert. At Siding Two, the loaded train would meet an empty whose engineer took charge of the detached two-unit Pusher consist, ran back out across the mainline switch, and attached them to his head-end, thus returning them to the mine whence they formed the Lead consist of the next loaded and the erstwhile head-end power became the next remote-controlled Pusher consist. Secondly, the test train was a success, and this isn't always the case (remember, Deno considers a demonstration test train to be a 'success' only if it runs its planned distance without mishap… and he does

Figure 187: In 1994 a Robe River Iron Associates iron ore train with two head-end units and two manned pushers out-of-sight at the hind-end departs the Mesa J loadout yard and crosses the Robe River to commence its 200-km run to the coast. RRIA

not distinguish between long and short journeys). This Robe River demo train operated non-stop over its 200 km (124 mile) run, taking about 4½ hours at an average speed of 40 km/h (25 mph).

Thirdly, the project included the installation of Tower Control for the remote operation of the train during loading, and fourthly, it was the first application of both Tower Control and LOCOTROL III in Australia.

Robe River ore trains at this time consisted of two 3900 hp GE Dash 8s and 200 cars, with two more GEs pushing for the first half of the journey. On 18 March 1995, with the project essentially complete, Deno and Robe River locomotive engineer Glen Davies—who had acquired previous DP experience in his home state of Queensland—ran a loaded train with 350 cars and seven Dash 8 units. The train measured 3800 metres (12 470 ft), grossed-in at 45 465 tonnes (50 120 short US tons) with 35 350 tonnes of payload, and operated in a 2 + 120 + 2 + 115 + 2 + 115 + 1 configuration.

BHP Billiton Iron Ore (Australia)

Originally as the Mount Newman Mining Co Pty Ltd, (Mount Newman Joint Venture) this railroad has progressed its use of remote distributed power from the original LOCOTROL 105 equipment through LOCOTROL II to LOCOTROL III, and now LOCOTROL XA. On 28 May 1996, Australia achieved the tonnage record over 408 km with ten GE Dash 8s pulling 540 carloads (train length: 5892 meters or 3.7 miles) at an average 56 km/h (35 mph). The payload was 72 191 gross tonnes (57 309 tons) and the train was configured 3 + 135 + 2 + 135 + 2+ 135 + 2+ 135 + 1.

On 21 June 2001, following the completion of a GE LOCOTROL III installation integrated with the Knorr-Bremse/NYAB CCBII air brake system, BHPBIO (now BHPIO) decided to run another record-breaking train.

Figure 188: Travelling away from the photographer in the left-hand image, the 350-car train described on the previous page begins its climb out of the Robe River valley near Deepdale. The right-hand image shows the train climbing through the Chichester Range towards Siding Two. RRIA

They built a 682-car consist and ran it 275 km (170 miles) to their Yandi mine to load. Returning to Port Hedland the loaded train weighed 99 734 tonnes (110 000 short US tons), conveyed 82 100 tonnes of ore, measured 7.3 km (4.5 miles) long, and was configured 2 + 166 + 2 + 168 + 1 + 180 + 1, the six units all being 6000 hp GE AC6000CWs. The train eventually made it to the port, but by Milt Deno's standards (he was not involved with this installation or this run) this would not have been a successful test since the train suffered a separation en route. Company personnel paralleling the train on the adjacent access road were on hand to assist with installing a replacement knuckle and getting the behemoth on the move.

Figure 189: The 682-car BHPBIO demo train approaches the Shaw passing track as it climbs through the Chichester Range. In the far distance the hind-end is just leaving the Hesta passing track location. In the extreme far distance, the railroad easement can be seen crossing the Fortescue River flood plain between the Chichesters and the Opthalmia Range on the horizon. BHPBIO

Burlington Northern Santa Fe Railroad (USA)

On 10 July 2009, BNSF operated a double-stacked 3.6 km (2.24 mile) LOCOTROL container train over 1151 miles from Los Angeles, California, to Clovis, New Mexico as the first of a series of test runs seeking greater operating efficiencies. The train was powered by seven units in a 3-2-2 configuration and conveyed 458 containers, 11 256 US tons.

Union Pacific Railroad (USA)

On 8 January 2010, UPRR despatched IDILB-08, a LOCOTROL intermodal train comprising 295 doublestack wells (approximately 86 multi-unit cars) from Dallas, TX, and ran it to the port of Long Beach in Los Angeles, CA, arriving 2 days later. Measuring 5.5 km (3.42 miles/18 061 ft) and grossing 14 061 tonnes (15 500 tons), the train was powered by nine GE ES44AC (4400 hp) locomotives, was configured 3 + 99 platforms (well-cars) + 2 + 98 + 2 + 99 + 2, and conveyed 618 containers.

Figure 190: With the Chocolate Mountains in the background, the IDILB-08 is passing Bombay Beach, California, on the shores of the Salton Sea. The hind-end of the train is far distant at the extreme right-hand side of the image. Joe Perry, ChasingSteel.com - with permission.

Figure 191: The IDILB-08 is approaching Cabazon CA—west of Palm Springs—as it traverses the San Gorgonio Pass (also known as Beaumont Hill) around 90 miles from the Port of Long Beach. 33 of the train's 296 wells have already passed the photographer as this image is captured. Two Remote units are just visible in the middle distance, directly above the 'V' in the gravel track. The rear-ward portion of the train is faintly visible descending a slight grade immediately below the right-hand-most group of wind generators. Joe Perry, ChasingSteel.com - Ontario, CA.

Figure 192: The IDILB-80 has just crested the grade as it approaches the Beaumont Ave overpass at Beaumont, CA. The train is about 100 miles from the Port of Long Beach. Ahead lies a 25-mile, 1.4% (1-in-70) descent through the San Timoteo Canyon to Colton, then a long traverse of the Los Angeles Basin reaching almost to Downtown LA, before an abrupt turn south to Vernon and and a direct 15-mile run down the Alameda Corridor to the port.
Bob Hanggie - Upland, CA.

7

THE Locotrol MODEL SERIES AND ASSOCIATED TECHNOLOGY

The content of this history will be deliberately limited to LOCOTROL III. In my view, that was the threshold of development of the 'original style' of the technology. Since then, evolutionary versions of the product have emerged that—while delivering improvements in the human interface and perhaps also the means of transmission (for example, wireline rather than wireless)—nevertheless continue to accomplish exactly the same task in much the same way.

This chapter will show the reader how the early product was presented to the market, some extra detail about how it operated, and reveal some associated technologies that were born of the developments in LOCOTROL, but that were themselves rendered irrelevant by it.

Throughout North Electric's early work developing their locomotive remote-control system, there was no consideration given to affording the technology a product or marketing name.[104] It seems that at the time, this requirement never occurred to the engineers who conceived this technology and those later to be engaged on its development.

As their early efforts on the nascent system evolved, it eventually came to be referred to within the plant as a system for the *Automatic Remote Control of Locomotives* then *Locomotive Remote Control* and *Remote Control System*, for a brief time *Multiple Consist Control*, and finally, 'LOCOTROL'. However, as successive iterations of the system were incrementally developed and tested in operation it became necessary to differentiate between them. From this requirement arose the eventual model designatory continuum, commencing with Model 100 and culminating—for the original LOCOTROL I series—with the Model 105SS system (utilising solid-state electronics).

It is perhaps understandable that the early terminology for this equipment would be couched in language that might sound unsophisticated by today's standards. In searching for ways—within the context of the era—by which to refer to the head-end or *controlling* locomotive (and any extra locomotive units coupled in multiple with it), the term 'Master' was coined. It followed, inevitably, that the *controlled* locomotive (and any extra locomotive units coupled in multiple with it) inserted into the body of the train, would become the 'Slave(s)'. This appears to have been acceptable to the prototype customer, Southern Railway, but would later become the subject of anxiety for other customers who were more sensitive to the negative—and increasingly politically-incorrect—connotations of these terms (not to mention the possibility of the moniker 'slave driver' being applied to the hapless locomotive engineer). This customer concern resulted in eventual adoption of the terms 'Lead' and 'Remote' that would prevail into the future.

The following early descriptive information was provided by the manufacturer for prospective customers and is depicted herein as it was presented in Radiation Incorporated and Harris Corporation pamphlets. Knowledgeable readers will be able to discern the degree by which the technical writing matures throughout the evolution of the technology.

---

104   The origin of the LOCOTROL name is described in Chapter 1.

LOCOMOTIVE REMOTE CONTROL

Figure 193: Early Radiation Incorporated 'Control Systems' pamphlet cover.

LOCOMOTIVE REMOTE CONTROL SYSTEM DESCRIPTION

1. SYSTEM ADVANTAGES

In all normal instances the locomotives, which power a train, are located at the front or lead end of the train. When long, heavily loaded trains are operated, it is necessary to use several locomotives to provide sufficient power. These units are arranged such that the lead unit is the controlling locomotive and all additional units are slaves. The slave units are grouped together and controlled by a cable called the "train line", and also by several air lines which are used mainly for braking. As many as six or eight units may be slaved in this manner. In railroad terminology this grouping of locomotives is called a consist.

Several advantages are gained when locomotive power is split into sections, one section of locomotives is placed at the lead end, and another approximately 2/3 of the length of the train toward the rear end. Some of these advantages are given below.

1. **RAPID AIR BRAKE RELEASE.**

    Air passages can be filled from both the master and slave units; this decreases the distance through which the air must move and results in more rapid brake release.

2. **SAFER EMERGENCY STOPS.**

    The air is expelled and brakes applied at both ends of the train, thus decreasing the "run-in" of slack.

3. **BETTER HANDLING IN MOUNTAINOUS AREAS.**

    It is possible for the rear power section to push over the tops of hills and also around the "zig-zags" of mountainous terrain. This increases the overall applied force in the forward direction.

4. **LONGER TRAINS ARE POSSIBLE.**

    When very long trains are pulled up long grades, where the entire length of the train must be pulled by the lead end, excessive loading

Figure 194: Radiation Incorporated, 'Locomotive Remote Control' system description (1).

occurs, causing the couplers to break. When power is supplied at
the rear, both push and pull forces are applied and the effective length
and overall strain is noticeably reduced.

Of course, some method must be devised for the control of the rear units. A radio control system designed to control the rear locomotives by signals transmitted over a radio link from the lead or controlling locomotives to the slave is described in this system description. The radio control system was designed and used for the following parameters.

2. SYSTEM PARAMETERS

The system operates within a 3kc bandwidth on FM half-duplex carrier. This allows similar systems to operate on a single radio frequency. An address or identity code is used to allow multiple systems to operate in the same area and on the same frequency.

The system provides fast-acting commands and high-speed feedback response to the operator and does so without any degradation of the signal.

The system is fail-safe and provides for protection against all false operations.

Each transmitted message contains an address, the status of all functions, and code checking data necessary to insure integrity of the data. The lead locomotive logic subsystem transmits data at regular intervals and, in addition, immediately transmits any control change. The slave locomotive logic subsystem receives and verifys these transmissions and carries out the control changes as instructed. As a reply, following a transmission from the lead unit, the slave unit will transmit all controlled functions. Whenever a change in the status of the control functions or alarms are detected, the slave unit will transmit to the lead unit.

The lead subsystem receives all transmissions from the slave subsystem and compares the called for control functions. If the status of any functions at the slave do not compare with those called for at the lead, the lead subsystem continues transmitting at rapid intervals until they do match. When all functions match at both ends, the lead unit transmits to the slave and receives a reply only once every thirty seconds. This transmission informs both the slave sub-

Figure 195: Radiation Incorporated, 'Locomotive Remote Control' system description (2).

system and the lead operator that everything is operating properly. If the slave fails to receive a transmission from the lead unit for sixty seconds, an alarm condition is activated and the slave unit goes to idle if an air brake application is detected. In the lead unit the status of the slave is displayed to the operator for verification and the operator is alerted to the loss of control continuity by an alarm light.

The system is capable of handling 40 different functions.

The mechanical packaging of the system is designed to withstand transverse coupling shocks of at least 3g and vibrations of at least .032 inches double amplitude at approximately 14 cycles per second.

3. GENERAL SYSTEM DESCRIPTION

The lead locomotive contains the lead or master subsystem. This subsystem consists of sub-units as illustrated in Figure I.

The input to the subsystem is provided by two primary sources: The control and display console; and the locomotive TRAIN LINE. Inputs from the TRAIN LINE enter via the isolating sense circuits, while inputs from the console are applied directly to the logic. Data from both of these sources are stored in the TRANSMIT REGISTER for analysis. The data thus stored is compared to data in the RECEIVE AND OPERATE REGISTER. The result of this comparison is sent to the COMMON CONTROL LOGIC. The data in the RECEIVE AND OPERATE REGISTER is received from the slave subsystem. This data consists of the latest transmitted status of all controls in the slave locomotive and any alarms which are detected by the slave. The COMMON CONTROL LOGIC is the overall control and analysis unit. This unit controls the signal analysis, checking, and storage of all incoming data. It also makes decisions as to when it is necessary to transmit to the slave and it governs the priority and timing of all events within the subsystem as well. In essence, it is a fixed-type program consisting of routines and subroutines for performing all of the required analysis and control.

The CONTROL AND DISPLAY CONSOLE is a remotely located switch and indicator panel. The data in the RECEIVE AND OPERATE REGISTER lights display lamps on this console. The lamps inform the operator of the status of all controls in the slave, such as: throttle position; amount of dynamic braking.

Figure 196: Radiation Incorporated, 'Locomotive Remote Control' system description (3).

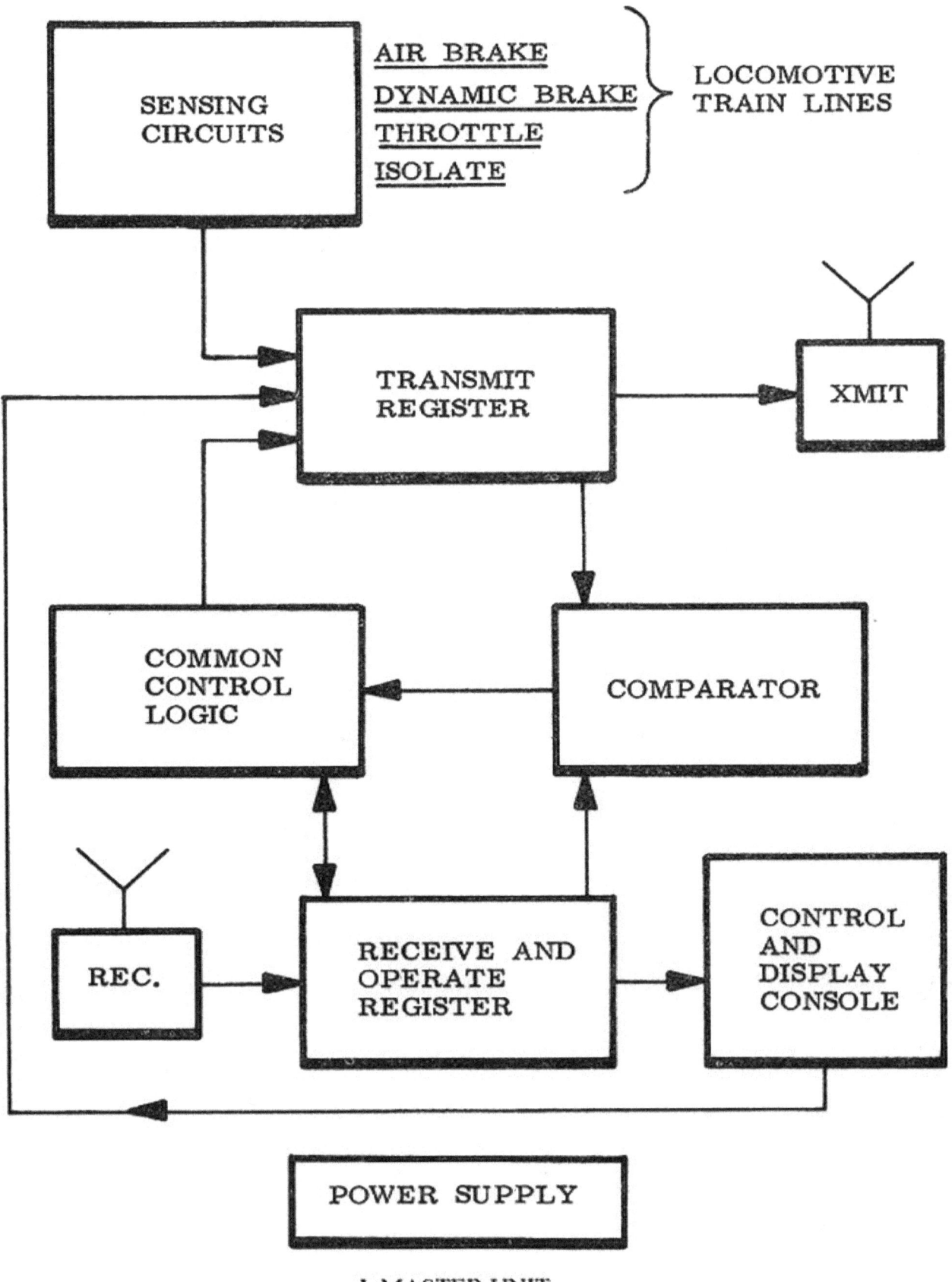

Figure 197: Radiation Incorporated, 'Locomotive Remote Control' system description (4).

Several lamps are used to inform the operator of alarm conditions in the slave locomotive consist, such as: HOT ENGINE; LOW OIL; and NO POWER. An alarm bell may also be sounded, if this is desireable.

Under supervision of the COMMON CONTROL LOGIC unit, the incoming data is stored in a serial shift register called the receive register. When the incoming data has been validated by code and address check, it is transferred in parallel to another storage register called the operate register. These two registers are shown as the RECEIVE AND OPERATE REGISTER. The modulation and demodulation of data at each end of the radio link is explained in detail in Section 5, COMMUNICATIONS CHANNELS.

The organization of the slave subsystem shown, is very similar to that of the lead subsystem and is illustrated in Figure II. The primary differences are that the control program is different, and the input-output is changed. The data in the RECEIVE AND OPERATE REGISTERS block drives the slave locomotive power contactors via the isolating DRIVER circuits. These power contactors, very heavy duty solonoid operated relays, are used to drive the TRAIN LINE and braking system in the slave locomotive and thus in all other tandem locomotives. The input to the slave subsystem is via isolating sense circuits connected to the TRAIN LINE and braking controls. They provide positive feedback with respect to what controls are actually being operated. This data is stored in the transmit register for analysis. When necessary the data is transmitted to the lead subsystem for comparison and display.

As in the lead subsystem, the received data is validated before it is used. In addition, some of the control functions are analyzed to see if they are logical. This adds a check features to the system.

In the event that a loss of control continuity develops, the slave unit COMMON CONTROL LOGIC looks for the detection of air flow in the main brake pipe. This indication signifies that the lead unit is applying the main brakes. The slave subsystem then "throttles down" and isolates the slave locomotive air brake system. The slave will remain in this status until continuity is regained. The slave will then return to the control settings that are indicated by the data received.

Figure 198: Radiation Incorporated, 'Locomotive Remote Control' system description (5).

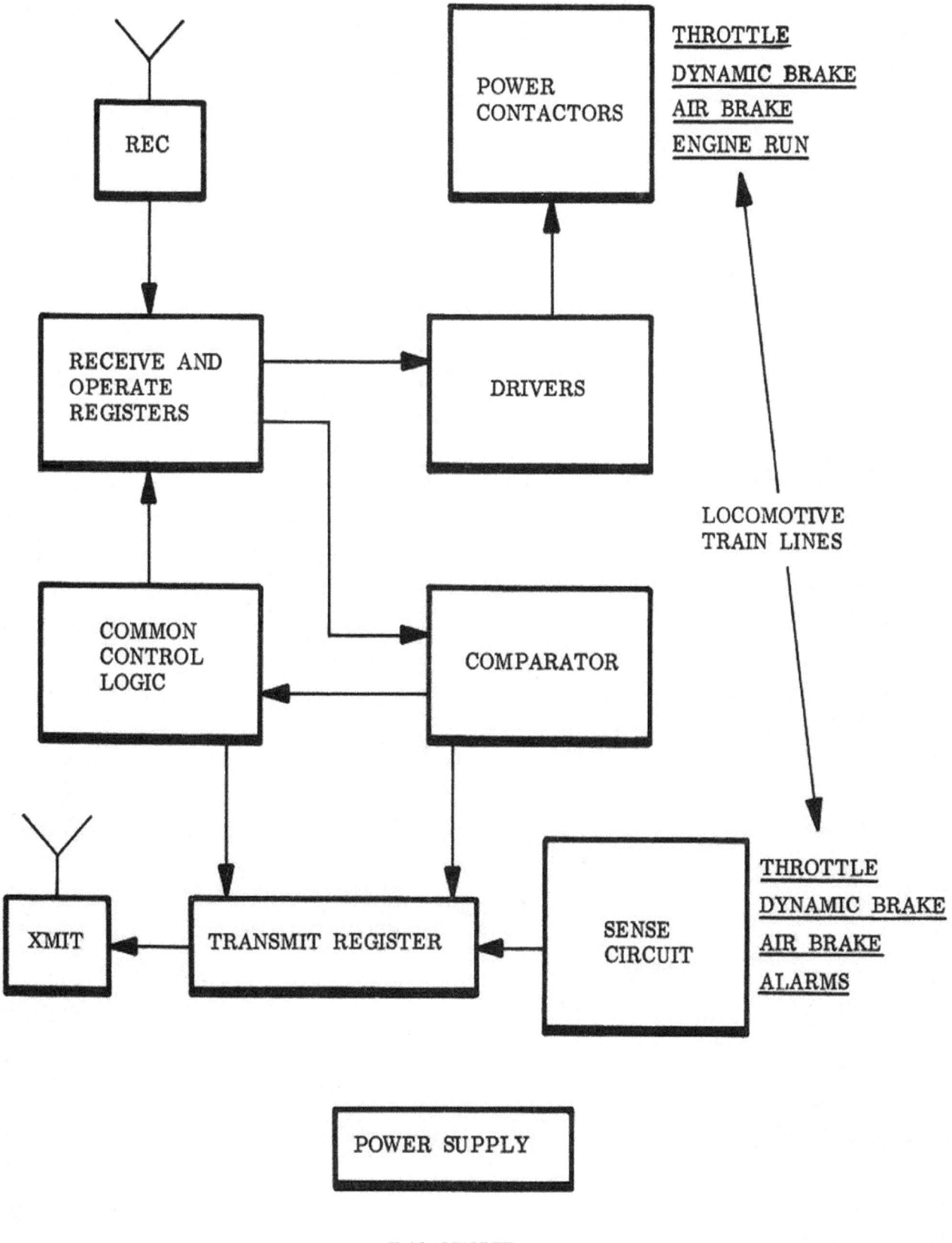

Figure 199: Radiation Incorporated, 'Locomotive Remote Control' system description (6).

4. LOGIC TO LOCOMOTIVE INTERFACING

The system to locomotive interface has been very carefully designed. All electrical circuits in a locomotive are isolated from the frame. The locomotive circuits are very noisy and sometimes contain transients ranging up to 5 KV. It is therefore desirable to keep the system isolated galvanically from the locomotive power and control circuits. For this reason, special AC coupled interface circuits have been designed. All control signals transmitted from the system and all feed back signals received by the system are by way of these special interface circuits.

The system power is derived from the locomotive power. The system power supplies receive their inputs via a filter from the locomotive 72 VDC power. The power supplies consist of a DC/DC converter followed by series regulators. These regulators provide +12VDC and -12VDC to the subsystem. The -10VDC for the Modems is obtained by a voltage dropping diode in the logic circuitry. The power is completely isolated from the locomotive power by the AC coupling of the DC/DC converter.

5. COMMUNICATION CHANNELS

The communications channel consists of a pair of narrow band FM radios (excluding standby radios). These radios operate in the 152-174 MC industrial band. The bandwidth of the channel is 3.5 kc having a frequency response variation of ± 3 DB. The radios are operated on a half duplex basis, (either transmit or receive, but not at the same time).

Selection of the data modulation scheme is based on system response time, channel characteristics, reliability, error checking, and economics. The scheme chosen was that of medium speed (600 baud) frequency shift keying. The center frequency is 2200 cps. This scheme is identical with that of a normal FSK teletype channel except that the speed is considerable higher.

A code checking scheme is used to assure security in the system. This scheme is based on W. Wesley Peterson's book, "Error Correcting Codes", MIT Press, Copywright 1961, Chapter 4 and 6.

The chance of undetected single and multiple bit errors or bursts is remote, and for all practical purposes, is nonexistent.

Figure 200: Radiation Incorporated, 'Locomotive Remote Control' system description (7).

III. MASTER UNIT WITH COVER REMOVED

Figure 201: Radiation Incorporated, 'Locomotive Remote Control' system description (8).

A complete description of the code check system is published by E. P. Wiesner in ELECTRICAL DESIGN NEWS, August, 1963.

6. MAINTAINABILITY & RELIABILITY

The equipment has been designed so that any circuit failure in the equipment can be easily corrected by isolating the inoperative circuit by trouble shooting techniques and then replacing the defective card. A reliability study has been performed by the North Electric Company and indications are that the mean time between failure of the logic portion of the system is 3.7 months while the total system reliability is calculated to be 3 months before a system failure occurs. A complete reliability report on this system is available. All frame equipments are modularized and can be removed from their mountings simply by decoupling the connectors and removing the units for maintenance.

7. MODIFICATIONS OTHER THAN NORTH EQUIPMENT REQUIRED TO BE MADE BY THE RAILROADS

Due to the remote location of slave locomotives, it is desirable to enable the lead locomotive to reset the ground relay (no power). Diodes are added to the ground relays and connected to a ground relay reset button in the locomotives.

Push button brake operation is necessary for best interfacing of air brake operation with slave operation. A unit in the lead locomotive changes electrical commands to pneumatic brake operation and gives direct commands to the control system. A second unit is located in a special radio control car (RCC) and converts electrical commands from the slave control unit to pneumatic brake operation. The RCC allows all locomotives to be used in a regular consist mode.

It is necessary to place some brake piping in the RCC to make the airbrakes appear the same as the airbrakes on a locomotive. Directional sand control piping should also be added to the RCC.

The RCC unit must have a train line cable connector so that it may be joined to the slave locomotives. A power connector must be added to the slave locomotives to furnish power for the control system.

The trainlines are operated by power contactors with safety interlocks in much the same way as the control stand is interlocked on a regular locomotive.

Figure 202: Radiation Incorporated, 'Locomotive Remote Control' system description (9).

Figure 204: Radiation Incorporated, 'Locotrol I' pamphlet cover.

INTRODUCTION

LOCOTROL is a solid-state electronic system used for the optimum distribution of locomotive power in long heavily laden trains. It provides precise automatic remote control from the lead consist to a second or third locomotive or consists of locomotives connected at optimum load points within a train build-up.

Digital logic and air brake units are installed in the lead and remote locomotives and automatic remote control of the remote units is obtained utilizing a radio link between the lead and remote units.

Division of locomotive power is not a new concept. Previous attempts have been made, using only voice communication between the manned control positions of the lead and rearward consists. The precise timing necessary for coordinated operation could not be achieved in this manner. The synchronous, automatic control of both consists made possible by the solid-state logic system employed in LOCOTROL solves this problem. The division of locomotive power into separate units, which operate in unison, results in increased power efficiency and faster, smoother and safer stops and starts.

CONVENIENT OPERATION

The LOCOTROL system was designed for simplicity of operation. In normal operating modes, it functions automatically. Just as the locomotives are made as safe as humanly possible with various built-in safety devices, so is LOCOTROL. Redundant features are incorporated where necessary to insure safety. Any decisions the operator makes will not make the system less safe.

The system can be employed in either of two modes: Synchronous automatic operation or independent operation. When the synchronous mode is employed, the lead locomotive operational functions, controlled from the engineer's position, are sensed, signals are transmitted to the remote consist and translated instantly into command signals to the remote consist control equipment.

When operating conditions require independent operation of the lead and remote consists, and "inhibit" capability (disabling the sensing circuits in the lead consist) is available to the train engineer. The remote consist can then be independently controlled by command signals from the lead locomotive. A rotating type deck switch on the control console in the lead locomotive provides this control. When required, the remote consist equipment can be isolated by controls on the console, making a "dummy" of the remote consist.

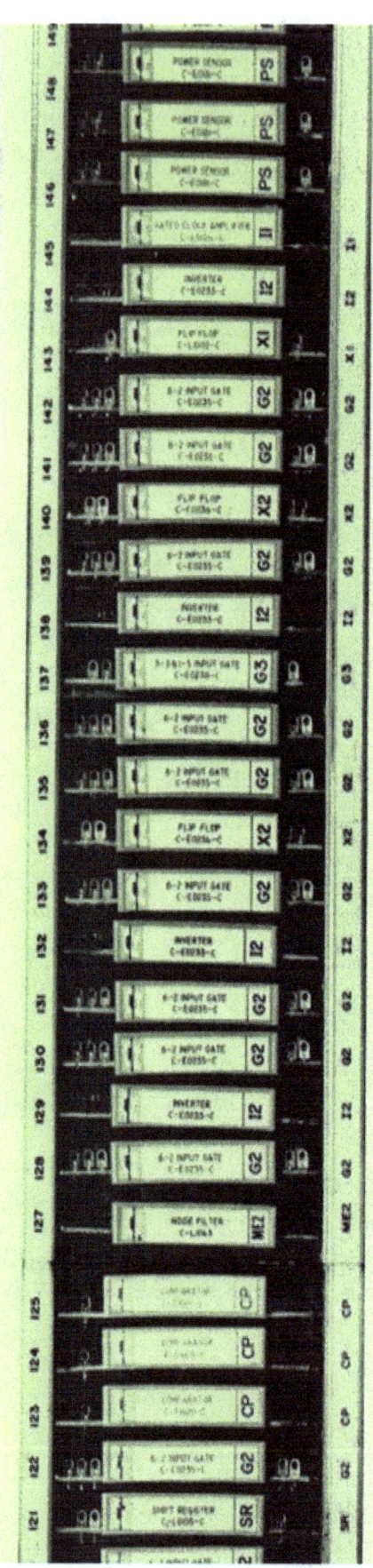

Figure 205: Radiation Incorporated, 'Locotrol I' system description (1)

WHY LOCOTROL

The use of LOCOTROL in effect shortens the length of a train. Extensive tests conducted by one of the largest train air brake manufacturers and actual in-train operation by one of the largest railroads have shown the following distinct advantages when LOCOTROL is employed.

MORE RAPID AIR BRAKE RELEASE — The air brake passages of the pneumatic brake system can be charged more rapidly since the distance through which the air must travel is effectively decreased, i.e., charging is accomplished from two separate points. Release time is improved by a factor of five.

SAFER EMERGENCY STOPS — The pneumatic brake system is divided into two units which operate simultaneously to reduce the "run-in" of slack. Application time is improved by a factor of three.

BETTER TRAIN HANDLING — The more even distribution of locomotive power results in increased and smoother overall applied force and is especially useful when used over mountainous terrain.

FASTER ACCELERATION — The train can gain operating speed more rapidly without undue strain on lead couplings.

LONGER TRAINS — When very long trains are pulled up long grades where the entire length of the train must be pulled by the lead end, excessive loading occurs sometimes causing the couplers to break. However, when power is supplied at the rear, as with LOCOTROL, both push-and-pull forces are applied and the effective length and over-all strain is noticeably reduced.

SHORTER SCHEDULING — Less Freight damage and associated monetary savings are all end products associated with these advantages.

BETTER TRAIN CONTROL — The capability of dynamic braking on the remote consist either synchronously with the lead consist or independently provides better train control and more rapid and smoother deceleration.

Status of the remote consist major operating functions is continuously displayed at the engineer's position in the lead unit. Alarms immediately apprise the engineer of a locomotive malfunction.

No minimum proximity limitations exist with LOCOTROL. The use of an Address, or identity, code, ensures the integrity of any LOCOTROL system and permits the use of the same radio frequency band width by multiple systems operating in the same area. Trains equipped with LOCOTROL can operate in the same area with no danger of interference to one another. The command signals transmitted by the control unit of each train are acted upon only by the remote unit or units within the same train.

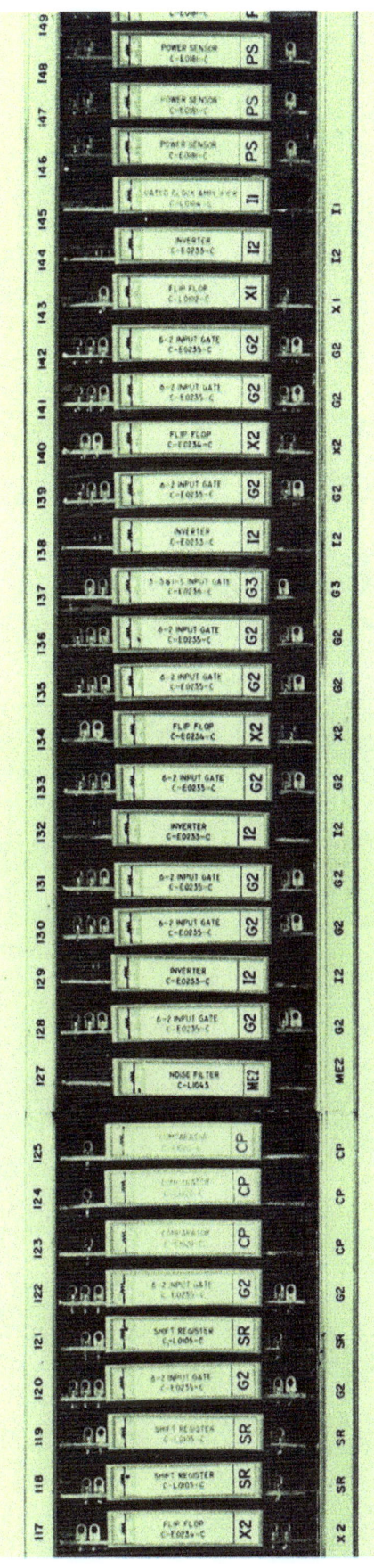

Figure 206: Radiation Incorporated, 'Locotrol I' system description (2)

EQUIPMENT DESCRIPTION

A LOCOTROL system consists of a lead station and one or more remote stations. In some systems more than one remote station is used. In this case, a message repeater is incorporated in the remote station between the lead station and the rear remote station.

A lead station is composed of seven major units:
1. Logic Cabinet No. 1
2. Logic Cabinet No. 2
3. Power Supply
4. Control Console
5. Air Brake Control Unit
6. Radio Equipment
7. Relay Interface Cabinet

A remote station is composed of six major units:
1. Logic Cabinet No. 1
2. Logic Cabinet No. 2
3. Power Supply
4. Air Brake Manifold
5. Radio Equipment
6. Relay Interface Cabinet

CONTROL CONSOLE – is placed at the engineer's position so that remote consist status and alarms are clearly displayed. In addition, the console also contains the switches necessary to isolate and independently operate the remote unit.

LOGIC UNITS – contain the train line sensing circuits and the control logic. The cabinets contain logic printed circuit broads.

INTERFACE RELAYS – provide the electrically isolated connection between the LOCOTROL logic and the consist train line.

RADIO EQUIPMENT – is standard FM half-duplex carrier with the LOCOTROL utilizing a 3 KC bandwidth.

AIR BRAKE MANIFOLD – Electric-to-pneumatic control of air brakes on both the lead and remote units.

Figure 207: Radiation Incorporated, 'Locotrol I' system description (3)

PUSHBUTTON AIR BRAKE CONTROL – electrifies the control of the braking system. It is necessary in order to provide an electrical control signal to the lead consist LOCOTROL unit and for operation of the remote consist air brake.

POWER SUPPLIES – are DC/DC Converters followed by series regulators. They are cabinet mounted and provide power to the LOCOTROL equipment.

RELIABILITY

The mechanical assembly of the equipment is designed to withstand the transverse shocks of train operation (especially those resulting during coupling operations). Operational stability of the equipment is maintained when subjected to shocks in excess of 1g over the frequency range encountered in locomotive operation.

MAINTAINABILITY

The equipment has been designed so that any circuit failure in the system can be quickly isolated by standard troubleshooting techniques. The inoperative circuit is then restored to service by replacement of the defective circuit card. All frame equipments are modularized and can be removed from their mountings simply by decoupling the connectors and removing the units for maintenance.

SPACE REQUIREMENT

Equipment installation in the locomotive units is based on the available space as determined by a study of the newer type locomotives. The LOCOTROL equipment can be mounted in a locomotive of the remote consist or in a radio control (RC) car. When the remote equipment is installed in an RC car, any number of locomotives can be placed in the remote consist. The remote control capability is then enabled by the inclusion of this RC car in the remote consist. The train line and the air brake lines required for control of the consist are then attached to the locomotive units in the standard manner. A power connector must be installed between a locomotive unit and the RC car to furnish power for the control system. The train lines are energized by interface relays with safety interlocks in much the same way as the control stand is interlocked on a regular locomotive.

Figure 208: Radiation Incorporated, 'Locotrol I' system description (4)

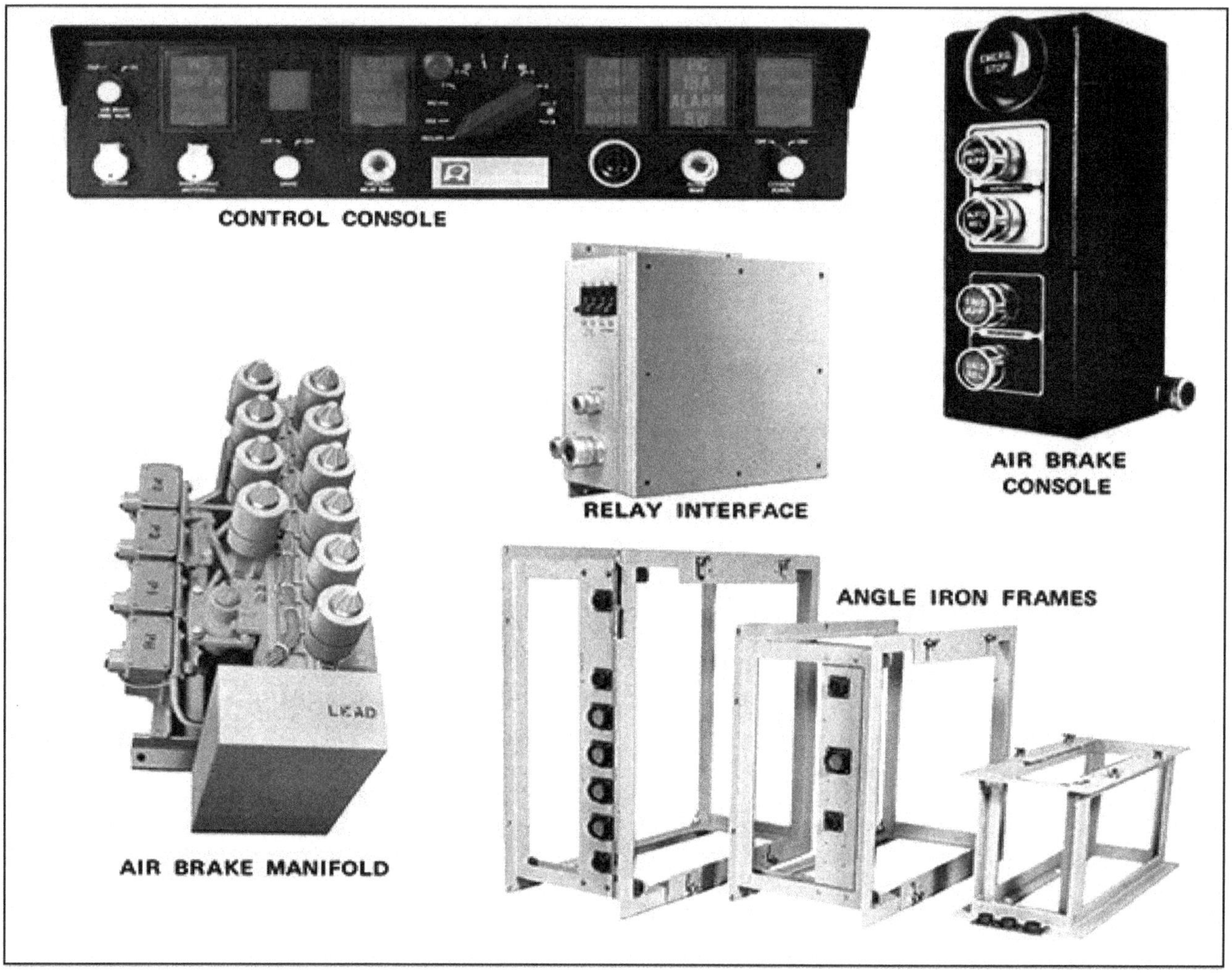

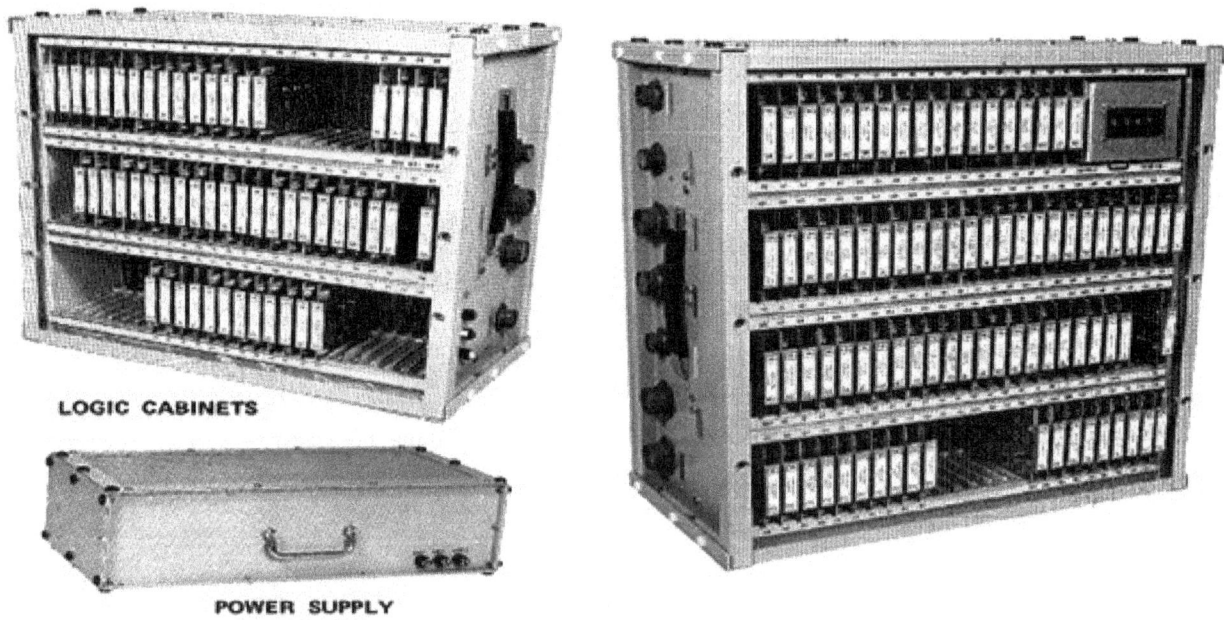

Figure 209: Locotrol 105 components. The Control and Air Brake consoles were located about the engineer's control stand.

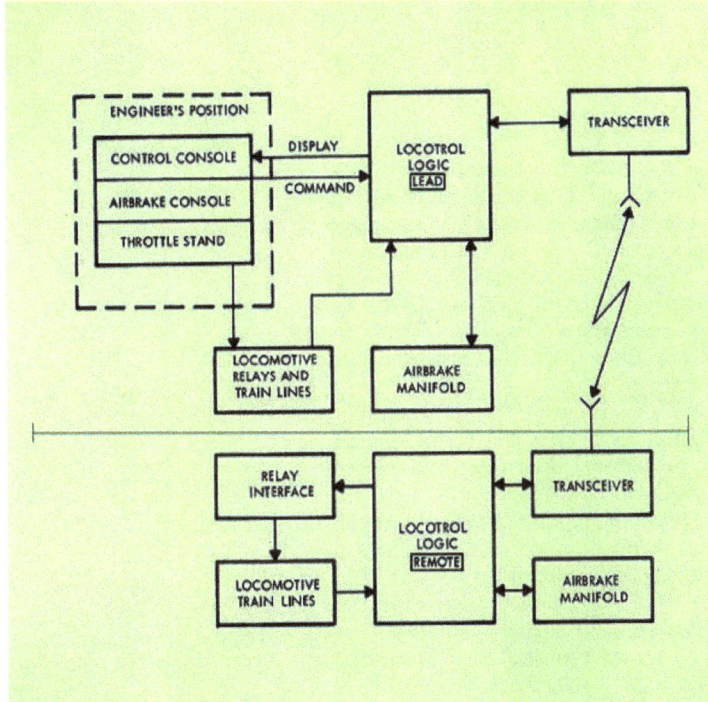

FUNCTIONAL DESCRIPTION

TRAIN OPERATION

The lead consist is controlled from the engineer's position over the "train line" by the throttle stand controls. The air brake system of the lead consist is controlled by the pushbutton air brake control unit at this position. When synchronous control of both consists is employed, the remote consist is controlled automatically. The command signals provided to the lead consist train line and air brake system are sensed, processed by the logic equipment, and transmitted by radio to the remote consist. These signals are received by the remote consist, processed by the logic equipment and applied to the train line and air brake control circuits. The remote consist thus operates in synchronism with the lead consist.

When operating conditions require individual control of the two consists, the remote consist can be operated from the control console in the lead locomotive independently of the lead consist. The sensing circuits from the lead train line are disabled. The control of operating signals transmitted to the remote consist is transferred to the control console. Command signals to the remote consist are then initiated by use of a rotating type deck switch on the console.

Isolation of the remote consist is accomplished by operation of the control switch on the console. The air brake feed valve can be cut off by operating a separate switch on the console.

The status of the remote train line and any alarm conditions are displayed on the console. Status displays include such information as throttle position or amount of dynamic braking. Alarm displays are the result of remote consist malfunctions such as "hot engine," "low oil," and "no power." Data in the LOCOTROL logic drive the display lamps in this console.

Figure 210: Radiation Incorporated, 'Locotrol I' system description (5)

LOGIC OPERATION

The principle of the LOCOTROL system is demonstrated by the functional block diagram of the lead unit, located in the lead locomotive. Inputs to the transmit register come from the sensing circuits, control and display console and the common control logic unit. The input to the operate register is received from the remote unit and consists of status data and alarm data which keep the lead unit informed of the conditions in the remote unit. The data in the operate register are compared to that in the transmit register by the comparator. The output of the comparator is one input to the common control logic.

When a control change is indicated by the inputs to the transmit register, a command signal is immediately transmitted to the remote unit, where it is verified and the command executed. The remote unit replies by transmitting all function status signals for display and comparison at the lead unit. Whenever a change in the status of the control functions or alarms are detected, the remote unit will transmit to the lead unit. If the status signals received by the lead unit fail to compare with the transmitted command signals, the lead unit continues to transmit these signals at short intervals until a comparison is obtained. When all functions compare, the lead consist transmits data to the remote consist every 21 seconds and immediately receives a reply as a continuity check.

Should the remote unit fail to receive a transmission from the lead unit for 45 seconds the feed valve is dropped out and the remote unit stepped down to idle.

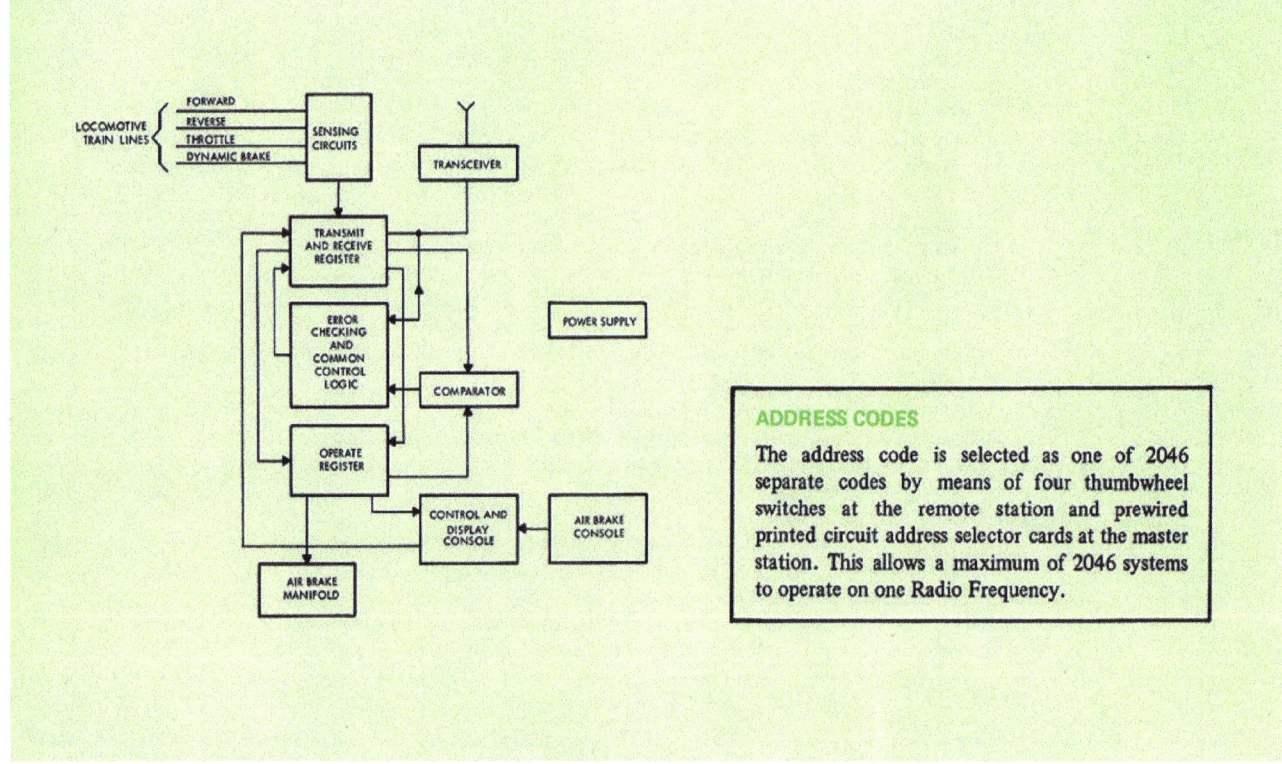

Figure 211: Radiation Incorporated, 'Locotrol I' system description (6)

EQUIPMENT SPECIFICATIONS

ENVIRONMENTAL

The equipment is mechanically designed to withstand, with a comfortable safety margin, the shock and vibration encountered in locomotive operation. It has been subjected to environmental tests similar to those given train locomotives. The logic cabinets and power supplies are furnished with angle iron frames which are bolted or welded into position in the locomotive. The LOCOTROL cabinets are then inserted into frames and secured in place by steel plates and wing nuts. Electrical connection is made through Bendix Pigmy connectors (or equivalent). The control and air brake consoles are designed for mounting to the locomotive throttle stand. The associated air brake equipment is also designed for standard locomotive installation. The system is sealed against dust, locomotive gases and oily fumes. It is not weather-proofed and must be located inside a locomotive or radio control car.

ELECTRICAL

Power Requirement	72 VDC ungrounded, 5 amps
Operating Voltage	0, -12, +12 VDC
Type of Transmission	FM Half-Duplex Carrier with 3 KC bandwidth frequency shift keyed
Transmission Frequency	152 - 174 MC Industrial Band
Temperature Range	-30ºC to +65ºC
Display	Front Panel mounted indicators

MECHANICAL

Dimensions:
Master Unit No. 1 Frame	20.32" high x 23.2" wide x 10.3" deep
Master Unit No. 2 Frame	25.6" high x 23.2" wide x 10.3" deep
Remote Unit No. 1 Frame	20.3" high x 23.2" wide x 10.3" deep
Remote Unit No. 2 Frame	25.6" high x 23.2" wide x 10.3" deep
Control Console	5.5" high x 22.0" wide x 9.0" deep
Power Supply Frame	9" high x 25" wide x 12" deep

Mounting:
Logic Cabinets	Angle Iron Frames
Control Console	Quick Disconnect
Power Supply Converter	Angle Iron Frames

1-L-2M BP571 PRINTED IN U.S.A.

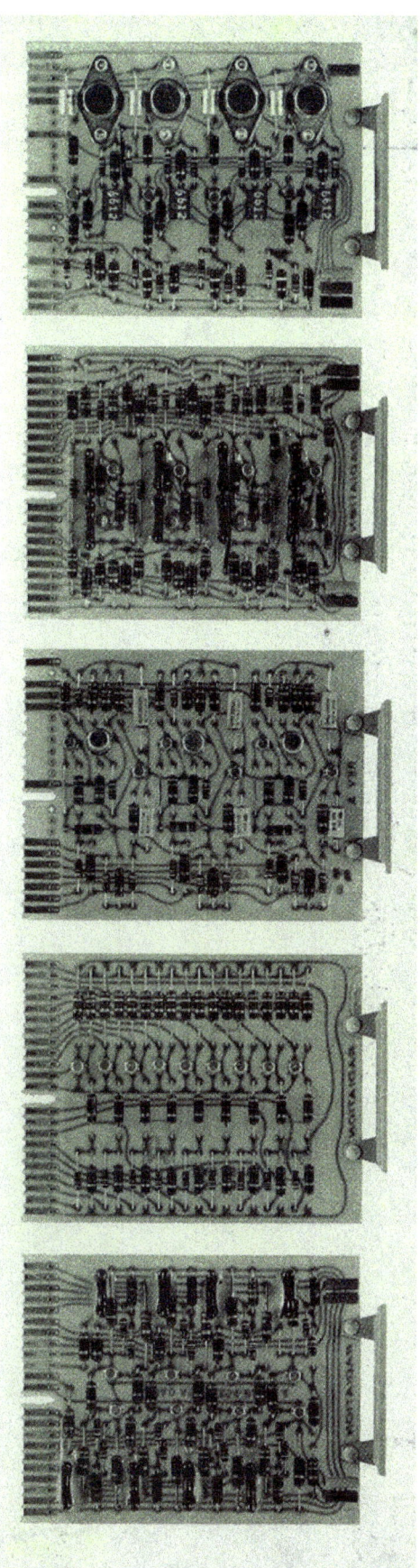

Figure 212: Radiation Incorporated, 'Locotrol I' system description (7)

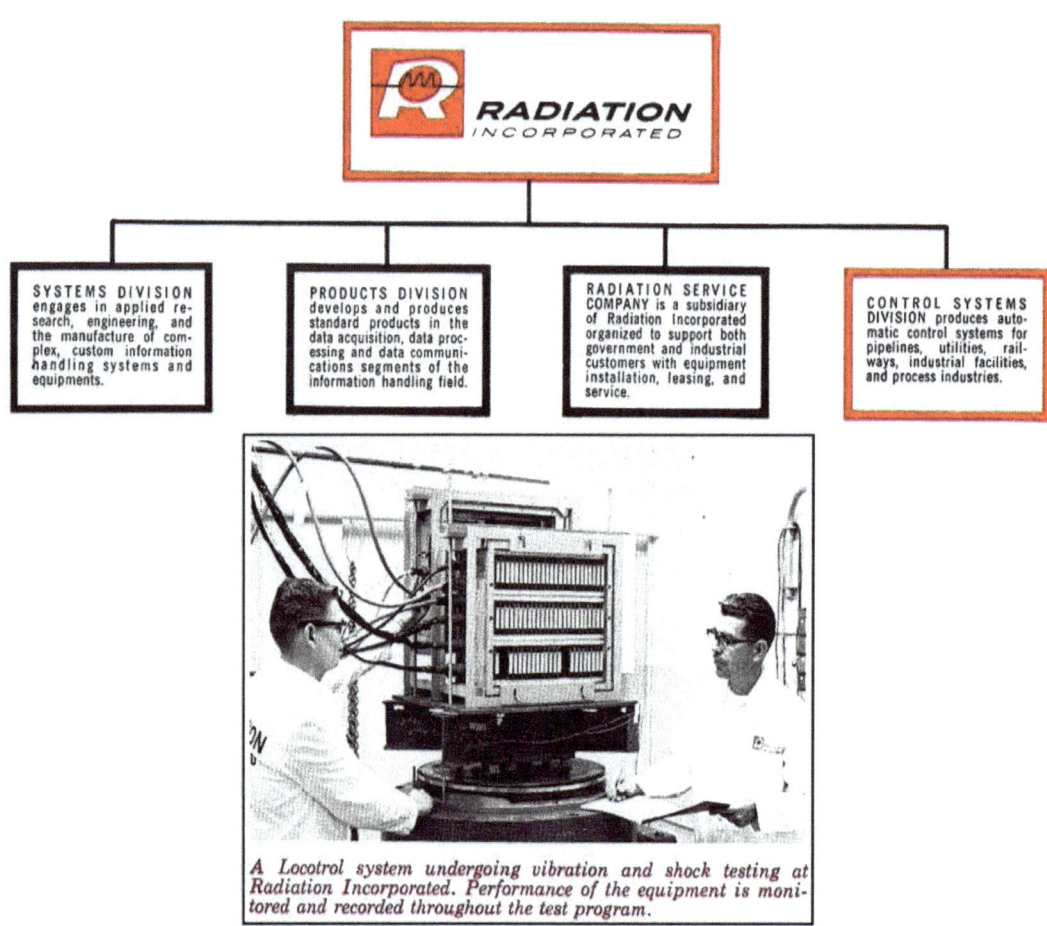

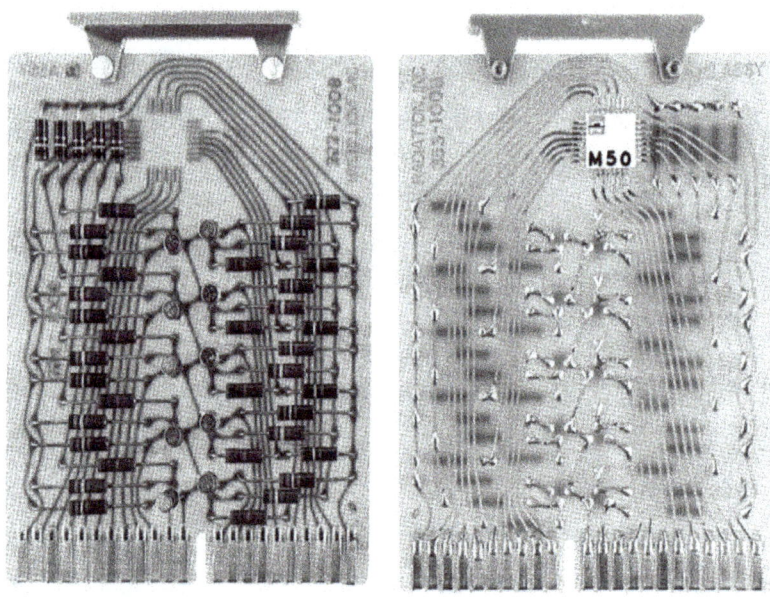

Figure 213: Radiation Incorporated, 'Locotrol I' system description (8)

Locotrol I

The indicators and functionality of the Control Console could—to a degree—be customised for the individual railroad; for instance, the Mt Newman Mining LOCOTROL I systems had Automatic Override. For this reason, on MNM consoles, the 'Override' push-button as depicted was redundant (and therefore non-functional) and the OVERRIDE indicator was not required and was available to be used for a feed valve FLOW indication.

The Lead and Remote Radio 2 indicators would illuminate to indicate that the associated logic on the respective unit had selected that system as the operating radio. The absence of an illuminated Radio 2 display indicated that the radio/s were operating on System 1, as was normally the case.—

Operating LOCOTROL

The following series of diagrams—based on a LOCOTROL I console—illustrate aspects of how a locomotive engineer operates this DP system. The Feed Valve is a device on the air brake manifold on the LOCOTROL-equipped Remote unit (or the remote-control car) that supplies air into the brake pipe at the Remote unit. It is controlled by the locomotive engineer to do what the locomotive brake valve assembly on the Lead unit does: to charge the train brake pipe. The instructions should be interpreted with reference to the illustrations of the engineer's Control Console.

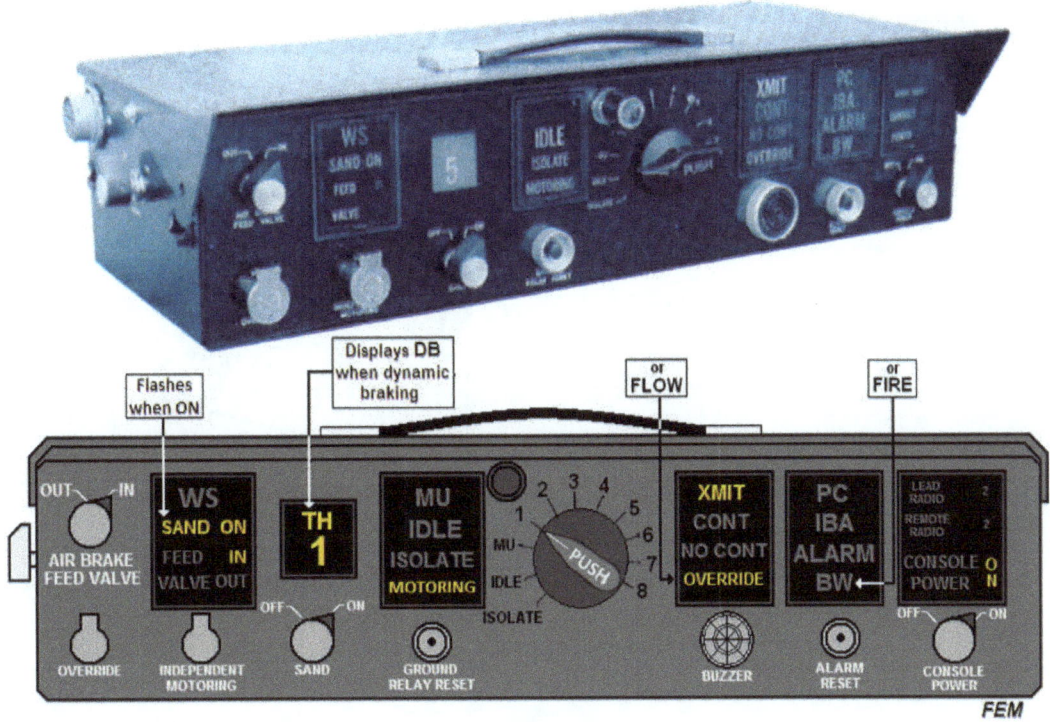

Figure 214: Locotrol 105 Control Console detail. The clarity of some of the indicators depicted in the photo above the drawing does not necessarily mean they are illuminated (i.e. showing a status condition). This manner of presentation was a peculiarity of this model console. Some indicators looked as if they were illuminated (i.e. active) when they were not. Note that conditions shown in the photo (upper portion of illustration) and the author's drawing (lower portion) are different. The photo of the console depicts the Mode Selector in 'Idle' and yet the status indicator is reporting the Remote unit at 'Notch 5'. This is an apparent anomaly, and the author assumes that when the photo was taken to illustrate the pamphlet, the Mode Selector had momentarily been placed to Idle but the Remote unit had not yet transmitted the updated status. It would normally do this within three or four seconds. The author's diagram below the photo depicts a state of 'Independent Control'. The Mode Selector has been placed to the Throttle 1 position, showing MOTORING as distinct from (dynamic) BRAKING, and this condition at the Remote unit is confirmed by the console tedicators.

LOCOTROL I
Operating Processes

Figure 215: Harris Corporation Locotrol I – Operating processes – Cover page

REMOTE UNIT FEED VALVE CONTROL

FEED VALVE 'IN':

1. Turn **Mode Selector** switch out of **ISOLATE** position.
2. Turn air brake **Feed Valve** switch to **IN**.
3. Depress the Auto Release pushbutton on the air brake control console.
4. Observe **Feed Valve IN** [indication].

FEED VALVE 'OUT':

1. Turn air brake **Feed Valve** switch to **OUT**
2. Turn **Mode Selector** switch to **ISOLATE**
3. Observe **Feed Valve OUT** [indication].

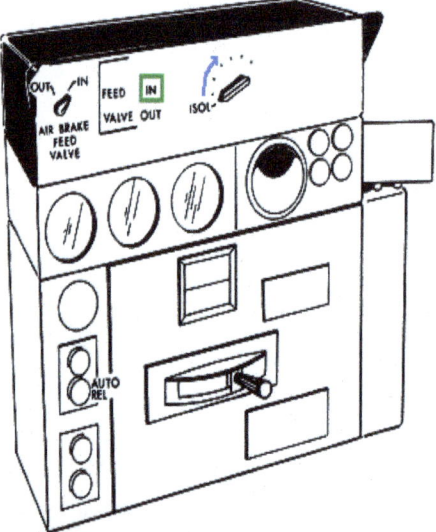

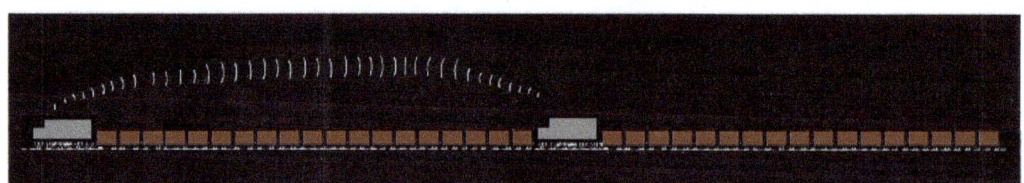

Figure 216: Locotrol I – Operating processes (1) – Remote unit Feed Valve control

TERMINAL AIR BRAKE AND LEAKAGE TEST

1. Turn the **Mode Selector** switch out of the **ISOLATE** position.
2. Turn the air brake **Feed Valve** switch **IN**.
3. Depress the **Auto Brake RELEASE** pushbutton.
4. Observe **Feed Valve IN** [indication].
   (Note: Lead unit **Feed Valve** must be at the **IN** position)
5. Once the desired brake pipe pressure is attained, make the desired brake pipe reduction then wait <u>a minimum of 12 minutes</u>.
6. Turn the air brake **Feed Valve** switch to **OUT**.
7. Turn the **Mode Selector** switch to **ISOLATE**.
8. Observe **Feed Valve OUT** [indication].
9. Turn the Lead unit **Feed Valve** to **OUT**.
10. Test for brake pipe leakage.
11. When the leakage test is complete, repeat steps 1-4 (except Lead unit **Feed Valve** is **OUT**).
12. When an increase in pressure on the Lead unit brake pipe pressure gauge is observed (indicating the brake pipe between Remote and Lead units is continuous), place the Lead unit **Feed Valve** to **IN**.
13. When the proper authority indicates the brakes have released, the train is ready to depart.

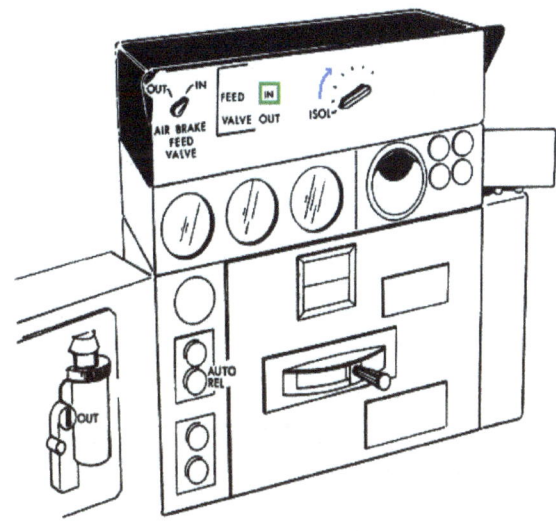

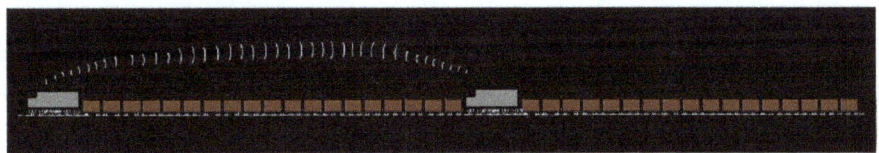

Figure 217: Locotrol I – Operating processes (2) – Terminal air brake and leakage test

WHEELSLIP

> **Wheelslip** indicator illuminates.
> Audible alarm sounds.

1. Turn **Sander switch ON** to sand Remote unit (**SAND ON** light flashes).
2. Audible alarm ceases.
3. **WHEELSLIP** indicator extinguishes when wheel-slip condition ceases.
4. Turn **Sander switch OFF** to end Remote unit sanding (**SAND ON** light is extinguished).

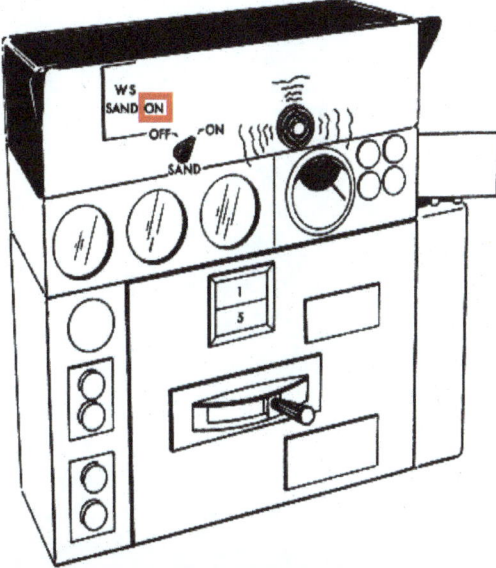

Figure 218: Locotrol I – Operating processes (3) - Wheelslip

OVERRIDE

For use when approaching areas of expected or known radio 'No Continuity'.

1. Depress **OVERRIDE** button (Override timer is activated when the button is released).
2. Oberve **OVERRIDE** indication is illuminated.

   Note: 'Override' condition remains in effect until one of the following occurs:
   (a) Override timing relay times out.
   (b) Radio CONTINUITY is regained.
   (c) An Automatic brake application is made.

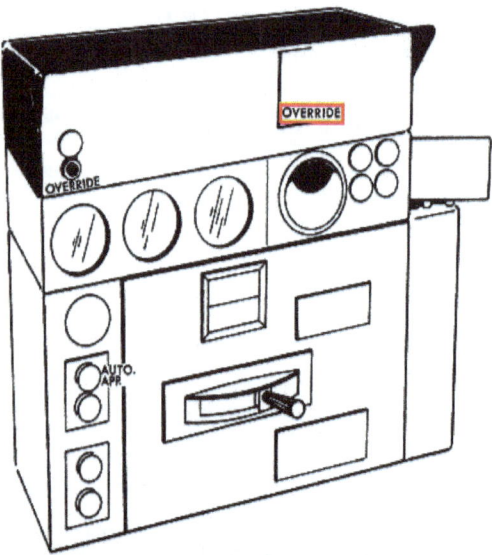

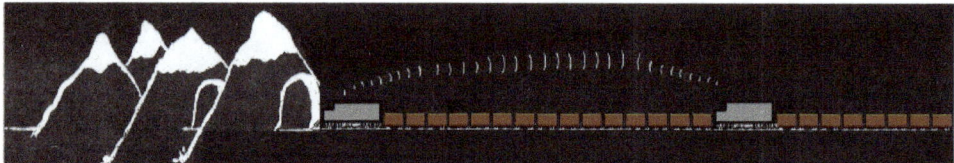

Figure 219: Locotrol I – Operating processes (4) - Override

RECOVERY FROM A *PENALTY* APPLICATION

A Penalty brake application results from train overspeed, release of the 'dead-man pedal' (where fitted), or use of the conventional automatic brake valve handle.

1. Turn the **Feed Valve** switch to **OUT**.
2. Turn **Mode Selector** to **ISOL** (Isolate).
3. Observe **Feed Valve OUT** indication.
4. Note: Throttle must be in **IDLE**.
5. Unlatch Auto Brake Valve handle and place to **SUPPRESSION** position until the locomotive PC is recovered.
6. Return ABV handle to **RELEASE**.
7. Turn the **Mode Selector** away from **ISOL**.
8. Turn **Feed Valve** switch to **IN**.
9. Depress **Auto Rel** pushbutton.
10. Observe **Feed Valve IN** indication.

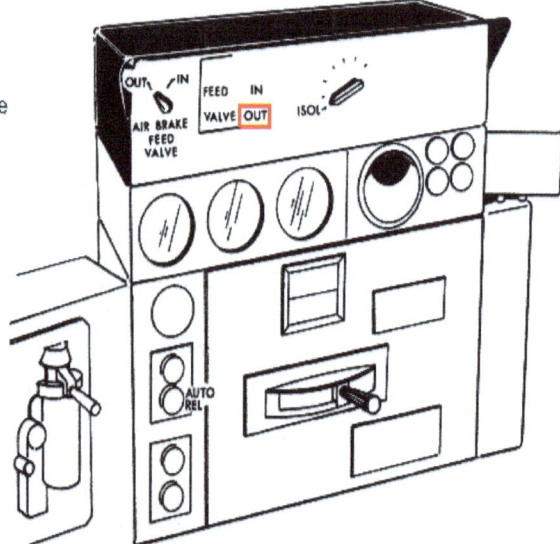

Figure 220: Locotrol I – Operating processes (5) – Recovery from a Penalty application

RECOVERY FROM AN *EMERGENCY* APPLICATION

After the train has stopped:

1. Turn **Feed Valve** switch to **OUT**
2. Turn **Mode Selector** to **ISOL** (ISOLATE) position
3. Observe **Feed Valve OUT** indication
4. Determine cause of Emergency application
5. If cause was a Break-in-Two, the angle cock on the rear portion must be left open until the train has been re-coupled
6. Once the cause has been determined and corrected, depress the **Auto Release** pushbutton and observe that the equalising reservoir and brake pipe air pressures are restoring

When brake pipe pressure is rising on the end of the train:

7. Turn **Mode Selector** out of **ISOL** position
8. Turn **Feed Valve** switch to **IN**
9. Depress the **Auto Release** pushbutton
10. Observe **Feed Valve IN**

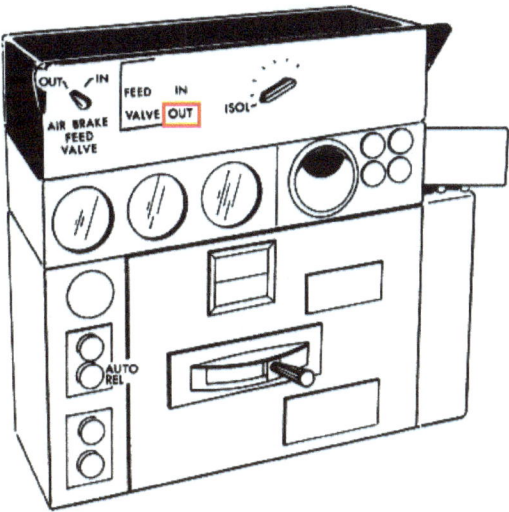

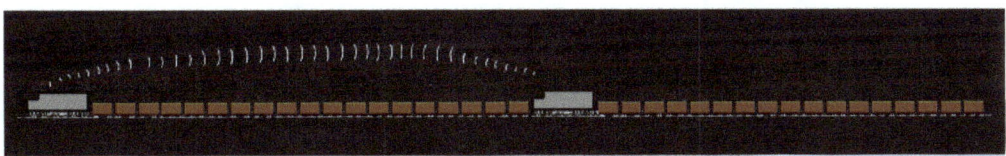

Figure 221: Locotrol I – Operating processes (6) - Recovery from an Emergency application

ALARM

(Ground Relay, Hot Engine, Low Oil, No Power on Remote unit)

> **ALARM** indicator illuminates
> Engine alarm bell sounds

1. Depress **ALARM RESET** button (silences Alarm bell - Alarm indicator remains **ON** as long as alarm exists).
2. Depress **GROUND RELAY RESET** button (if the units are equipped with electrical reset, and if this relay has tripped out, it will reset and the Alarm indicator will extinguish).
3. If the Alarm indicator remains **ON**, follow established company procedures to locat and correct the alarm condition.

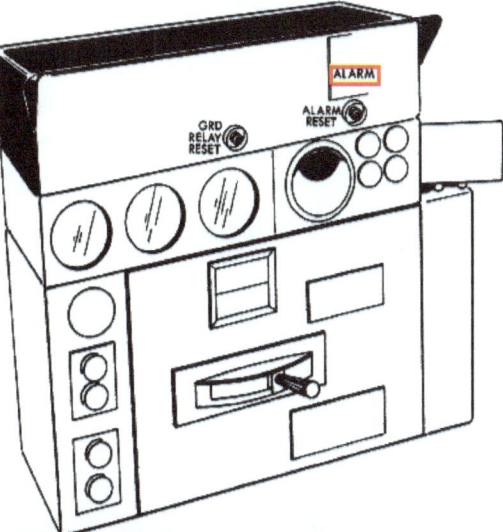

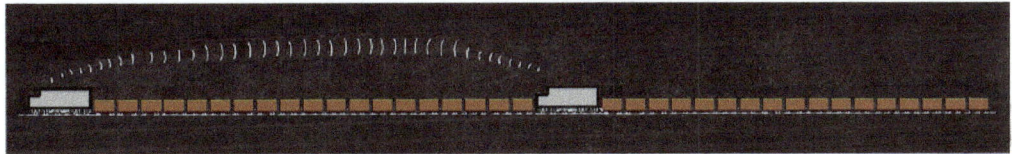

Figure 222: Locotrol I – Operating processes (7) - Alarm

MAKING A CUT BETWEEN *LEAD* AND *REMOTE* UNITS
(to pick up or set out cars, etc.)

1. Turn **Feed Valve** switch to **OUT**.
2. Turn **Mode Selector** to **ISOL** (Isolate).
3. Observe **Feed Valve OUT** indication.
4. Close the angle cock at the rear of the last car of the FRONT section of the train.
5. The angle cock at the front of the lead car in the REAR section of the train must be left open.

CAUTION: Remember that the Remote unit will still respond to commands from the Lead unit even when the Lead unit is not physically coupled to the rest of the train. It is absolutely essential that the above process be followed to ensure that the REAR portion of the train is in the following condition:

   (a) Remote unit **Feed Valve OUT**.
   (b) Remote unit **ISOLATED**.
   (c) An Emergency air brake application has been made on the REAR portion of the train.

6. Make the cut.
7. After the train is recoupled, depress the **Auto Rel** pushbutton and observe that the equalising reservoir and brake pipe are being restored.
8. Turn the **Mode Selector** switch away from **ISOL**.
9. Turn the **Feed Valve** switch to **IN**.
10. Depress the **Auto Rel** pushbutton.
11. Observe **Feed Valve IN** indication.

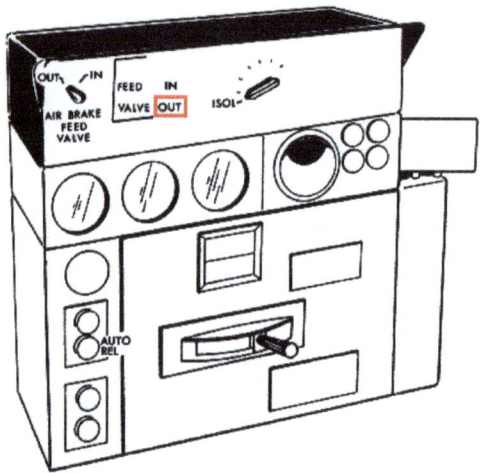

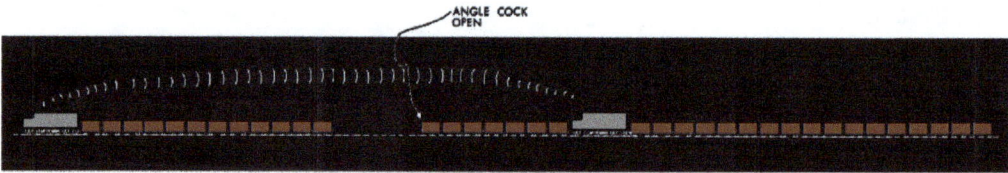

Figure 223: Locotrol I – Operating processes (8) – Making a cut between Lead and Remote units (to pick up or set out cars

LOCOTROL II and III

LOCOTROL II

Some commonalities of LOCOTROL II and III

LOCOTROL II is a technically advanced and improved version of the original LOCOTROL 103 and 105SS products which have been proven during nearly 20 years of use on major railroads of the world.

The new system uses the advantages of microprocessor-based control and monitoring to gain further security, ease of operation and maintenance, and added capabilities when compared to the original design. Using an "intelligent" processor, which can be programmed, now means that various parameters and limits can be configured to match a specific type of equipment and operating condition. Telemetry functions, additional operating interlocks, memory, self diagnostics, and

Figure 224: Harris Controls 'Locotrol II' pamphlet cover and system description (1)

a logger function are included in the newer, more efficient design. Locomotive applications have been broadened to include the newest models of diesel-electric and electric units.

Among the added features made possible by these improvements are:

Faster data transmission speeds. This allows faster updating of data, adds to functional capacity, provides extra communications security, and increases precision of measured data. System response times are generally improved.

Digital data displays. Actual Brake Pipe, Brake Cylinder, and Equalizing Reservoir pressures and Brake Pipe Flow rates are displayed in real time on the Engineer's console.

Improved security. A unique addressing system using locomotive serial numbers mandates correct system setup. LOCOTROL II can become operational only after correct completion of a standard Brake Pipe Continuity test during which the actual response to pressure change within the pipe physically confirms proper lead/remote address setting and radio linkage.

Continuous system self-checking features. At initial system turn-on and continuously during operation, a series of checks and comparisons is made according to a programmed sequence in order to detect any equipment malfunction or communication error.

Wider temperature range. More rugged and reliable components,

LOCOTROL II Console and Electronics Module (covers removed).

Figure 225: Harris Controls 'Locotrol II' system description (2)

lower power consumption, and a wider range of operating temperatures insure that LOCOTROL II will comfortably withstand the rigorous environment of constant locomotive operation.

Introduction

LOCOTROL is a solid-state electronic system for the control of distributed locomotive power in trains. It provides precise, automatic remote control from the Lead consist to a second consist of locomotives, and reports the status of those locomotives back to the operator at the head end of the train. Therefore locomotives can be connected at optimum load points within a train make-up.

Digital control logic and air brake units are installed in the Lead and Remote locomotives, and automatic remote control of the Remote units is obtained utilizing a radio link between the Lead and Remote consists.

Division of locomotive power is not a new concept. Previous attempts have been made, using only voice communication between the manned control positions of the Lead and rearward consists; but the precise timing necessary for coordinated operation cannot be obtained in this manner. The synchronous, automatic control of both consists made possible by the solid-state logic system employed in LOCOTROL solves this problem. The division of locomotive power into separate consists, with coordinated operation, results in increased power efficiency and faster, smoother, and safer stops and starts.

Convenient Operation

The LOCOTROL system was designed for simplicity of operation. In normal operating modes, it functions automatically. Just as locomotives are made as safe as possible with various built-in safety devices, so is LOCOTROL. Redundant features are incorporated where necessary to insure safety.

There are four operating modes. The first two modes, *Synchronous Automatic* ("Multiple Unit") and *Independent*, allow all system functions to be employed. When synchronous operation is used, the Lead locomotive operational functions, as controlled from the engineer's position, are sensed and signals are transmitted to the Remote consist for instant translation into corresponding command signals. When operating conditions require independent operation of the Lead and Remote consists, an "inhibit" capability (disabling the sensing circuits in the lead consist) is available to the train engineer. The Remote consist can then be independently controlled by command signals from the Lead locomotive. A rotating type deck switch on the control console in the Lead locomotive provides this separate control.

A third mode, *Idle*, allows only the operation of all air brake functions. The Remote engine is continuously idled. This mode thus allows the Remote consist to act as an air brake repeater.

The fourth mode, *Isolate*, disables all control functions except the Emergency Air Brake command, effectively making a "dummy" of the Remote consist.

Advantages of Locotrol Operation

Better Train Control — The capabilities of *motoring*, *dynamic braking*, and *air brake control* on the Remote consist — either synchronously with the Lead or independently — provide better control of the train under all conditions.

Reduced Drawbar Stress — Through the insertion of a consist of locomotives toward the middle or rear of a train, drawbar (coupler) forces are distributed more evenly within the train.

Longer Train Operation — With this method of distributing drawbar stress, longer trains can be operated. Theoretically, LOCOTROL trains can be operated with up to three times the tonnage of a conventional train.

Reduced Traffic Density — Through the operation of longer trains, more tonnage can be handled by fewer trains. This reduces the number of trains required to haul a specific tonnage.

Increased Hauling Capacity — As railroads increase train length and tonnages hauled, the railroad effectively increases its hauling capacity.

Precise Control of Slack Action — With the capability of controlling a consist at an optimum point within a train, slack action (interaction between cars within a train) can be reduced. This results in reduced lading damage and reduced equipment damage.

Faster Air Brake Response — The concept of dispersed motive power within a train allows the operator to control his air brakes (through the feed valve) at two points within the train. This allows the train to be charged from two sources ... with the rear source charging in two sections. Air brakes are thus charged and vented from three different directions in a LOCOTROL train. Air brake response times are improved by a factor of approximately three.

Safer Operation — The improved motoring and air brake features available in LOCOTROL operation result in safer overall train handling.

Reduced Operating Costs — The above factors combined result in reduced operating costs and more efficient operation for today's railroads.

Status Reporting

Status of the Remote consist's major operating functions is continuously displayed at the engineer's position in the Lead unit. Alarms immediately apprise the engineer of a locomotive malfunction.

Use of Multiple Systems

Even when using the same radio frequency band for more than one system, no minimum proximity limitations exist with LOCOTROL.

Each LOCOTROL system is insensitive to spurious signals from any source including any other LOCOTROL equipment operating in the same area. This is because an indi-

Figure 226: Harris Controls 'Locotrol II' system description (3)

vidual "address" code, which is a part of every transmission, identifies each LOCOTROL system. Thus, two LOCOTROL-equipped trains, even though using the same radio frequency band in the same area, could not receive or respond to each other's signals. A remote can respond only to the Lead equipment transmitting the correct individual address codes; and in turn, that Lead can receive data returned only from the addressed Remote equipment. Furthermore, the two units must be within the same train and must have completed a brake pipe test sequence.

A LOCOTROL II System consists of Lead Station and Remote Station equipment. The lead equipment is installed in a locomotive which is to function as the controlling unit, and the remote equipment is installed in another locomotive (or control car) which is to function as the remote unit.

The LOCOTROL II System's lead unit hardware consists of the following components:

   a. Electronics Module
   b. Interface Module
   c. Input/Output Module
   d. Control Console
   e. Air Brake Console
   f. Radio Module
   g. Air Brake Control Unit

The LOCOTROL II System's remote unit hardware consists of the following components:

   a. Electronics Module
   b. Interface Module
   c. Input/Output Module
   d. Radio Module
   e. Air Brake Control Unit

The equipment is mechanically designed to withstand, with a comfortable safety margin, the shock and vibration encountered in locomotive operation. It has been subjected to environmental tests similar to those given locomotives. The Electronics Module, radios and converters are furnished with mounting frames which are to be installed in the locomotive. The Locotrol modules are then inserted into the frames and secured in place. The Control Console is designed for mounting to the locomotive throttle stand. Other mounting systems can be supplied. For instance, lead and remote unit cabinets can be furnished to contain all electronics and electrical interface equipment in a single enclosure.

The system is sealed against dust, locomotive gases and oily fumes. It is not weather-proofed and must be sheltered from the elements.

Major components of the control system are briefly described as follows:

   a. **Control Console** — Contains remote status indicators and alarms. Also contains the switches necessary to isolate and independently operate the remote unit.
   b. **Electronics Module** — Contains the system electronics which perform the functions necessary to control overall system operation.
   c. **Input/Output Module** — Provides electrical isolation between the Logic Module input circuitry and the locomotive train line signals. Provides the binary control signals between the Logic Module output circuitry and the Interface Module. Also provides the electrically isolated analog control signals between the Logic Module output circuitry and the locomotive train lines.
   d. **Interface Module** — Provides the binary control voltages to the locomotive train lines.
   e. **Radio Module** — FM, half-duplex carrier having a 3 khz band width. May be single or dual transceiver.
   f. **Air Brake Console** — Provides electrical control of the lead and remote unit air brakes.

Power Requirements. The Locotrol System is designed to operate from the standard locomotive battery voltage source of 72 volts DC (+8V, −15V), ungrounded. Maximum current drain (less radios) is 5 amperes. Current drain for the radios is 1.6 amps. in the Receive mode and 3 amps. in the Transmit mode. Other power sources, such as 110 volts DC, can also be accommodated.

Interface Requirements. Both input and output circuits have interfacing which provides isolation between locomotive and Locotrol electrical systems with transient protection devices incorporated to prevent spurious operation or damage to the equipment.

The input interfacing is composed of electrical sensing circuits connected to the locomotive train lines and pressure sensors connected to the air brake system. The electrical sensing circuits detect the presence, absence, or levels of battery voltage on the locomotive train lines. The pressure sensing circuits detect air pressures and air flow in the air brake system. Together, these electrical and pressure sensing circuits detect, condition, isolate, and input into the Locotrol System the operational status of the locomotive.

The output interfacing consists of relay driver circuitry and associated relays and magnet valves. These control the locomotive train lines and air brake functions while providing D.C. isolation between the Locotrol System and the locomotive battery.

Communication Link. Communications between lead and remote units is by VHF and UHF radio. The use of this type of equipment requires operating licenses issued by the Federal Communications Commission. Data transmission by the radio is in the form of a frequency-shift-keyed (FSK) signal. A 3 kHz bandwidth is used with modulation frequencies of 1300 Hz and 2100 Hz.

In conventional trains operating without a LOCOTROL II System, the locomotive is controlled from the engineer's position in the cab by manipulation of levers and controls. These controls are connected through the train lines and control the applied voltages to the respective locomotive control functions.

Figure 227: Harris Controls 'Locotrol II' system description (4)

With a LOCOTROL II System, the division of locomotive power by the placement of locomotive units at the front and some other point in the train's buildup, makes available the full potential of locomotive power. However, there are no control lines between the separated units. The LOCOTROL II System employs digital control, utilizing radio equipment as a telemetry link, to provide the precise timing necessary for coordinated operation of the lead and remote locomotive units. This permits division of locomotive power and control into the separate units operating in unison, resulting in the following advantages:

Initial charging time of the train air brake system is significantly reduced.

Brake applications are quicker than in conventional trains.

Air brake releases are also quicker.

Power efficiency and overall tonnage capabilities are enhanced through synchronous control of the locomotive power at two locations.

More rapid acceleration and deceleration of the train is possible.

Stopping distance is reduced by faster application of brakes.

Locotrol provides improved control of slack action, which reduces damage to cars, couplings, and lading.

Locotrol enables long trains to achieve and maintain higher speeds for faster, smoother runs and improved scheduling.

Low speed train handling is improved by the ability to make brake applications and running releases.

Each function of the locomotive is controlled by the presence or absence of a battery voltage. When the control voltage for a specific function is detected by the respective sensor circuit, an output is generated which indicates that the function has been activated.

All control functions, actuated or not, are transmitted on each transmission. This control information is transmitted serially in a specific message format and with a specific message protocol. Each control function is designated as a specific bit position or a group of bit positions in the message format. With each function control code occupying a specific position in the transmitted message format, the remote locomotive unit can determine which function is actuated and perform that function in unison with the transmitting locomotive.

Communication protocol dictates that the master station in the LOCOTROL II System (lead locomotive) transmit a Command message under the following conditions:

a. *When an operator command change is sensed.*

LOCOTROL II Consoles in locomotive cab.

Figure 228: Harris Controls 'Locotrol II' system description (5). Cab control equipment - typical installation in EMD locomotive. A mounting bracket is fixed to the top of the brake stand and the horizontal Control Console is clipped to this. The vertical Air Brake Console is installed into the control stand to the left of the throttle handle

b. *When the remote unit does not respond to a Command message.*

c. *In response to a remote initiated communications check.*

d. *A minimum of every 20 seconds.*

Therefore, the remote unit is always immediately notified of a control change and the lead locomotive is immediately notified if the commanded change has taken place.

Each message received by the remote unit causes generation of a Status message back to the lead locomotive. The remote's status message is then decoded and the remote status is displayed on the lead's Control Console. The LOCOTROL II System performs data integrity checks to prevent undesired operation and improper displays as part of its Communications Security function.

The remote unit also generates a message to the lead unit under the following conditions:

a. *When a status change or alarm condition is detected at the remote.*

b. *A remote initiated communications check, if a significant unexpected change in brake pipe charging flow rate is detected.*

If communication is lost between the lead and remote locomotive units for more than a preset period of time, usually 45 seconds after the last successful communication check, or only five seconds after the transmission of an automatic or emergency brake application message, a Communication Interruption condition exists. If this occurs, the radio changeover logic is automatically activated. If one of the combinations of radios results in reestablishing communications, the system continues to operate with the combination of radios. If all radios fail to communicate, the remote locomotive begins the idle-down procedure. This leads to several other operating changes which occur after specific time periods and under certain conditions which are more fully described in technical documentation.

Equipment Specifications

Environmental — The equipment is mechanically designed to withstand, with a comfortable safety margin, the shock and vibration encountered in locomotive operation. It has been subjected to environmental tests similar to those given train locomotives. The logic cabinets and power supplies are furnished with angle iron frames which are bolted or welded into position in the locomotive. The LOCOTROL cabinets are then inserted into frames and secured in place by steel plates and wing nuts. Electrical connection is made through Bendix Pigmy connectors (or equivalent). The control and air brake consoles are designed for mounting to the locomotive throttle stand. The associated air brake equipment is also designed for standard locomotive installation. The system is sealed against dust, locomotive gases and oily fumes. It is not weather-proofed and must be located inside a locomotive or radio control car.

Electrical

Power requirement (less radios)
  72 VDC ungrounded, 5 amps
Operating Voltage
  0, −12, +12 VDC
Type of Transmission
  FM Half-Duplex Carrier with 3 KC bandwidth frequency shift keyed
Transmission Frequency
  152-174 MHz VHF; 450-474 MHz UHF
Temperature Range
  −30°C to +65°C
Display
  Front Panel mounted indicators

Mechanical

Electronics Module
  24" high, 20" wide, 15" deep, 30 lbs.
Control Console
  4" high, 18" wide, 8" deep, 15 lbs.
Air Brake Console
  15" high, 7" wide, 6" deep, 15 lbs.
Interface Module
  24" high, 36" wide, 10" deep, 60 lbs.

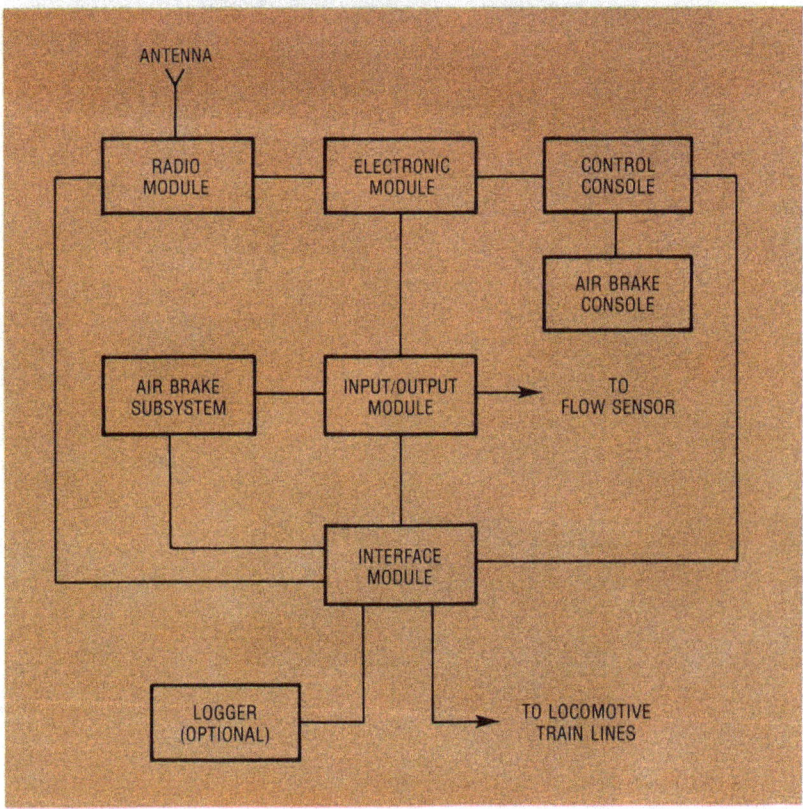

Simplified Block Diagram, LOCOTROL II System

Figure 229: Harris Controls 'Locotrol Harris Controls 'Locotrol II' system description (5)

Operating Modes — LOCOTROL II or LOCOTROL III

Several modes of operation can be selected from the system's Control Console.

Multiple Unit

All remote consists perform the same traction, dynamic brake and air brake functions as the lead unit.

Idle

All remote consists perform the same air brake functions as the lead unit. All remote throttles remain in idle.

Feed Valve Out

All remote consists perform the same traction and dynamic brake functions as the lead unit. The remote train air brake control is disabled.

Isolate

All remote consists remain in idle with air brake functions disabled (except emergency application).

Independent Throttle (LOCOTROL II)

All remote consists perform the same air brake functions as the lead unit. The remote throttles are controlled via the Control Console. To permit operation through grade changes, the remote consist may be operated in traction while the lead is in the dynamic brake mode.

Speed Control (option)

The lead consist and all remote consists independently control engine speed and generator excitation to obtain the selected speed for train loading operations.

Tower Control (option)

The remote consist (LOCOTROL II) or all consists (LOCOTROL III) can be controlled from an off-board location. By using limited traction and air brake functions, the train can be moved through an indexer/dumper facility.

Individual Remote Control (LOCOTROL III)

Remote consists can be set to synchronous, idle, feed valve out, isolate or engine stop modes.

Harris Controls Division, P.O. Box 430, Melbourne, Florida 32902-0430, Telephone (407) 242-4121, Fax (407) 242-4073

Figure 230: Harris Controls 'Locotrol II and III' – Operating modes

Operating Configuration—LOCOTROL II

The basic LOCOTROL II system consists of two identically equipped locomotives...one used as a lead unit and one used as a remote. This system can be operated in either the Multiple Unit Mode (MU) or the Independent Mode. When the MU mode is used, the train operation signals from the engineman's position are transmitted by radio to the remote station for motive power and air brake control. The Independent Mode permits different degrees of throttle in the two consists.

LOCOTROL II Station Equipment:

- Control Console
- Air Brake Console
- System Module
- Radio Equipment
- Air Brake Control Unit
- Air Flow Sensor Module

Major Component Functions of the LOCOTROL II System are:

Control Console contains remote status and alarm indicators. The microprocessor-based console also provides the switching necessary to select remote control modes and independent throttle positions.

Air Brake Console provides control of lead and remote air brake functions. Emergency, automatic and independent brake functions are included.

System Module contains the system's microprocessor and associated electronics which control overall system operation. It provides the control voltages to the locomotive train lines. It also includes electrical isolation and transient protection of input signals.

Radio Equipment includes one or two data radios and associated power converter which provide the communication between the lead and remote locomotives. The communication is half duplex (single frequency) and may be in the VHF or UHF bands.

Air Brake Control Unit contains the electro-pneumatic valves, pressure switches and pressure transducers used for controlling and monitoring all the air brake functions.

Air Flow Sensor monitors the train's air line charging rate.

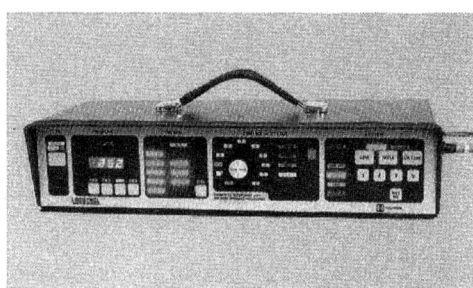

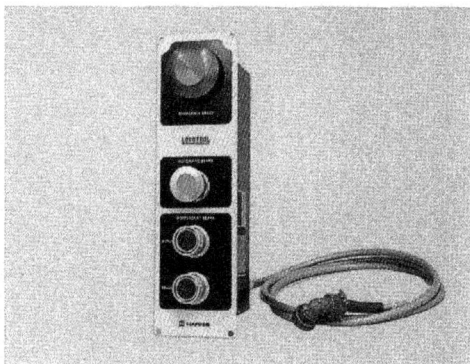

Harris Controls Division, P.O. Box 430, Melbourne, Florida 32902-0430, Telephone (407) 242-4121, Fax (407) 242-4073

Figure 231: Harris Controls 'Locotrol II' – Operating configuration

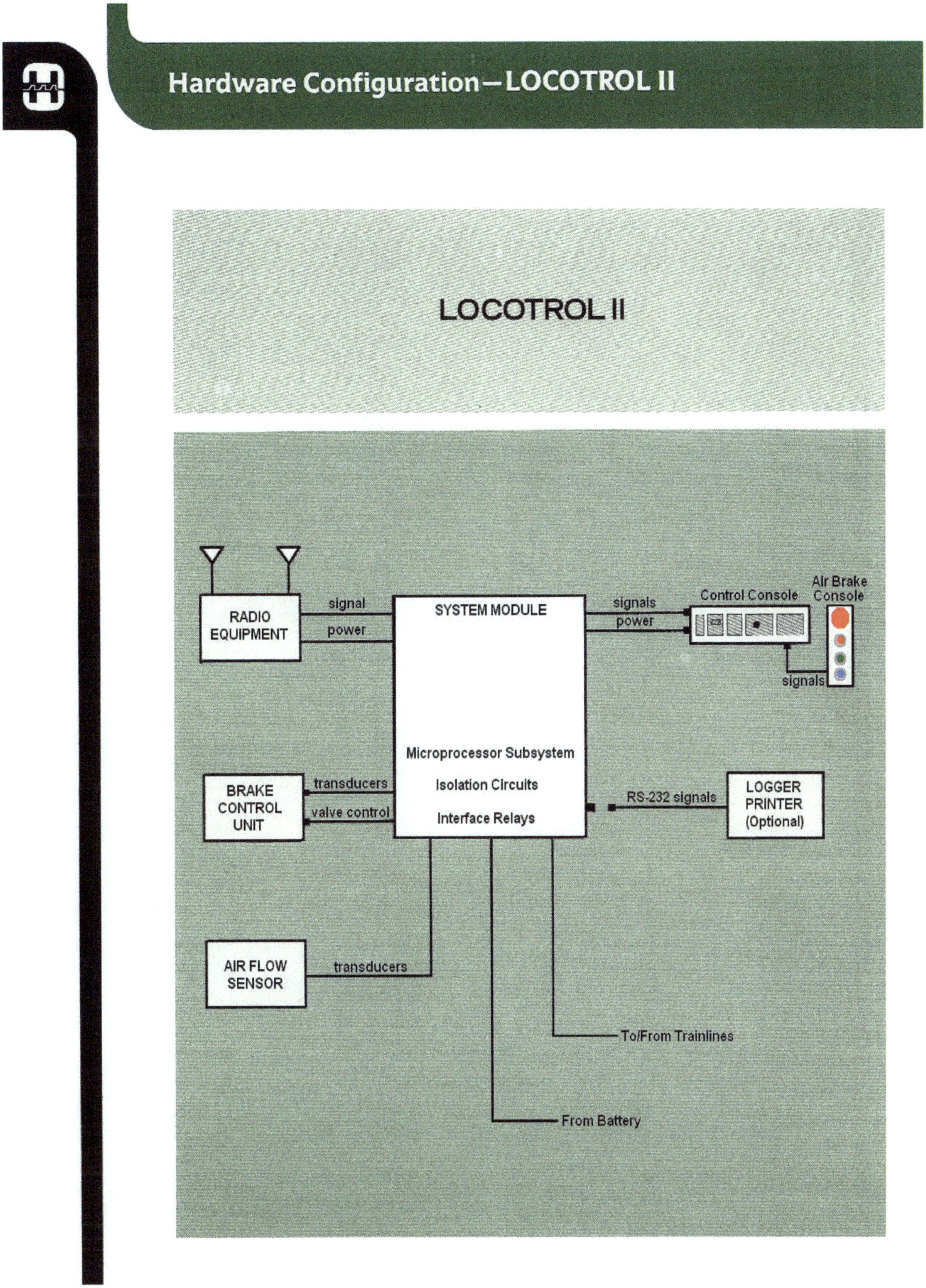

Hardware Configuration—LOCOTROL II

Figure 232: Harris Controls 'Locotrol II' – Hardware configuration

Operating Configuration—LOCOTROL III

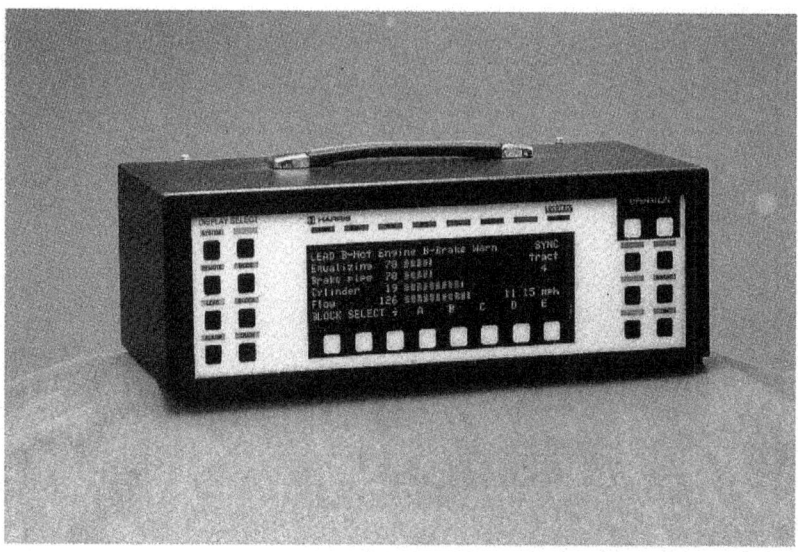

LOCOTROL III provides multiple remote consist operation, slow speed control and event recording. Each equipped unit can be operated as a lead unit or as any of the remote units. This system makes it possible to run long trains on high traffic track between large terminals and then separate them into shorter trains to be used, for example, in intermodal and general freight service. This provides an important competitive edge for short hauls as well as heavy haul operations.

LOCOTROL III Universal Station Equipment:
- Control Console
- Air Brake Console
- System Module with Event Recorder
- Radio Equipment
- Air Brake Control Unit
- Air Flow Sensor Module

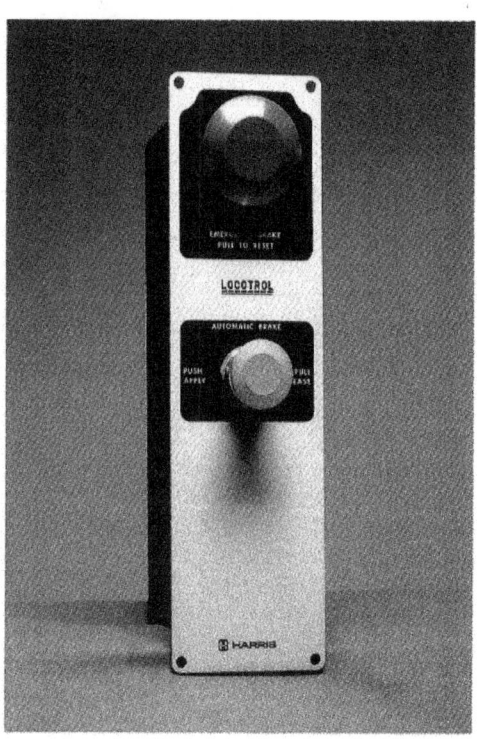

Figure 233: Harris Controls 'Locotrol III' – Operating configuration

Major Component Functions for the LOCOTROL III System are:

Control Console displays the remote status, alarms and other operating information. In addition, it provides the switching necessary to select remote control modes. Its noteworthy features include a dedicated microprocessor, a daylight readable alphanumeric display, high reliability lightbar alarm displays and flat panel switch technology.

Air Brake Console controls the train's automatic brakes by a single push/pull operator. Emergency brake applications can be initiated via an electro-pneumatic operator.

System Module contains the system's microprocessor and associated electronics which control the overall system operation. It also includes the electrical interface circuits, including isolation and transient protection.

The Event Recorder performs functions including the recording of specific events. This time-tagged event and summary data is accumulated on board for transfer to the logger unit. This optional logger unit is a portable computer which facilitates the collection of data from the on-board unit, the preview of data while on board the locomotive and the printing of data reports off board.

Radio Equipment includes one or two data radios and associated power converter that provide the communication between the lead and remote consists. The communication is half-duplex and may be in the VHF or UHF bands.

Air Brake Control Unit contains the electro-pneumatic valves, pressure switches and pressure transducers used for controlling and monitoring all the air brake functions.

Air Flow Sensor Module monitors the train's air line charging rate.

Figure 234: Harris Controls 'Locotrol III' – Major component functions

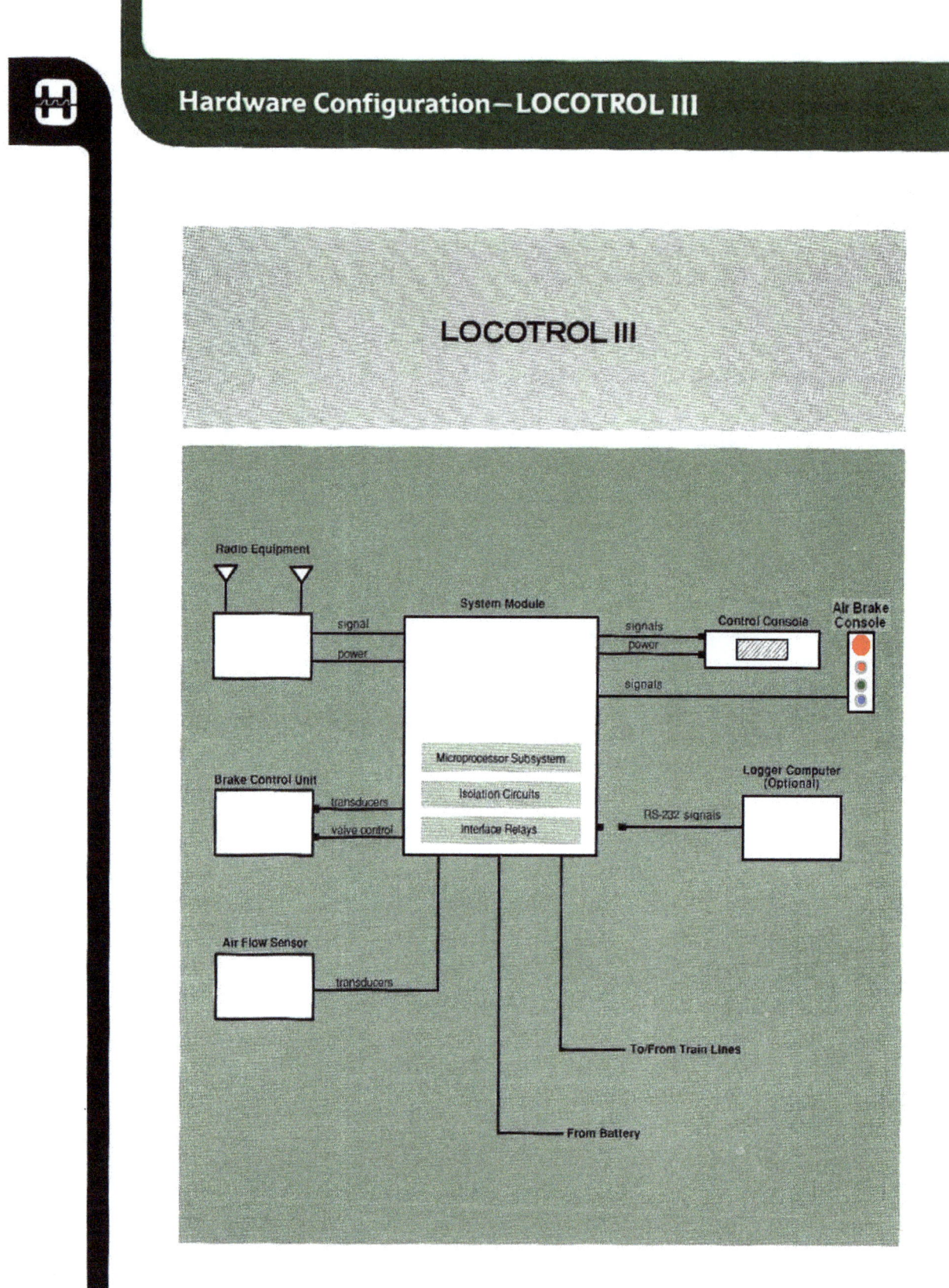

Figure 235: Harris Controls 'Locotrol III' – Hardware configuration

Technical Data—LOCOTROL II or LOCOTROL III

Equipment Stability

All the equipment included in the LOCOTROL product line is mechanically designed to withstand the shock and vibration encountered in locomotive operation. The Electronics Module, radios and Control Console are furnished with mounting bases which are permanently installed on the locomotive. The removable modules are then secured to the bases and connected via circular electrical connectors. The low-profile control console is typically mounted on the locomotive throttle stand. The air brake console is also typically installed in the throttle stand. Optionally, an equipment cabinet can be furnished to house the Electronics Module, Interface Module and radio equipment.

Control Functions

Each function of the locomotive is controlled by the presence or absence of a battery voltage. When the control voltage for a specific function is detected by the respective sensor circuit, a signal is generated which indicates that the function has been activated. The control functions are transmitted serially in a specific message protocol. Each control function is assigned a specific bit or a group of bits in the message format. Since each function control code occupies a specific position in the message format, the remote station can determine which function is activated and perform that function in unison with the transmitting lead station. Status information is similarly transmitted by the remote to the lead unit.

Data Transmission

The LOCOTROL system does not accept or process radio signals from any non-LOCOTROL source. It also does not accept data transmitted by any other LOCOTROL system operating in the same area. This is because there is a unique "address" code, which is part of every transmission, to identify each LOCOTROL system. Thus, two LOCOTROL-equipped trains, using the same radio frequency in the same area, could not respond to each other's signals. A remote can respond only to the lead station transmitting the correct individual address codes. In turn, that lead can receive data returned only from the addressed remote station. Furthermore, train movement is permitted only when the units are within the same train and have completed a brake pipe continuity test.

System Interface

The input and output circuits provide isolation between the electrical systems of the locomotive and the LOCOTROL microprocessors. Transient protection devices are incorporated to prevent spurious operation or damage to the equipment.

The input interfacing is composed of electrical sensing circuits connected to the locomotive train lines and pressure sensors connected within the air brake system. The electrical sensing circuits detect the presence, absence or level of battery voltage on the locomotive train lines. The pressure sensing circuits detect air pressure and air flow in the air brake system. Together, these electrical and pressure sensing circuits detect the operational status of the locomotive.

The output interfacing consists of relay driver circuitry, analog driver circuitry and associated relays and solenoid valves. These control the locomotive's train lines and air brake functions while providing electrical isolation between the LOCOTROL system and the locomotive's battery.

Figure 236: Harris Controls 'Locotrol II and III' – Technical data

Communication Protocol

Communication protocol dictates that the lead station in the LOCOTROL system transmit a command message under the following conditions:

- When an operator command change is sensed.
- When a remote unit does not respond to a command message.
- In response to a remote initiated communications check.
- A minimum of every 20 seconds.

In this manner, each remote station is immediately notified of a control change and the lead station is immediately notified if the commanded change has taken place.

Each valid message received by a remote station causes transmission of a status message to the lead station. This message is then decoded and the status data is displayed on the Control Console. The LOCOTROL system performs data integrity checks to prevent undesired operation and improper displays as part of its communications security function.

The remote station also generates a message to the lead station under the following conditions:

- When a significant status change or alarm condition is detected at the remote.
- A remote initiated communications check, if a significant unexpected change in train air line charging rate is detected.

If radio communication is lost between the lead and remote stations for more than 25 seconds, a communication interruption condition exists. If this occurs, the radio change-over function is automatically activated at all affected stations. If any combination of lead and remote radios results in re-establishing communications, the system continues to operate using that combination of radios. If radio communications are not re-established within 45 seconds, the remote stations prepare for the idle-down sequence. If a remote station subsequently senses a brake application, it reduces traction and air brake control to a passive status.

Figure 237: Harris Controls 'Locotrol II and III' – Communication protocol

Technical Specifications

Power Input

72 VDC nominal battery:	56-82 VDC
Electronics Subsystem:	2 Amps
Radio Equipment:	3 Amps
Control Outputs (Typical):	15 Amps

Communication

Single or Dual Data Transceivers
Radio Frequency: UHF or VHF Half-Duplex
Modulating Signal: FSK per CCITT V.23

Overall Dimensions

	LOCOTROL II	LOCOTROL III
System Module:	29" × 32" × 15"	29" × 32" × 15"
Control Console:	19" × 4" × 10"	17" × 6" × 9"
Air Brake Console:	4" × 14" × 5"	4" × 14" × 5"
Air Brake Control Unit:	25" × 16" × 13"	25" × 16" × 13"
Air Flow Sensor Module:	11" × 9" × 3"	11" × 9" × 3"

Controlled and Monitored Functions

Engine Control:

Engine Run	Direction
Generator Field	Throttle Valves

Dynamic Brake:

Brake Mode and Set-Up   Brake Control[1]

Air Braking:

Equalizing Reservoir[1]	Feed Valve
Emergency Application and Reset	
Independent Application, Release and Bail	

Miscellaneous:

Ground Reset[2]	Vigilance Disable[2]
Manual Sand[2]	

Speed Control:

Speed Mode[2]   Generator Excitation[1, 2]

Monitored Functions

Alarms:

General Alarm	Brake Warning
Wheel Slip	Ground Fault
Pinion Slip[2]	Power Cutout

Air Brake:

Penalty	Brake Handle Movement
Train Separation	Main Reservoir[1, 2]

Telemetry:

Main Generator Voltage[2]	Main Generator Current[2]
Train Speed[1, 2]	

NOTES:
[1] Analog Signal
[2] Optional

Figure 238: Harris Controls 'Locotrol II and III' – Technical specifications

Support Equipment and Services

Harris offers the following optional equipment to facilitate the testing and maintenance of the LOCOTROL systems.

- Simulators for testing and troubleshooting
- Testers to permit PC board diagnosis and repair
- Specialized test aids

Harris also offers the following optional support services to expedite system installation and check-out as well as to optimize operation and maintenance.

- On-site application engineering
- Installation consultation and system commissioning
- On-site operator training
- On-site or in-factory maintenance training
- Warranty and replacement parts through the customer service department

Supporting Products

Several related products are available to provide expanded operational capabilities and to improve performance.

- Event recorder logger unit for data display and printing
- Tower master unit for train control during unloading operations
- Message repeater unit to improve data radio communications in tunnels, deep cuts or where RF interference, multipath or shadowing degrade communications.

Documentation

Each LOCOTROL system includes complete operation and service documentation.

- Installation manual
- Operation manual
- Train driver's handbook
- Service manual
- Technical drawing manual

Figure 239: Harris Controls 'Locotrol II and III' – Support equipment and services. Harris software engineers Lewis Cox (left) and Don Johnson work on test equipment.

ASSOCIATED TECHNOLOGY – Early evolutionary uses for air brake control technology

The work conducted at Radiation (later Harris Corp) to evolve a credible radio remote-control distributed power product led to the development of several associated products intended and designed to facilitate the operation of long trains without the expense and complexity of DP. These technologies became part of the Railroad Product Line: however, the success of the LOCOTROL product itself appears to have worked against either product being adopted by the rail industry.

As they worked through the 1990s and into the new decade on further development of the LOCOTROL product, Milt Deno and Harris engineers Dale Delaruelle, Gene Smith, and others realised the technology could be applied specifically to refining air brake functionality and improving the safety of train operations generally.

As the sole member of Harris' Railroad Product Line development team possessing a railroad technical and operations career background, perhaps Milt Deno knew intuitively that air brake skills in train handling generally were on a subtle but inevitable decline with the coming generations of locomotive engine service personnel, as the purity of the 'old-time' corporate knowledge and competencies declined. And perhaps he comprehended that this was due to contemporary air brake systems and equipment that tended to render the train handling task less onerous than it had been in the past. And perhaps he pondered the train wrecks he'd attended—some resulting from the locomotive engineer having permitted the train to 'run short of air'[105] for downhill braking—across the mountainous Canadian provinces in both fine weather and foul throughout his years as a divisional master mechanic. He might not, though, have foreseen the degree to which ECP braking would eventually reduce the judgement chore for locomotive engineers regarding the exercise of anticipatory skills in air brake manipulation.

The two concepts that are described in this section were never produced commercially and are little known. They are recorded here [1] as unexplored proposals that arose from the development of radio remote control and/or the application of technology generated by that development, and [2] so that their existence will not pass unacknowledged.

SOAR

This system's name—the **S**imultaneous **O**perating **A**irbrake **R**epeater—reveals it as a child of the evolution of LOCOTROL. No SOAR systems were ever sold, the concept being overtaken and rendered redundant by LOCOTROL itself.

A sales pamphlet is reproduced on the following pages.

---

105 A critical aspect of safe train operations—especially in heavy grade territory—is the maintenance of adequate air pressure in car reservoirs. Where locomotive engineers sometimes make successive air brake applications without allowing sufficient time for the proper recharge of the brake pipe, car reservoir pressure can be depleted to the extent that the train cannot be controlled on a descending grade. From these circumstances, a runaway will often result. See also, paragraph 2 on page 256.

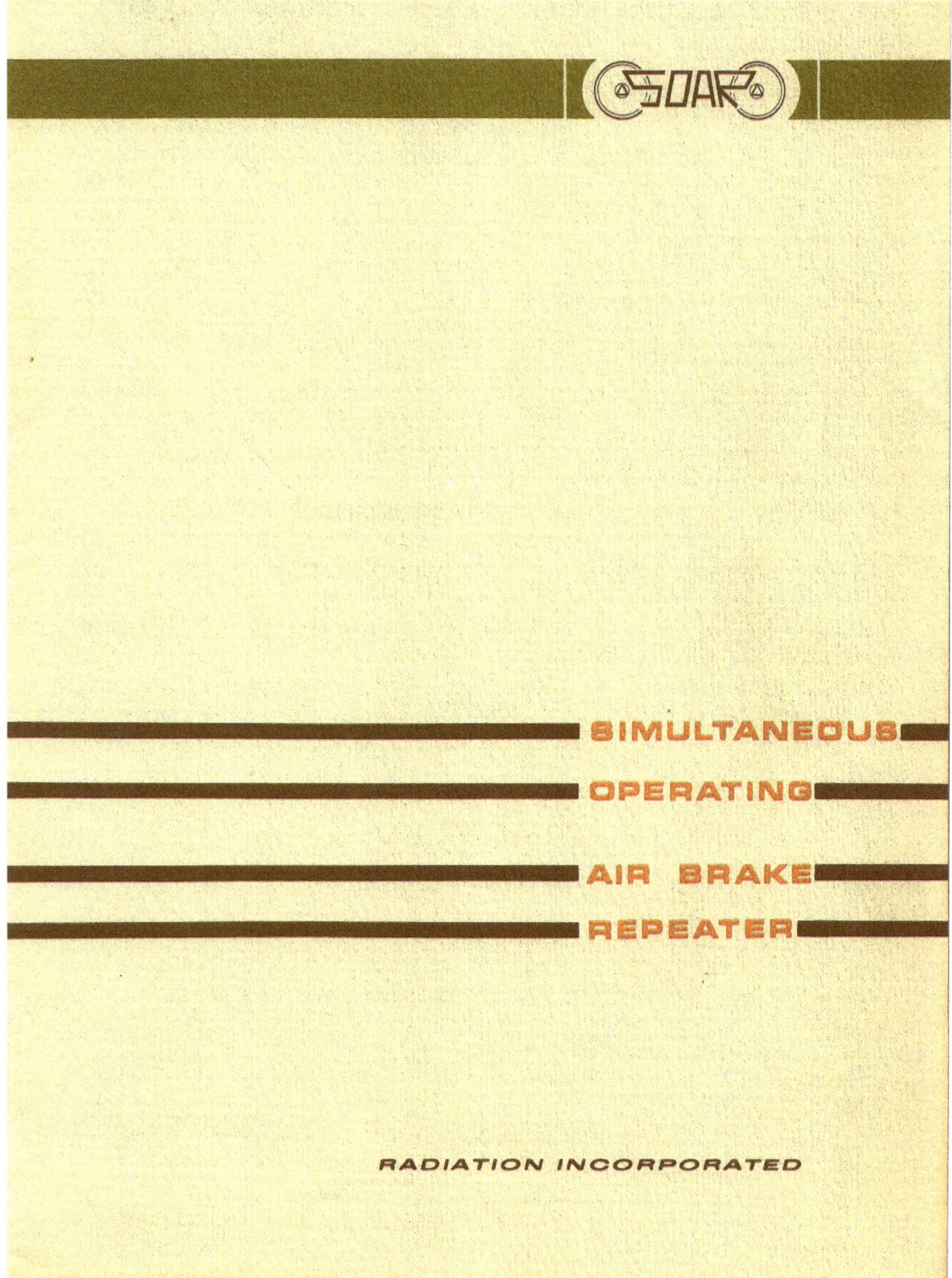

Figure 240: SOAR (1) - Pamphlet cover

INTRODUCTION

SOAR (Simultaneous Operating Air Repeater) is a solid state electronic system developed as a remote control air brake repeater to improve the braking operation of a train. Digital logic and air brake units are installed in the lead locomotive and in one or two remote repeater cars. The repeater cars are placed within the train at points where the brake pressure has dropped below desired operating levels to re-establish the brake pipe setting to that of the lead locomotive. Figure 1 illustrates the use of SOAR. Normal brake operation is not altered by the inclusion of SOAR equipment. The brake pipe remains continuous throughout the train allowing the brake pipe to be exhausted only at the lead locomotive if a failure in continuity occurs. Instant Communication between the lead locomotive and remote cars is maintained by VHF or UHF radio to provide simultaneous brake application and release.

The system can be employed in either a synchronous or cut-out mode. When operated synchronously, the lead locomotive air brake pressure is sensed by transducers, coded into a message format, transmitted to the repeater car and translated instantly into command signals to synchronize the air brake pressure setting.

When operating in the cut-out mode (SOAR inoperative), the automatic charging capability and application are inhibited at the repeater car. However, the lead locomotive continues to operate as a normal unit.

Unique break-in-two protection is afforded the railroad by A1 Charging Cutoff Pilot Valves and transmitted emergency brake applications. (A break-in-two is sensed by the nearest A1 Charging Cutoff Pilot Valve and initiates an emergency brake application to all trailing units.)

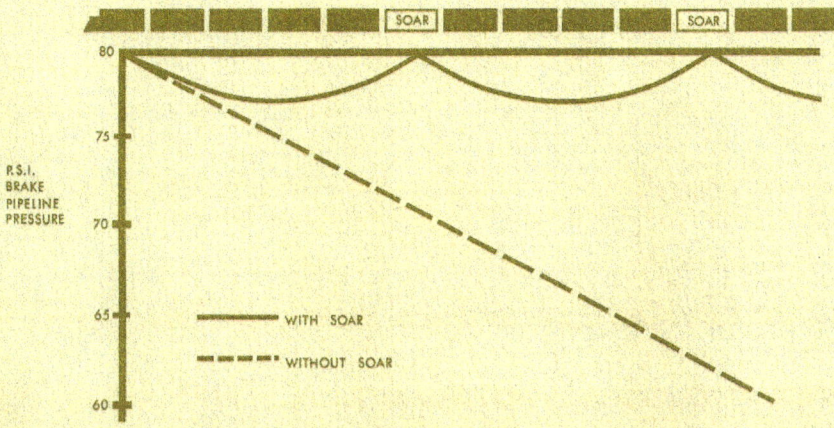

SOAR SYSTEM USED TO REESTABLISH BRAKE PRESSURE

EQUIPMENT DESCRIPTION

The SOAR system consists of a lead unit and one or two remote repeater cars.

A Lead unit is composed of seven major pieces of equipment:

1. Logic Cabinet
2. Power Supply
3. Power Cabinet
4. Control and Display Console
5. Pneumatic Interface
6. Frame Weldments
7. Radio with Converter and Antenna

A remote repeater car is composed of nine major pieces of equipment:

1. Logic Cabinet
2. Power Supply
3. Power Cabinet
4. Frame Weldments
5. Radio with Converter and Antenna
6. Air Brake Equipment Rack
7. Compressor
8. Generator with Batteries
9. Diesel Engine

Figure 241: SOAR (2) - Introduction

ADVANTAGES

SOAR in effect shortens the length of a train by boosting the brake pipe pressure at strategic points to allow the system to handle more cars. In addition, SOAR provides these advantages.

MORE RAPID AIR BRAKE RELEASE — The air brake passages of the pneumatic brake system can be charged more rapidly since the distance through which the air must travel is effectively decreased, i.e., charging is accomplished from two separate points. Release time is improved by a factor of five.

RAPID DRY CHARGE — The air brake equipment can be dry charged more rapidly since charging is accomplished from multiple points.

FASTER APPLICATION — The brake pipe is discharged to atmosphere at multiple points, resulting in faster application of train brakes.

LONGER TRAINS — Multiple sources of air with faster applications and releases allow more cars to be added to train as long as draw bar pull is not exceeded.

SAFER EMERGENCY STOPS — The pneumatic brake system is divided into two units which operate simultaneously to reduce the "run-in" of slack. Application time is improved by a factor of three.

COLD WEATHER OPERATION — Addition of air source overcomes leakage during cold weather operation allowing more cars in the train.

SHORTER SCHEDULING — The faster application, release and charging times allows reduction in terminal-to-terminal time.

BREAK-IN-TWO PROTECTION — To minimize slack "run-in" following a break-in-two near the lead unit, any emergency application sensed at the lead unit will be transmitted to the remote unit.

BRAKE PIPE CONTINUITY — To ensure brake pipe continuity, inter-locking is provided in the brake pipe charging control function. Brake pipe pressure at each remote must be above 15 psi before charging can occur by the lead locomotive. When dual remotes are used, the 15 psi for the second remote is supplied primarily from the first remote.

DISPLAYS — Status of the remote repeater cars major operating functions are continuously displayed at the engineer's position in the lead locomotive. Alarms immediately apprise the engineer of brake system malfunctions.

SECURITY — The use of an address or identity code ensures the integrity of the SOAR system and permits the use of the same radio frequency by multiple systems with no danger of interference to one another. A switch locking feature is provided to prevent address code changes by unauthorized persons.

LOGIC CABINETS — contain the air brake sensing circuits and control logic printed circuit cards.
POWER SUPPLIES — are DC to DC Converters followed by series regulators.
POWER CABINETS — provide for power interface and switching.
CONTROL and DISPLAY CONSOLE — is placed at engineer's position so that the status of the repeater car is clearly displayed and controls are accessible.
PNEUMATIC INTERFACE — provides the pneumatic-to-electric transducer for sensing at the lead unit.
FRAME WELDMENTS — houses and secures the Logic Cabinets and Power Supplies.
RADIO with CONVERTER AND ANTENNA — communication equipment is either UHF-FM or VHF-FM half-duplex carrier.
AIR BRAKE EQUIPMENT RACK — contains the electric-to-pneumatic operation and pneumatic-to-electric sensing transducers at the repeater car.
COMPRESSOR, GENERATOR with BATTERIES, DIESEL ENGINE — may be supplied by Radiation Incorporated or the customer.

Figure 242: SOAR (3) - Advantages

TRAIN OPERATION

The SOAR system has been designed to operate with one or two remote repeater cars. Selection of single or dual remote operation is made by a Control Switch on the Control Console. When operating conditions require independent movement of the lead consist, deactivation of the repeater cars is accomplished by operation of the Control Switch in the "Off" position.

The lead unit air braking is controlled using standard air brake controls from the engineman's position. When synchronous control is employed, the remote repeater car is controlled automatically. The air brake pressure is sensed at the lead unit whenever a change occurs in the air brake system. This change is encoded into a digital command by the logic equipment and transmitted by radio to the repeater car. The remote repeater car decodes the command which results in the application of a corrective signal to the air brake control circuits.

Status and Alarms of the SOAR repeater cars is continuously displayed on the Control Console. Status displays include selected information such as continuity, charging and application. Alarms are displayed when changes in the brake system indicate low main reservoir and low brake pipe pressure.

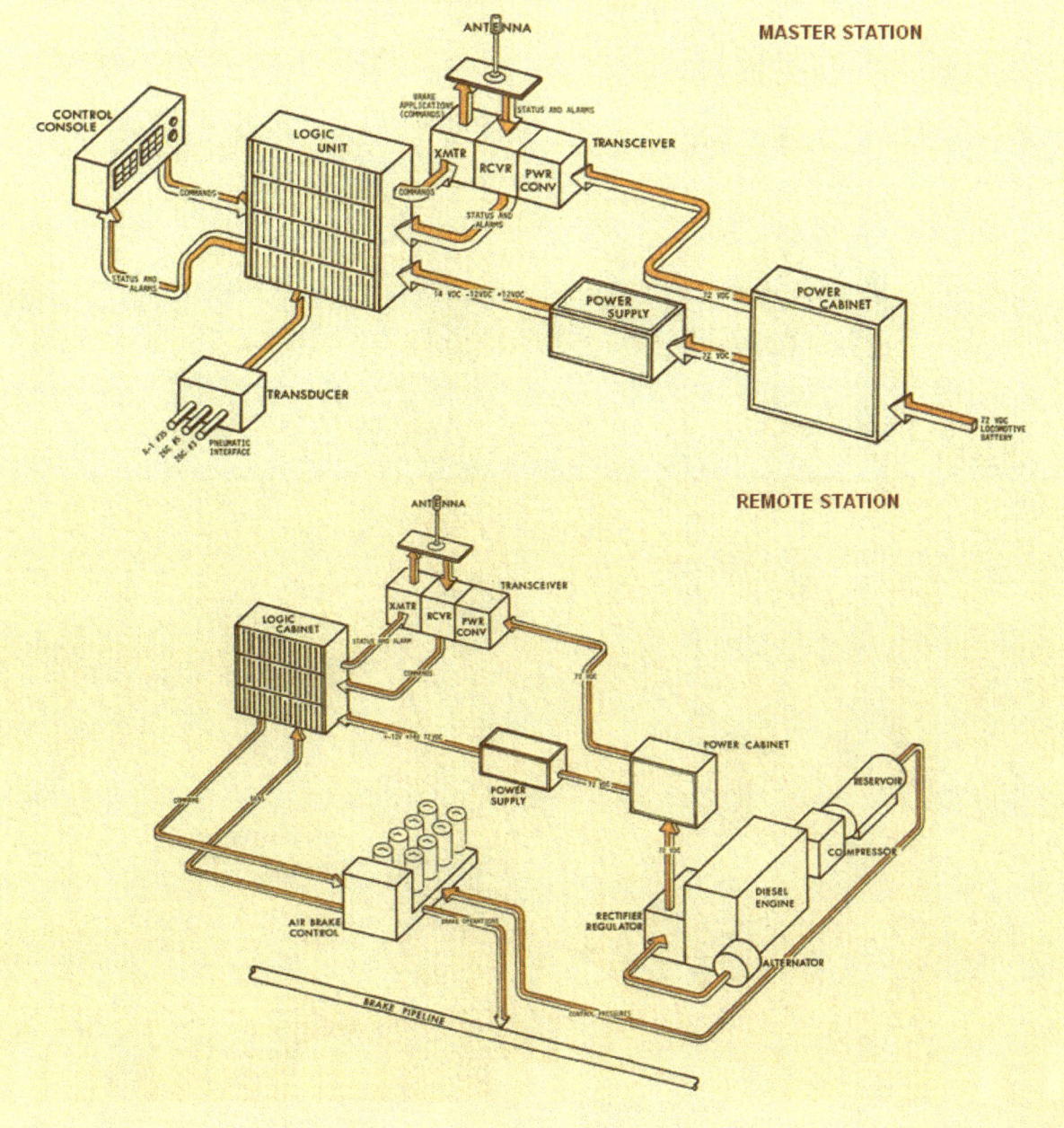

Figure 243: SOAR (4) - Train operation

LOGIC OPERATION

The principle of the SOAR system is demonstrated by the flow diagram of the lead unit.

Inputs to the transmit circuitry originate from the Transducer Pressure Switches and Control Console. When a control change is indicated by the inputs, a command signal is immediately transmitted to the repeater car where it is verified and the command executed. Whenever a change in the status of the control functions or alarms is detected, the repeater car will transmit all the status and alarm points to the lead locomotive. The status is then compared with the original transmitted signals. A correct comparison will cause the transmission of data to the repeater car every 20 seconds and immediately receive a reply as a continuity check. An incorrect comparison will cause the lead unit to transmit the command signals at 4 to 6 second intervals until a comparison is obtained.

If the lead unit receives a reply to each continuity check and status change, from each remote, a state of "continuity" exists and the console 'CONT' display will be illuminated. If the lead fails to receive a valid reply to any transmission from either remote within 2 seconds, a continuity search mode will be declared. This will extinguish the console "CONT" display.

A state of continuity will exist at the remote if transmissions are received regularly. Should the remote fail to receive a valid message for a period of 30 seconds, a state of "no continuity" will be established. The remote unit will then respond to sensed brake applications and losses of continuity as follows:

- Sensed first service brake application <u>with</u> continuity will cause the remote to become passive.

- Sensed first service brake application <u>with no</u> continuity will cause the remote to become passive.

- Loss of continuity with brakes applied will cause the remote to become passive.

- Loss of continuity with brakes released and no sensed brake application will cause the remote to remain in operation for a maximum period of 0 to 40 minutes (adjustable). After that period, the remote will become passive.

In marginal communication areas or in tunnels, a Message Repeater can be used. It will temporarily store then retransmit received SOAR digital messages.

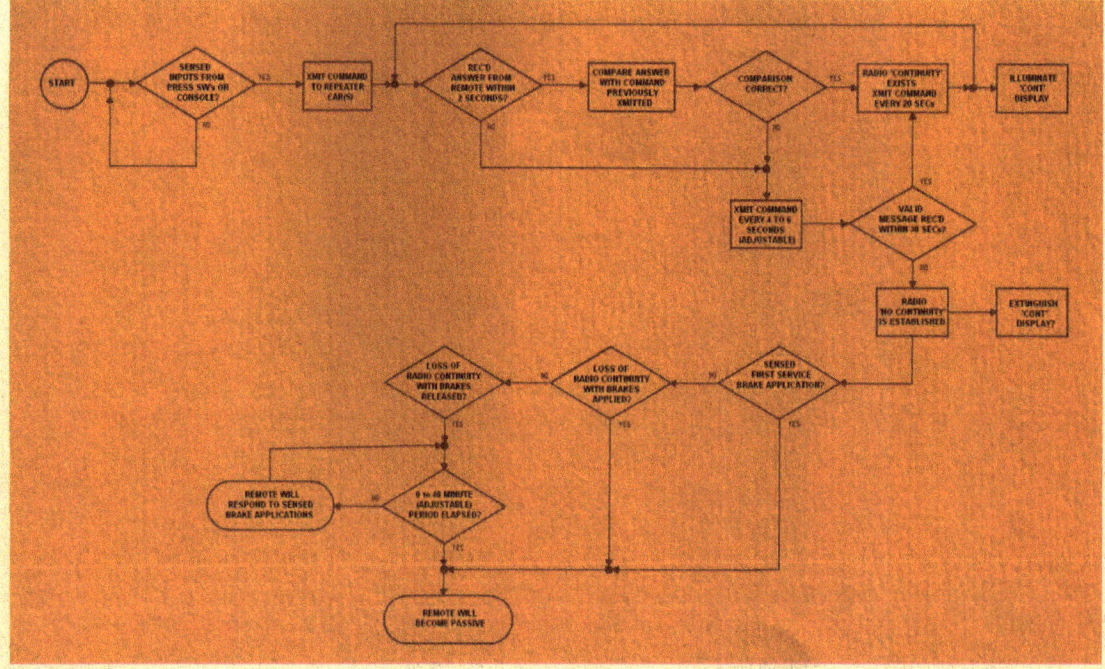

Figure 244: SOAR (5) - Logic operation

BASIC SPECIFICATIONS

ELECTRICAL

Power Requirements .... 72 vdc nominal, ungrounded, 5 Amperes

Operating Voltage .... 0 vdc, -12 vdc, +12 vdc (Radiation Supplied)

Type of Transmission .... FM Half-Duplex Carrier Frequency Shift Keyed

Transmitter Frequency .... 450 MHz or 160 MHz Band

Display .... Console backlighted indicators

MECHANICAL

Dimensions:

Lead Unit and Remote Unit Logic Frame .... 25.6-inches high, 23.2-inches wide, 10.3-inches deep

Power Supply Frame .... 9-inches high, 25-inches wide, 12-inches deep

Control Console .... 5.5-inches high, 10-inches wide, 9-inches deep

MOUNTING

Logic Cabinets .... Angle Iron Frames

Control Console .... Quick Disconnect

Power Supply Converter .... Angle Iron Frames

ENVIRONMENTAL

Temperature Range .... -30°C to +65°C

Relative Humidity .... Above 96 per cent

RADIATION INCORPORATED
SUBSIDIARY OF HARRIS-INTERTYPE CORPORATION

CONTROL DIVISION

P.O. Box 430   Melbourne, Florida 32901   (305) 727-5600

Figure 245: SOAR (6) - Basic specifications

THE ADAPTIVE AIR BRAKE

Milt Deno had accrued comprehensive experience of rugged-country railroading as both a diesel inspector and master mechanic with CP Rail on their mountainous operating divisions throughout Alberta and British Colombia. He knew that trains were becoming longer and heavier. The performance of the new car control valve types greatly exceeded that of older versions but could be less forgiving in some respects. For instance, engineers could be blindsided by the 49 psi (338 kPa) value below which ABD and ABDW valves would not go into Emergency. Some train wrecks had resulted from the locomotive engineer permitting the train to 'run short of air' during sustained downhill braking and thus being unable to control it (see Footnote 57 and 58).

As the divisional boss of locomotive engineers, and after attending and investigating several train runaway incidents, Milt had originally conceived the idea of a train Air Brake Control System that would adapt itself to the actions of the locomotive engineman and prevent deficient air brake manipulation from resulting in an insufficient supply of car brake cylinder pressure.

When he eventually found himself in the balmy climes of Florida and in the innovative embrace of electronics giant, Harris Controls, Deno saw the opportunity to fulfil this goal. It did not harm his resolve that his boss, John Boucher, encouraged he and his colleagues liberally to pursue patents using Harris's resources.

Deno might not, though, have foreseen the rapidity with which the technology of Electronically-Controlled Pneumatic braking (ECP) would be taken up by the railroads. And he may not have anticipated the degree to which ECP would revolutionise—even simplify—train handling, and the degree to which it would eventually reduce the judgement chore for enginemen regarding the exercise of anticipatory skills in air brake manipulation. And did he appreciate that ECP braking might render his very Adaptive Air Brake concept redundant? Doubtless he did not.

Herein, then, is offered as an exposé of an obscure but clever concept that held great promise for improving safety in the operation of long 'conventional' freight trains but that was utterly overtaken by the enhanced control of air brake management afforded by LOCOTROL.

※

The section describes the Adaptive Air Brake Control System—invented and patented by Deno, Gene Smith, and Dale Delaruelle—and its functionality. Anyone who has looked at patent applications will know that their descriptions of complex systems are inevitably generic, verbose, and often bewilderingly legalistic. Where practicable, the narrative here—as used by the patent attorneys in the drafting of these United States Patents filed in 1988 and 1989—has been adapted by the author to render it more readable and, hopefully, more comprehensible. The author has added explanatory comment where he thinks it might be useful, and no technical detail has been modified. Readers will most benefit who are among those described so characteristically by the patent application narrative as 'skilled in the art [the discipline of railway air braking]'.

Patent Title: *THE ADAPTIVE AIR BRAKE CONTROL SYSTEM*
This invention relates in general to air brake systems for railroad trains, being particularly intended as a control mechanism for ensuring effective braking action in accordance with prescribed operational characteristics of the brake system. Please refer to the diagrams throughout.

Abstract
An Adaptive Air Brake Control System monitors various brake pipe parameters (such as fluid path volume and air flow rate) and controllably modifies action taken by the enginemen or performs emergency control

of the brakes, to continuously enable the braking system to adapt itself to dynamic operating conditions and anomalies in the integrity of the fluid path. In accordance with a pressure reduction modification mechanism [the engineer's brake valve or brake controller], the application of a pressure reduction to the equalising reservoir is precisely controlled by considering the actual state of the brake pipe, to ensure the requested brake application is effected as intended. The control mechanism also monitors the integrity of the fluid flow path of the brake pipe [on both locomotive/s and train] so that the engineman may be alerted, and a prescribed train safety measure effected in the event of a potentially hazardous anomaly [i.e. dangerously low auxiliary reservoir pressure] in the link. It also provides the engineer with a precise indication that the train's brakes have been fully released or applied.

Background to the invention

One of the most critical aspects of the operational control of railway vehicles—particularly freight trains—is ensuring the predictably successful operation of the air brake system. This system is subject to a variety of dynamic effects, not only because of the controlled application and release of brake pipe pressure in response to operational requirements, but because of the occurrence of unpredictable anomalies in the integrity of the brake system itself: for example, a deficiency in brake pipe recharge at the time a brake application is desired.

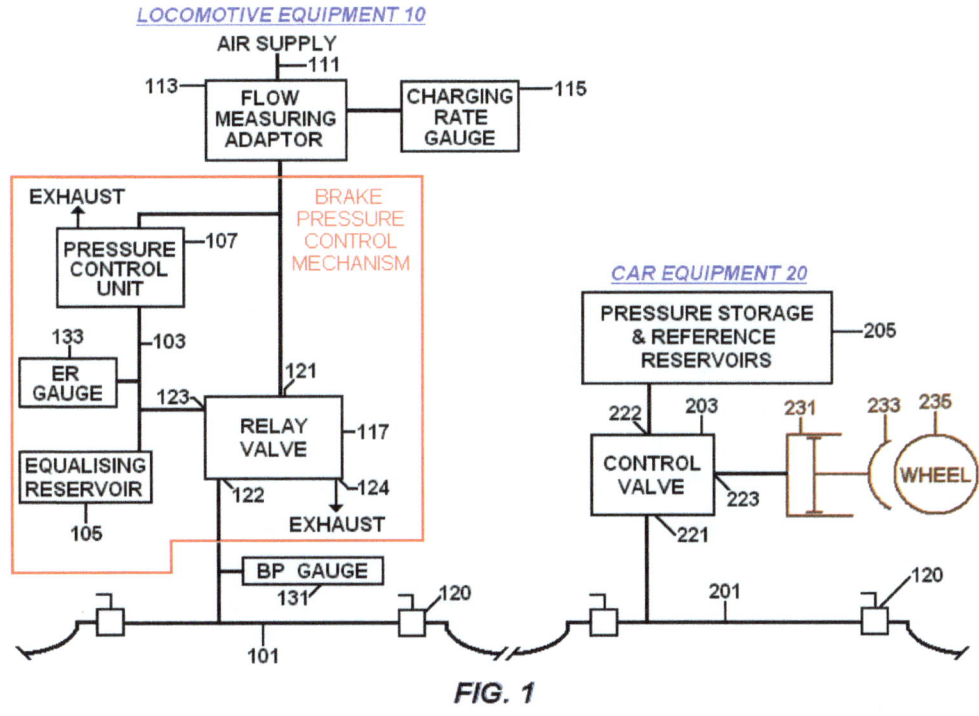

Figure 246: Adaptive Air Brake (1). FIG 1: A diagrammatic illustration of a conventional air brake control system employed on a freight train.

Referring to Figure 1, which diagrammatically illustrates a typical air brake system employed on a freight train, the application and release of braking action is generally controlled by the engineman within the grouping, *Locomotive Equipment 10*. This assemblage contains an Air Brake Control System including a controllably-pressurised brake pipe **101**, which is coupled (via **120**, one of a series of cut-out valves) to the train brake pipe **201**, through which air brake pressure is supplied for each of the cars (per grouping *Car Equipment 20*) of the train. The Air Brake Control System also includes an air pressure supply **111** [feed from main reservoir] to charge the locomotive and train brake pipe, for controlling the operation of the braking mechanisms.

Within air supply input link **111** are—for measuring the air supply charging rate into the brake pipe—an air flow measuring device **113** [flow adaptor] and an associated differential pressure gauge **115** [flow meter]. Air supply link **111** is coupled first to an input port **121** of a relay valve **117** and second, a bidirectional port **122** that intersects the brake pipe **101**. The relay valve also includes a third port **123** that is coupled through an air pressure control link **103** to an equalising reservoir **105** and a pressure control unit **107** [regulating valve] through which the link **103** and equalising reservoir **105** are controllably charged and discharged during a braking operation. A fourth port **124** of relay valve **117** is controllably vented to the atmosphere as an exhaust port. Coupled with brake pipe **101** and air pressure control link **103** are respective pressure measuring gauges **131** and **133** by which the equalising reservoir **103** and brake pipe **101** pressures are monitored.

The brake control unit within a typical car (*Car Equipment 20*) of the train includes an operating valve **203** [control/triple/distributor], a first port **221** coupled to the train brake pipe **201**, a second port **222** coupled to pressure storage and reference reservoirs **205** [auxiliary and control reservoirs], and a third port **223** which is coupled to car brake cylinder **231** that controls the movement of the brake shoes **233** relative to the wheels **235** of the car.

In operation, the cut-out valve **120** [headstock angle cock], through which the locomotive brake pipe **101** and successive segments of the train brake pipe **201** are coupled in serial fluid communication, is assumed to be fully open, so that there will be a continuous fluid path between the *Locomotive Equipment 10* and the *Car Equipment 20* of the train.

The brake system is initially pressurised by the operation of pressure control unit **107** [brake valve regulating valve], which fully charges equalising reservoir **105** via supply line **103**. Relay valve **117** then functions to connect its ports **121** and **122** so that the regulated pressure is supplied to the locomotive brake pipe **101** and thereby to the [train brake pipe] **201** to charge the fluid path to its maximum regulated pressure as established by the pressure in the equalising reservoir **105**.

The pressure within fluid path **101/201** is determined to have reached the maximum—as established by the fully-charged equalising reservoir—when the pressure at port **122** equals that at port **123**. Through control valve **203** on each of the cars (*20*) in the train, combined pressure storage and reference reservoirs **205** are fully charged, thereby establishing a reference pressure for maximum withdrawal of the piston of each brake cylinder **231** and thus complete release of the brakes on each car.

When the engineman desires to apply brakes to the wheels of the [trailing] cars, pressure control unit **107** is manually-operated—typically via a handle-operated pneumatic control valve which is coupled to air pressure control link **103**. This will cause a partial pressure reduction in air pressure control link **103** and thereby an equal reduction in the pressure within equalising reservoir **105**. The reduced equalising reservoir pressure is sensed by relay valve **117** (via port **123**) which, in turn, causes its bi-directional port **122** to be connected to exhaust port **124**.

Locomotive brake pipe **101** is thereby vented to atmosphere until the brake pipe pressure matches the reduced pressure in equalising reservoir **105**. Because of the considerably larger volume of the fluid path through the brake pipe and trainline **101/201**, the length of time required for air pressure within **101/201** to reduce to that in the equalising reservoir **105** is significantly longer than the time required to achieve a reduction in air pressure in the equalising reservoir **105**, which occurs rapidly in response to the operation of pressure control unit **107**.

As the pressure in locomotive brake pipe **101** and train air line **201** drops, the respective control valves **203** in each of the cars (*20*) sense that pressure reduction by comparing train air line **201** pressure with that in the combined storage and reference reservoir **205** and produce a corresponding increase in the pressure applied to the brake cylinders **231**. This results in an application of braking to the car wheels in proportion to the sensed pressure reduction in train air line **201**. Further pressure reductions made by the engineman to equalising reservoir **105** produce corresponding pressure reductions in train air line **201** and, thereby, additional braking effort by the braking mechanisms at each of the cars (*20*). In other words, for the intended operation of the brake system, the braking effort applied to each of the cars (*20*) is proportional to the reduction in pressure in the equalising reservoir **105** within the *Locomotive Equipment 10*.

A brake release is accomplished by the engineman operating pressure control unit **107** [per the Automatic brake valve handle] towards the Release position to effect restoration of equalising reservoir pressure to its fully-charged state, as previously described. With equalising reservoir **105** recharged, a pressure differential once more exists (this time from an increase rather than a reduction) in line **103** between ports **122** and **123**, and this is sensed by relay valve **117** which connects air supply link **111** to locomotive brake pipe **101** so as to recharge (thereby increasing the pressure in) both lines **101** and **102**. This pressure increase is sensed by the car control valves **203** in each of the cars (*20*) to connect brake cylinders to exhaust and cause the brakes to be released.

During normal operation, the application and release of brakes is controlled in accordance with the above-described sequence of events: however, there may be circumstances—dictated either by action taken by the engineman or by other unpredictable events—that create the potential for unsafe operation of the braking system. One of these conditions occurs from the engineman applying braking subsequent to the release of a previous brake application, but prior to the system having been fully recharged and pressure within the brake pipe having stabilised.

Specifically, when the engineman initiates a release of the train brakes after a previous application, air for recharging the brake system is supplied via input link **111** and relay valve **117** as described above. During this brake release and recharge interval, the pressure within brake pipe **101**—being supplied via relay valve **117**—rises slowly compared with the rapid rate-of-charge of equalising reservoir **105**, which has been recharged by the operation of regulating valve **107**. If, prior to the brake pipe becoming fully recharged, the engineman initiates a new brake application (reducing equalising reservoir **105** pressure by activation of the regulating valve **107** through manipulation of the brake valve handle), the pressure differential between the partially-charged train brake pipe **201** and pressure storage reservoirs **205** in each of cars (*20*) will be different to that intended by the pressure reduction newly-applied to equalising reservoir **105**. Consequently, each of the operating valves **203** will sense a smaller pressure differential between ports **221** and **222** than the reduction applied to equalising reservoir **105**, so that the braking effort imparted to brake cylinders **231** in each of the cars will be less than what the engineman has requested (and intended) via brake valve handle movement. If not immediately comprehended by the engineman, this 'reduced-effort' brake application can place the train in a potentially unsafe condition.

Still, even when they might recognise the insufficiency of the new braking application, the engineman often attempts to remedy the problem by a further incremental reduction in the pressure in equalising reservoir **105**. Again, however, the application of only a partial braking effort as described above will take place, so that there may still be inadequate braking action applied by the cars of the train. Simply put, if the engineman tries to make up for insufficient braking of one pressure reduction request in a piecemeal fashion, and fails each time, it is possible that continuing efforts in this process will be unsuccessful and the originally-intended braking effort will never be accomplished. An experienced engineman (in terms of the train and conditions the train currently encounters), upon realising the unsafe condition, may apply a severe pressure reduction to make up for the original lack of braking response to the pressure reduction of an incompletely-charged system.

Yet, the action taken by the engineman is only a guess—even though possibly an educated one—as to whether a further braking effort will successfully brake the train, and this cannot be equated with safe train operation. Another circumstance in which a 'guesstimate' braking control procedure is often used is in the course of determining when the train brake pipe **101/202** is fully charged, so that the brakes are fully released, and the train may safely proceed. For this purpose, a practice commonly employed by enginemen is to interpret the sound of air passing through the brake relay valve to determine when the train brake pipe is fully charged, and the brakes therefore fully released.[106] In effect, this practice constitutes a 'seat of the pants' procedure that is not necessarily reliable.

As pointed out above, in addition to potential safety hazards that might arise from the application of insufficient brake control pressure reductions by the engineman (possibly due to an unexpected dynamic event—it should be noted—as much as bad judgement by an engineman), the integrity of the train brake pipe **101/201** is subject to unforeseeable changes (such as an end-of-car isolating valve/angle cock being accidentally hit or deliberately tampered with, resulting in a change in brake pipe continuity) that, if undetected, could permit the brakes to be applied normally to one part of the train but not applied or only partially applied to the remainder.

The need for qualifying the integrity of the brake pipe **101/201** is especially important when the locomotive is uncoupled and removed from the train at a destination, or for switching purposes. During this procedure, the engineman will normally initiate a brake application on the train to lock the wheels of the cars. After applying this input, the engineman will wait for some period until the brake pipe pressure reduction has propagated down the entire train (air line **201**) and the air exhaust port **124** of relay valve **117** has ceased venting brake pipe pressure to atmosphere. Once the engineman is satisfied that the air pressure has been completely reduced as required by listening to what he considers to be the last venting of air from the exhaust port of relay valve **117**—and has settled—a crew member may then proceed to close the brake pipe isolating valve **120** and disconnect from the train.

If the pressure in the train brake pipe **101/201** had not stabilised prior to the crew member disconnecting the locomotive from the train, an undesired release of the brakes may be caused by a reaction within the brake pipe to the abrupt obstruction to its exhausting air flow. In addition, if there is an obstruction in the brake pipe **101/201**, preventing proper exhaust of brake pipe pressure, the brakes on some or all the cars of the train may not have applied. Again, the engineman's reliance upon the audible cue as a control criterion for the operation of a freight train is a far-from-safe railway operational procedure.

Brief system description

In accordance with this invention, the above-described shortcomings of a conventional brake control system are obviated by a new *Air Brake Control System* that monitors numerous brake pipe parameters, including fluid path volume and air flow rate, in addition to conditions that are normally indicated as gauge readings to the engineman. With this data, the system controllably modifies action taken by the engineman or performs emergency control of the brakes to continuously enable the braking system to adapt itself to dynamic operating conditions and anomalies in the integrity of the fluid path.

First feature:
A control mechanism is provided for precisely controlling the application of a pressure reduction in the equalising reservoir, and thereby the pressure in the brake pipe, to ensure that the requested brake application is effected as intended.

---

106 Note that this comment refers specifically to the American environment where brake pipe recharge is read from a *flow indicator* gauge only, and is not 'heard' in quite the same vigorous fashion as it is in Australia and New Zealand where locomotive engineers have the additional benefit of a *flow meter* that provides more robust visual and audible cues.

To this end, the control mechanism—which it is intended would be implemented by way of a dedicated microprocessor and attendant signal interface components—in addition to being coupled to monitor the engineman's requested pressure reduction, also monitors the brake pipe to determine whether it is currently undergoing a change in air pressure (as would be the case during a recharge of the brake pipe after the release of a previous brake application). As the pressure within the equalising reservoir drops in response to the engineman's pressure reduction, the charging rate of the air flowing into the brake pipe through the relay valve will decrease as the reducing pressure value within the equalising reservoir and that slowly rising within the brake pipe approach each other. When the flow rate into the brake pipe becomes negligible (effectively equal to zero, indicating that brake pipe pressure is now the same as that within the equalising reservoir) the pressure value within the equalising reservoir at that time is stored as data. Using this stored value, the requested equalising reservoir pressure change is automatically modified (increased) by an amount equal to the difference between the maximum pressure of the equalising reservoir and the stored equalising reservoir pressure value. Therefore, the pressure in the brake pipe is accurately reduced by an amount that effectively achieves the engineman's requested pressure reduction.

Similarly, if the requested equalising reservoir pressure reduction is less than the brake pipe pressure (so that the brake pipe continues to draw a charging supply), the control mechanism imparts an additional pressure reduction to the equalising reservoir so that the charging air flow into the brake pipe will be forced to drop to zero. When this occurs, the pressure value within the equalising reservoir is again stored, and using this value, the requested equalising reservoir pressure change is modified (increased) by an amount equal to the difference between the maximum pressure of the equalising reservoir and the newly-stored equalising reservoir pressure value. Therefore, the pressure in the brake pipe is again accurately reduced by an amount that effectively achieves the engineman's requested pressure reduction.

Second feature:

The processor-based control mechanism monitors the integrity of the fluid flow path of the brake pipe so that the engineman may be alerted and a prescribed train safety measure be effected in the event of a potentially hazardous anomaly in the link (for example, inadvertent angle cock closure). For this purpose, whenever the train is newly-configured—thereby defining a new length of brake pipe—the volume of the newly-defined fluid flow path is calculated, and its value stored as a reference. Based upon continuously-monitored brake system parameter data, this volume is repeatedly checked. Should an anomaly occur in the physical configuration of the system (for example, the accidental closure of an angle cock) that changes the measured volume of the fluid flow path (that is to say, the locomotive now 'sees' a shorter brake pipe) and that if undetected might prevent the train's braking system from operating effectively, a prescribed pressure reduction will be applied to the equalising reservoir, so that the pressure in the fluid flow path is reduced by an amount sufficient to brake the train to a stop.

Third feature:

The control mechanism can provide the engineman with a precise indication that the train brakes have been fully released in accordance with a brake release action (initiated, for example, prior to the application of a locomotive tractive force). For this purpose, as in the brake pipe integrity checking scheme described above, whenever a brake application or release action is initiated, the volume of the fluid flow path through the brake pipe is measured at a subsequent brake release. Based upon this measured volume—which is indicative of the length of the fluid flow path (brake pipe)—a 'brakes released' signal is generated following the lapse of a period after the initiation of the brake release action. This period corresponds to that required to release all brakes

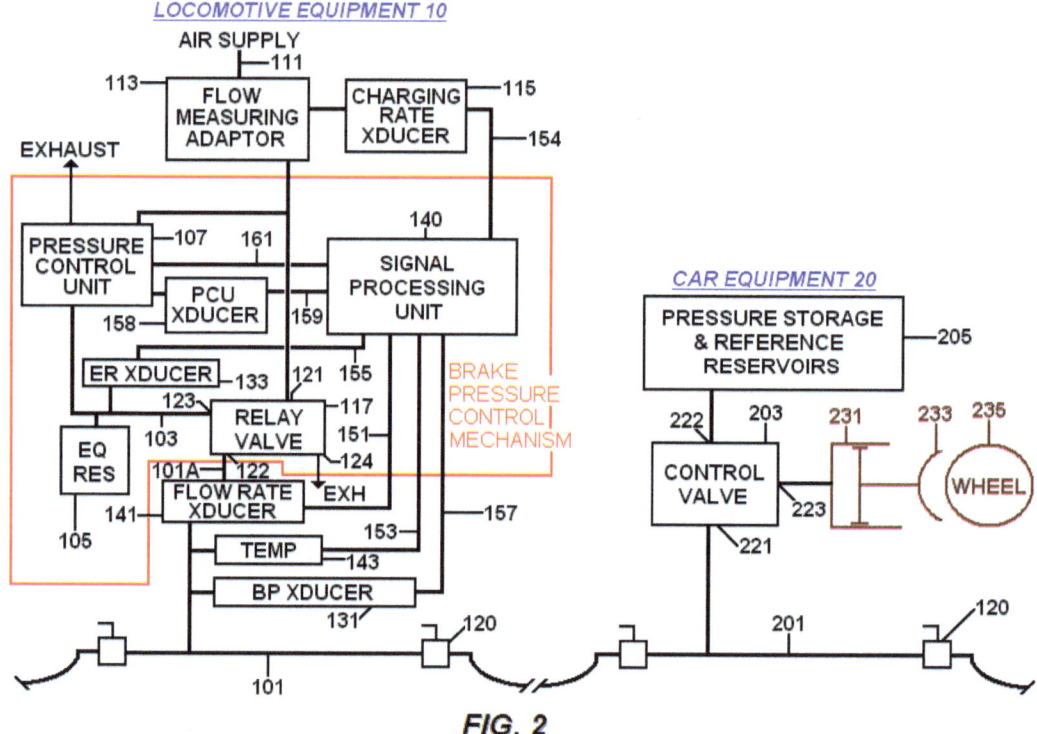

Figure 247: Adaptive Air Brake (2). FIG 2: How the conventional air brake control system is modified to provide the added adaptive control features.

on this train and is determined in accordance with the measured volume of the brake pipe. Until the 'brakes released' signal is generated, an 'inhibit' signal prevents the application of a tractive force to the locomotive.

Fourth feature:

Similarly to the above-described third feature, the control mechanism can provide the engineman with a precise indication that the fluid path has been properly discharged in response to a Full-Service brake application request (initiated, for example, prior to separating the locomotive from the remainder of the train). Again, in response to the engineman's requested action (this being the Full-Service equalising reservoir pressure reduction) the volume of the brake pipe is measured. As the brake pipe is vented through the relay valve exhaust port, brake pipe pressure, temperature and exhaust flow rate are monitored. If the measured volume matches that of the train when most recently configured, then when the pressure and the rate of exhaust flow drop to low and stable values (indicating that the brake pipe is effectively depressurised) an output signal is generated indicating that the brakes on the train are effectively fully-applied and that it is thereby safe to disconnect the locomotive. Until, this output signal is generated, the application of a tractive force to the locomotive is inhibited.

Brief description of the drawings

FIG 1 diagrammatically illustrates a typical air brake system employed by a railway freight train: the application and release of braking action being generally controlled by the engineman within the locomotive. This figure has been described previously. FIG 2 shows a modification of the Air Brake Control System depicted in FIG 1 that incorporates the control mechanism proposed in this Adaptive Air Brake Control System. FIGs 3–7 are flow charts of the control mechanism depicted in FIG 2.

Detailed system description

Before describing this Adaptive Air Brake Control System in detail, it should be noted that the invention resides primarily in a novel structural combination of conventional brake pipe parameter sensing circuits and signal processing components and not in their detailed configurations. Accordingly, the structure, control, and arrangement of these conventional circuits and components have been illustrated in the drawings by standard block diagrams, which show only those specific details that are pertinent to the invention so as not to complicate the disclosure with structural details that will be clear to those skilled in the discipline of railway air braking, having the benefit of this description. Thus, the block diagram illustrations of the figures do not necessarily represent the mechanical structural arrangement of the prototypical system but are primarily intended to illustrate its major structural components in a convenient functional grouping whereby it may be more readily understood.

FIG 2 effectively replicates FIG 1 and depicts a modification of the Air Brake Control System previously described. It shows how the conventional brake control system is modified to provide the added adaptive control features.

To enable a brake system control signal processing unit to monitor brake system parameters, the two gauges depicted in FIG 1 are replaced by corresponding sensing transducers. Also depicted is the addition of a bi-directional air flow transducer **141** and an air flow temperature transducer **143** coupled in a brake pipe connecting section **101A** between bi-directional port **122** of relay valve **117** and brake pipe **101**. Bi-directional air flow transducer **141** measures the flow of air (in either direction) between port **122** of relay valve **117** and the locomotive brake pipe **101** (and train brake pipe **201** connected thereto) and produces an output signal representing measured air flow rate. The output of air flow transducer **141** is conveyed over link **151** to a brake pipe parameter signal processing unit **140**. Similarly, temperature sensor **143** monitors the temperature of brake pipe air and provides a representative signal over link **153** to signal processing unit **140**. Signal processing unit **140** may be a conventional computer-control unit including microprocessor, memory and signal interface circuitry.

Additional inputs to signal processing unit **140** are equalising reservoir **105** pressure (as supplied by a signal from equalising reservoir transducer **133** over link **155**), brake pipe pressure (as monitored by brake pipe transducer **131** supplied over link **157**), and the pressure via transducer **158** coupled in line **159** at the output of pressure control unit **107** (this corresponding to the value of an equalising reservoir pressure requested by the engineman). Signal processing unit **140** is also coupled to receive additional inputs (such as by way of an attendant keyboard) representative of additional operational parameters, as is further described.

Signal processing unit **140** monitors brake pipe fluid pressure and flow parameters supplied by the outputs of the respective transducers described above, and produces control signals for application to a display and various train motion control devices (such as an equalising reservoir pressure reduction modification signal over link **161** to pressure control device **107**, to controllably modify a pressure reduction initiated by the engineman) to ensure safe operation of the train for a number of brake application and release conditions, including those initiated by the engineman, as is further described.

As previously pointed out, a potentially dangerous condition in train operation occurs when the engineman makes a brake application subsequent to a preceding brake application and release, but prior to the complete recharge and stabilisation of the pressure within the brake pipe system. Briefly readdressing the problem, when the engineman initiates a release of the train brakes after a previous brake application, air for charging the brake system is input through the air supply link **111** and relay valve **117**, to recharge the brake pipe. During this charging interval, the pressure within the brake pipe rises at a rate that is considerably slower than the rapid rate at which equalising reservoir **105** is charged. When the pressure within the brake pipe equals the pressure of link **103**, the relay valve interrupts the connection of the air supply link **111** to the brake pipe, since the train brake pipe **201** is now fully charged. If—prior to the brake pipe becoming fully recharged—the engineman operates pressure control unit **107** to reduce the pressure in equalising reservoir **105** and initiate a new brake application, the pressure differential between the partially-charged train brake pipe **201** and the reservoirs **205** on the cars (*20*) will be at variance with the pressure differential intended by the pressure reduction newly-applied to equalising reservoir **105**. Consequently, each of the operating valves **203** will sense a smaller pressure differential between ports **221** and **222** than that applied to equalising reservoir **105**, so that the braking effort imparted to the train will be less than what the engineman has intended. If not immediately recognised by the engineman, this 'reduced-effort' brake application can create a potentially unsafe operating condition.

In accordance with the control mechanism of this invention, any pressure change (reduction) in the equalising reservoir is controllably augmented by an amount that corrects for a shortage of recharge in the brake pipe at the time of the new brake application, to ensure that the brake application intended by the engineman is achieved. For this purpose, the control mechanism monitors the brake pipe pressure and charging rate to determine whether it is currently undergoing a change in air pressure, as would be the case during a recharge of the brake pipe after the release of a previous brake application. As the pressure within the equalising reservoir drops in response to the engineman's brake control actions, the charging flow rate into the brake pipe through the relay valve will decrease as the values of the now-reducing pressure within the equalising reservoir and the slowly-rising pressure within the brake pipe approach each other. When the flow rate into the brake pipe becomes negligible (effectively equal to zero, indicating that equalising reservoir pressure now equals that within the brake pipe), the value of the pressure within the equalising reservoir at that time is stored. Using this stored value, the equalising reservoir pressure change requested by the engineman (by the operation of pressure control valve **107**) is modified by an amount equal to the difference between the maximum charging pressure of the equalising reservoir and the stored equalising reservoir pressure value. Consequently, the brake pipe pressure is accurately reduced by an amount that effectively achieves the engineman's requested reduction.

The sequence of steps executed by signal processor **140** to implement this control mechanism are set forth in the processing flow diagram of FIGs 3–5.

Assuming the train—of some defined length—is currently subject to a brake application, at this point the engineman starts the process by initiating a brake release <START> (step **301**) by operating pressure control device **107** to cause equalising reservoir **105** to be charged, thereby advising signal processing unit **140** that a recharge has been requested. Ignoring for the present the <START RELEASE TIMER> (step **303**) and <RELEASE TIMER = FULL RELEASE?> (step **311**)—to be described below—the process proceeds to step **305** <MEASURE BRAKE PIPE PRESSURE, FLOW RATE TEMP & MAX ER PRESS>, wherein signal processing unit **140** measures brake pipe pressure, temperature, and flow rate. By monitoring these variables, signal processing unit **140** calculates, in step **307** <CALCULATE TOTAL AIR VOLUME (STP)>, the total quantity or volume of air that is supplied through the air supply link **111** and relay valve **117** during charging of the brake pipe. Brake pipe volume is calculated by signal processing unit **140** based on measured flow rate, temperature, and pressure of the air within the fluid flow path using standard gas law equations.

A stabilised and fully charged condition—a 'YES' condition of <IS FLOW RATE = STABLE & LOW?> (step **315**)—occurs when the charging air flow rate becomes negligible or is constant and below a prescribed threshold (for example, a flow rate around 5 cfm). Until the brake pipe becomes fully charged, however, the process proceeds to step **317** <OPERATOR REQUEST FOR BRAKE APP?>, where an inquiry is made as to whether the engineman has made a new brake application; namely whether the requested equalising reservoir pressure reduction requires modification to bring the brake pipe pressure down to a level that will cause the operating valves on each car to supply the intended and required brake cylinder air pressure. If there is still a significant air flow through the brake pipe link **101** (as monitored by transducer **141**, step **315**) and the engineman has initiated a brake reduction (step **317**), the process proceeds to step **401** <MEASURE FLOW RATE> in FIG 4.

Next, in step **403** <IS ER PRESSURE DECREASING?>, the output of equalising reservoir pressure transducer **133** is monitored to determine whether the equalising reservoir pressure is still decreasing or has reached the pressure requested by the engineman. If the equalising reservoir pressure is not decreasing, the process proceeds to FIG 5, step **501** <STORE EQUALISING RESERVOIR PRESSURE = OPERATOR REQUEST APP>, at which point the equalising reservoir pressure value is stored. For the purposes of the present description, however—given the operating condition in which the engineman has requested a further brake pipe reduction—then the pressure within equalising reservoir **105** will still be

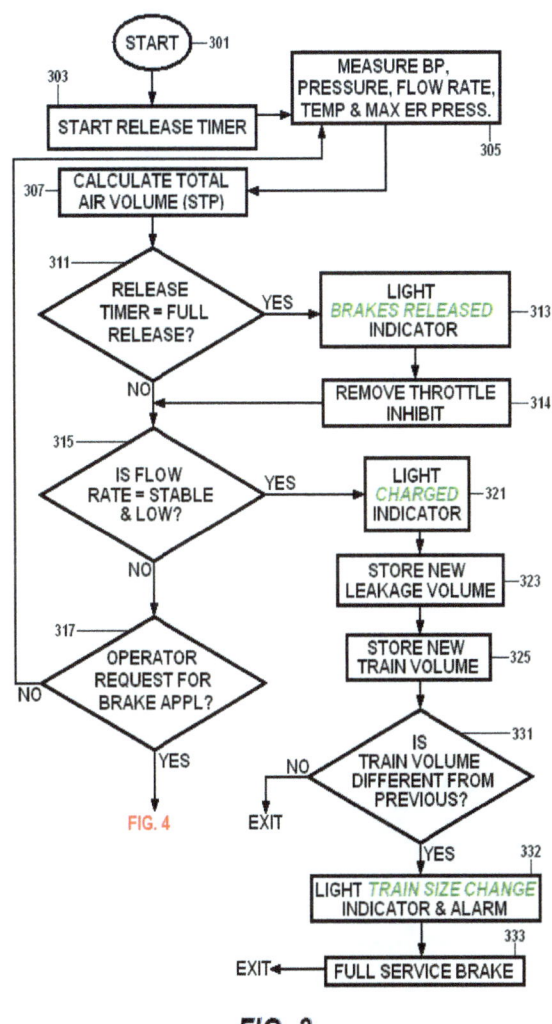

Figure 248: Adaptive Air Brake (3). FIG 3: Signal processing unit 140 monitors BP pressure and flow parameters as measured by transducers, and produces control signals for application to a display and various train motion control devices (such as an ER pressure reduction modification signal to [the engineman's brake controller], to controllably modify a pressure reduction initiated by the engineman, to ensure safe operation of the train for a number of brake application and release conditions, including those initiated by the engineman.

decreasing, so that the process proceeds to step **404** <FLOW RATE = 0?>. Until the flow rate drops to the above-referenced negligible value, the process loops back to step **401**. Eventually, however, as the pressure within brake pipe **101** continues to slowly increase and the pressure in the equalising reservoir continues to reduce, the flow rate will diminish until it reaches a value (as measured by flow transducer **141**) that causes the process to proceed to step **405** <Dpsi = MAX ER minus CURRENT ER>, in which a differential pressure value (Dpsi), equal to the maximum (fully charged) equalising reservoir pressure minus the output of equalising pressure transducer **133** measured at the time at which the brake pipe flow rate effectively drops to zero (brake pipe pressure matches that of the equalising reservoir), is stored.

Next, in step **406** <OUTPUT ER PRESSURE = OPERATOR REQUESTED PRESSURE minus Dpsi>, signal processing unit **140** governs pressure control unit **107** and equalising reservoir **105** to modify (further

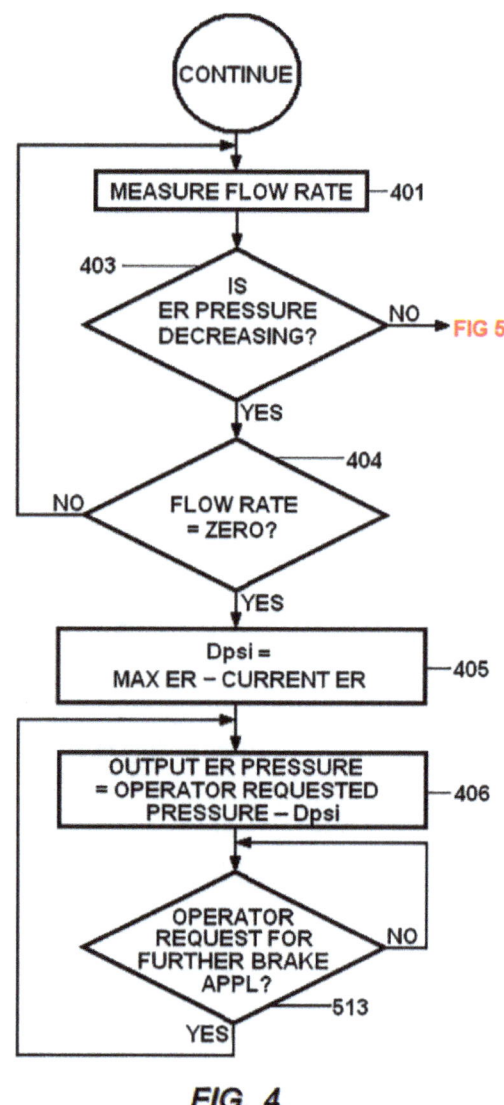

Figure 249: Adaptive Air Brake (4). Process flow diagram FIG 4: Sequence of steps executed by signal processor 140 to controllably modify a pressure reduction initiated by the engineman.

reduce) the requested pressure by the value determined in step **405**. Also, for any further brake application (prior to release), as indicated by step **407**, signal processing unit **140** again modifies the requested pressure reduction by the value determined in step **405**. In either event, the modification of the requested equalising reservoir pressure will ensure that the pressure in the brake pipe is at the correct value to cause the car control valves of the train to respond effectively to the requested brake pipe reduction.

As pointed out above, if the brake application requested by the engineman causes the pressure in the equalising reservoir to be higher than the brake pipe pressure ('NO' to step **403**), the process proceeds to step **501**, FIG 5, where the value of the current equalising reservoir pressure is stored.

Next, in step **503**, <CAUSE FURTHER ER REDUCTION>, the equalising reservoir pressure is reduced and then, in step **504** <FLOW RATE = 0?>, it is determined whether the air flow rate as monitored by flow rate transducer **141** has dropped to zero. Until the flow rate drops to zero, the process is looped back to step **503** resulting in a falling equalising reservoir pressure. When the brake pipe air flow rate reaches zero—indicating that brake pipe pressure matches that of the equalising reservoir—the process proceeds to steps **505**, **511** and **513**, which effects a subtraction of the sensed pressure differential Dpsi from the requested pressure, like that which takes place in steps **405**, **406**, **407**, previously described.

As noted above—in addition to potential safety hazards that may result from actions the engineman may take (for example, initiating a brake application prior to stabilisation of the charging pressure in the brake pipe)—the Adaptive Air Brake system control mechanism continuously monitors the integrity of the train brake pipe and takes corrective action should an anomaly occur. For example, in the event of an obstruction or blockage of the brake pipe resulting in a change in its effective volume, a warning indication is given to the engineman and a signal is generated that inhibits movement of the train until the obstruction is located and cleared.

One possible cause of a change in the volume of the brake pipe is an improper positioning of a brake pipe angle cock somewhere in the train. If this improper setting remains undetected, it is possible that the train may proceed with brakes controlled on part of it but disabled on the remainder. To prevent this from occurring, the system control mechanism employs the sequence of steps shown in FIG 6 (for brake application) and similarly in FIG 3 (for brake release).

In response to the engineman initiating a brake application, in step **601** <MEASURE BP PRESSURE, FLOW RATE, TEMP & MAX ER PRESS>, air pressure, temperature and flow rate from the brake pipe are measured. Next, in step **603** <CALCULATE AIR VOLUME (STP)>, the volume of the brake pipe is calculated. In steps

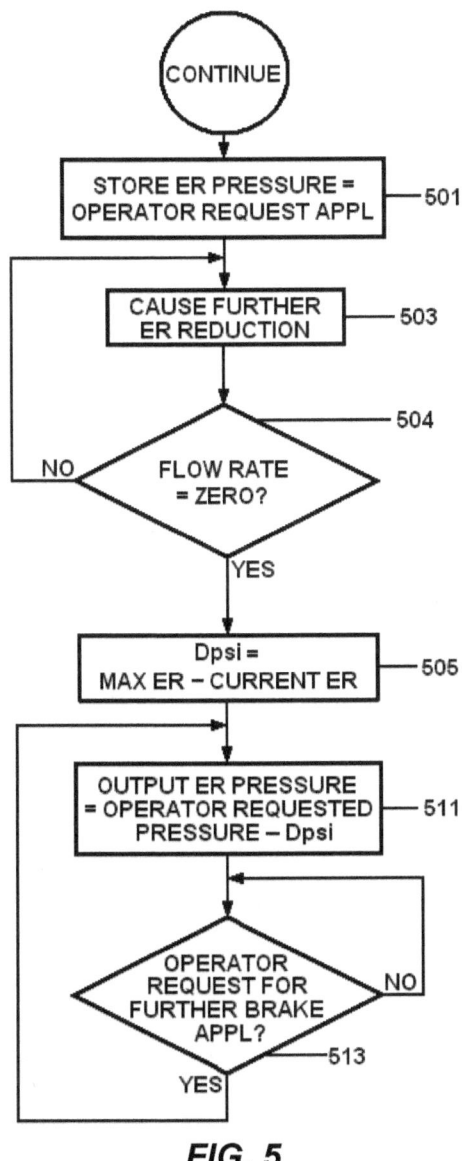

FIG. 5

Figure 250: Adaptive Air Brake (5). Process flow diagram FIG 5: System control mechanism sequence of steps executed by signal processor 140 to controllably modify a pressure reduction initiated by the engineman [continued from FIG 4].

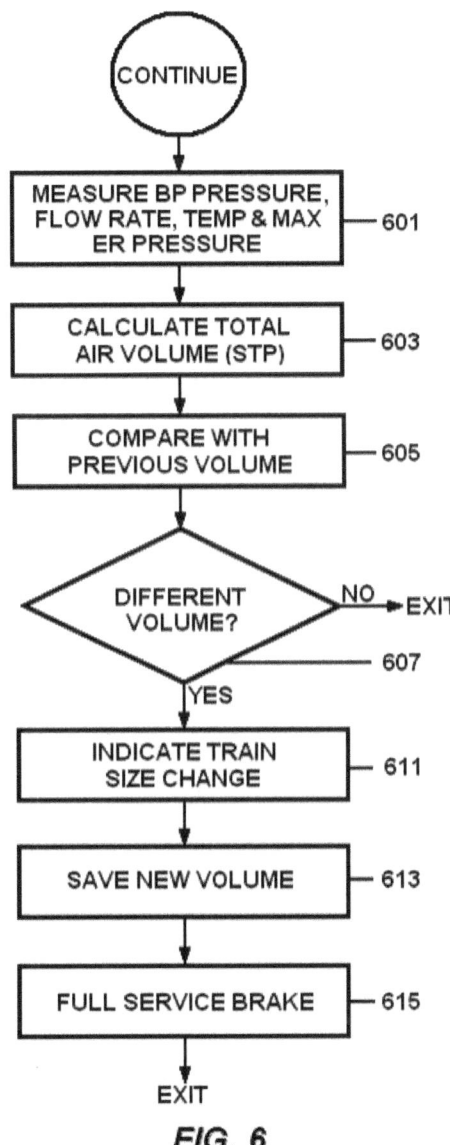

FIG. 6

Figure 251: Adaptive Air Brake (6). Process flow diagram FIG 6: System control mechanism sequence of steps for brake application (and similarly in Figure 3 for brake release) to prevent the train from proceeding with brakes controlled on part of it but disabled on the remainder.

605 and **607** the presently-calculated volume of the brake pipe is compared with a previously-determined reference value. If the newly-calculated brake pipe volume is the same as that of the reference, the program is exited. However, if the currently-calculated volume is different from its expected value, the process proceeds to step **611** <INDICATE TRAIN SIZE CHANGE> and step **613** <SAVE NEW VOLUME>. Consequently, a warning indication of a change in the size of the train is signalled via the engineman's control stand (step **611**), the new volume is stored (step **613**), and a Full-Service brake application is initiated (step **615**) to bring the train to a complete stop, so that the integrity of the brake pipe can be checked, and the problem corrected.

In conjunction with this capability of a brake pipe integrity check, it is important that the engineman knows when it is safe to disconnect the locomotive(s) from the remainder of the train, as—for example—at a destination or switching yard. Normal procedure requires the engineman to initiate a Full-Service brake application on the train in preparation for disconnecting. The engineman then waits until this brake application propagates through the brake pipe and the locomotive relay valve exhaust is no longer venting air. The crew member can then proceed to close the brake pipe angle cock and disconnect the locomotive from the train. As previously stated, a critical requirement of this process is that the outward flow of air from the brake pipe be stabilised, because if it is not, the abrupt restriction to the flow may result in a pressure wave in the brake pipe that can cause the release of the train brakes. Rather than have the engineman rely upon an audible cue (a very imprecise process), the system control mechanism employs the process shown in FIG 7.

At step **701** <MEASURE BP PRESSURE & FLOW RATE>, signal processing unit **140** measures brake pipe parameters and then inquires, in step **703** <FULL SERVICE BRAKE APPL?> whether the engineman has requested a Full-Service brake application (for maximum non-emergency braking effort). Assuming a Full-Service brake application, the process proceeds to step **705**, <FLOW RATE = 0?> and monitors the output of flow rate sensor **141** to determine when air flow has stopped (effectively zero flow rate), thus indicating that the exhaust flow from port **124** of relay valve **117** has stopped. Once the outflow from the brake pipe has stabilised (step **705**, 'YES'), the process proceeds to step **707** <LIGHT *READY TO CUT-OFF* INDICATOR> to cause a display on the control stand to indicate to the engineman that it is safe to disconnect the locomotive from the train.

Until the brake pipe outflow has stabilised, a negative answer to step **705** causes the process to continue to step **709** <INHIBIT MOTIVE FORCE> to cause a locomotive traction force inhibit signal to be applied to the locomotive so that no part of the train can be moved until the Full-Service brake application is complete and effective.

An additional aspect of the control mechanism of the Adaptive Air Brake Control System is to provide the engineman with a precise indication that the brakes of the train have been fully released in accordance with a brake release action (initiated, for example, prior to the application of a locomotive tractive force). For this purpose—as in the brake pipe integrity checking scheme described previously—whenever a brake application or release action is initiated, the volume of the fluid flow path through the brake pipe is measured. Based upon this measured volume—which is indicative of the length of the fluid path—a 'brakes released' signal is generated after the lapse of a period of time following the initiation of the brake release action: this period being determined in accordance with the measured volume of the fluid

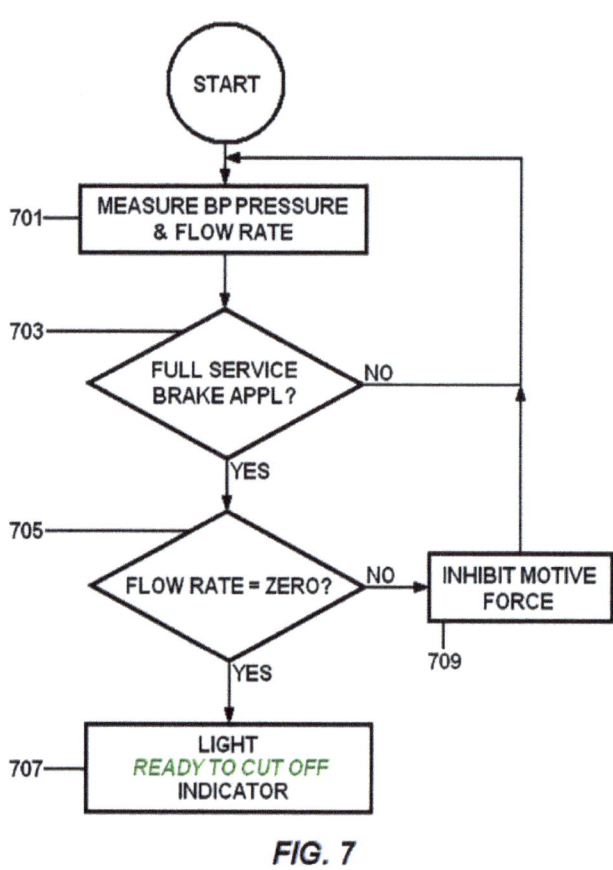

Figure 252: Adaptive Air Brake (7). Process flow diagram. FIG 7: System control mechanism sequence of steps to stabilise the outward flow of air to prevent its abrupt restriction that may result in a BP pressure wave and cause a brake release.

flow path. Until the 'brakes released' signal is generated, an inhibit signal prevents the application of a tractive force to the locomotive.

This timeout operation is effected in accordance with the sequence of steps within the flow diagram of FIG 3, in connection with the description of the pressure reduction modification. In response to the initiation of a pressure increase request, a soft-timer within signal processing unit **140** is started, (<START RELEASE TIMER>, step **303**). Subsequently, at step **311**, using the volume calculation performed in a previous execution of step **307** <RELEASE TIMER = FULL RELEASE?>, the process inquires whether the soft-timer has counted to a value that corresponds to the length of time required for the pressure increase to have propagated completely down the brake pipe. If not, the process proceeds to step **315**, previously described. If the answer to step **311** is 'YES', the process proceeds to step **313** <LIGHT *'BRAKES RELEASED'* INDICATOR> so that the display on the locomotive control stand will advise the engineman that the brakes are now fully released. Then, in step **314** <REMOVE THROTTLE INHIBIT>, a tractive force inhibit signal is removed from the locomotive to permit operation.

As a further feature of the brake pipe pressure monitoring and pressure reduction modification mechanism that is executed in response to the initiation of a brake release operation (FIG 3), the Adaptive Air Brake system provides a scheme for precisely indicating when the brake pipe has been completely recharged. For this purpose—as shown in FIG 3—upon the flow rate interpreted by sensor **141** reaching what is effectively a zero value (because of the pressure within the brake pipe increasing towards maximum charge as indicated by a 'YES' answer to step **315**) steps **321**, **323** and **325** are executed to provide an indication to the engineman that the brake pipe is fully charged. In addition, with the brake pipe being now fully charged, any leakage (for example, through hose couplings and pipe fittings) can be accurately determined from the output of flow sensor **141**, so that brake pipe volume can be accurately determined, taking the leakage value into account. With these new values stored, the volume is compared to that previously calculated in step **331** <IS TRAIN VOLUME DIFFERENT FROM PREVIOUS?>. If the volume is unchanged, the program is exited. If the processor calculates a different volume value, then it is concluded that there is an anomaly in the brake line that requires attention. Consequently—in addition to illuminating a display in step **332** <LIGHT *'TRAIN SIZE CHANGE'* INDICATOR & ALARM>—locomotive dynamic braking and a full braking pressure reduction is asserted in step **333**.

ෲ

The invention was subsequently modified by a further patent in October 1991. For this reason, this narrative will not further describe the original Adaptive Air Brake Control System, but will proceed to explain the modified scheme, which takes account of a basic flaw identified in the system logic of the original version.

Improved Adaptive Air Braking Control System

Patent Title: *ADAPTIVE AIR BRAKING SYSTEM WITH CORRECTION FOR SECOND ORDER TRANSIENT EFFECTS*

Abstract

Method and apparatus for compensating for transient effects in pneumatically-operated air brake system for a railway vehicle. A look-up table of experimentally-determined second-order correction values is stored in memory. Based on the magnitude of a previous application of the brakes, a correction value is selected from the look-up table having a magnitude to compensate for transient errors in the system. The correction value is then diminished by fixed decrements at fixed intervals. Upon a subsequent application of the brakes, the diminished correction value is used to adjust the pressure communicated to braking mechanisms [control valves] located on the train cars.

Background and summary of the modification (Refer to Figure 257)

In braking systems of the type described, the magnitude of the braking force applied to the wheels of the train cars by the braking mechanisms is directly proportional to the magnitude of the differential pressure reduction in the air line sensed by those respective braking mechanisms. However, due to the substantial volume of the air line—which extends the entire length of the train—an appreciable amount of time is required for the restoration of pressure in the air line to its original value following a braking event. A subsequent application of the brakes prior to complete restoration of pressure in the air line thus results in the communication of a smaller differential pressure reduction to the pneumatic braking mechanisms at the train car wheels, and therefore less braking force than requested and intended by the engineman via the air brake control system.

Even when he recognises the insufficiency of the new braking application, the engineman often attempts to remedy the problem by a further application of the brakes. Again, however, the application of a braking effort which is less than expected and desired will take place, so that there may still be inadequate braking action applied at the cars. Simply put, if the engineman tries to make up for insufficient braking of one pressure reduction request in a piecemeal fashion, and under-corrects each time, it is possible that continuing efforts in this process will be unsuccessful due to increasing train speed and that the originally-intended braking effort and train speed reduction will never be accomplished.

A method and apparatus to remedy this situation were disclosed in a further patent[107] in which the Air Brake Control System was modified to automatically augment the pressure reduction effected by the engineman whenever the brakes were applied prior to complete recharging of the train air line, so that the total amount of the pressure reduction communicated to the pneumatic braking mechanisms closely approximated that which would have been achieved had the train air line been fully charged.

The improved adaptive braking system [is therefore intended to increase] the accuracy of the adaptive braking function by considering, and compensating for, first-order transient effects in the charging and discharging of the train air line.

Deno and his colleagues discovered that improvements could be made to their Adaptive Air Braking Control System to render it capable of providing a total braking effort that more closely approximated that expected by the engineman. In particular, the sensing method described in US patent number 4 859 000 [to measure the decrease in the air line charging flow rate to signal the 'equivalent pressure' of the train airline] was characterised by inaccuracies due to hysteresis[108] and frictional effects in the brake system control mechanism. In addition, the presence of second-order transient effects in the propagation of pressure waves over the length of the air line during charging and discharging (often referred to as a 'pressure gradient'), provided a further source of error in the correction applied by the control system.

Accordingly, the modification was intended to provide an improved adaptive air braking system that was not subject to the previously-mentioned hysteresis and frictional effects, and that further compensated for inaccuracies in the system by considering the pressure-gradient effect.

The applicants claimed that this and other objects of their modified invention were achieved by directly

---

107 The applicant's commonly assigned US Patent № 4 859 000, filed 15 Aug 1988 and dated 22 Aug 1989.
108 See Glossary.

sensing the pressure in the brake line of the brake control system using a pressure transducer mounted thereon, and by providing a set of look-up tables containing experimentally determined pressure reduction factors to compensate for the pressure gradient and related effects. A central signal-processing unit was programmed to select a second-order correction value from the look-up tables (see Figure 255) based on the magnitude of the previous reduction requested by the engineman.

The correction value thus determined was then decremented over time to decay to zero when all transient effects in the brake control system had been eliminated. If the brakes were reapplied prior to complete recharging of the train air line following a braking event, an additional pressure reduction equal to the decremented correction value would be added to the pressure reduction effected by the engineman and the first-order correction referred to above.

Other objects, advantages, and novel features of the present invention will be apparent from the following detailed description of the invention when considered in conjunction with the accompanying drawings that are modified versions of those previously depicted.

Detailed description of the diagrams

FIG 1 (below) repeats Figure 247 on p262, and depicts a prior art adaptive air brake system of the type disclosed in US Pat. № 4 859 000. For the purpose of understanding the present invention, it is sufficient to note that signal processing unit **140** (a microprocessor) is coupled to monitor the engineman's requested pressure reductions

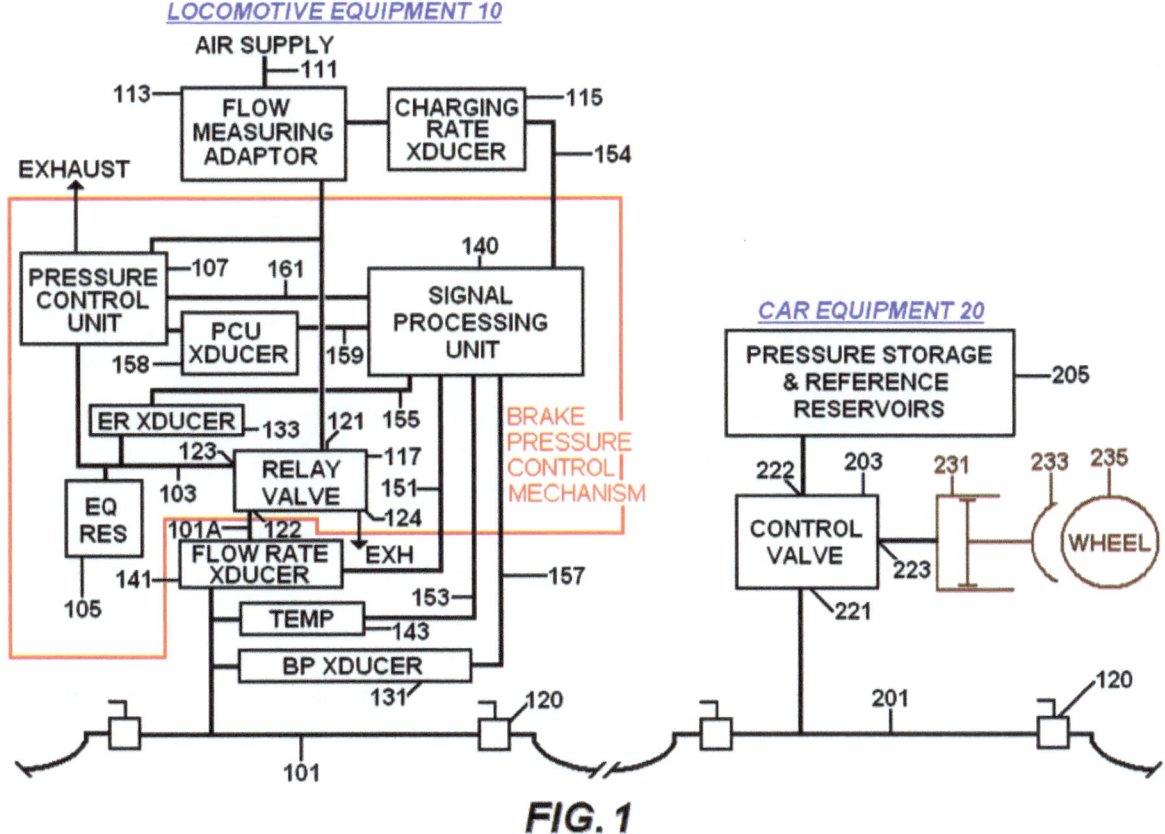

Figure 253: Adaptive Air Brake (8). With Correction for Second Order Transient Effects. FIG 1: Signal processing unit 140 monitors the engineman's requested pressure reductions and the BP to determine whether there is a change in air pressure, as would occur during recharge following the release of a previous brake application.

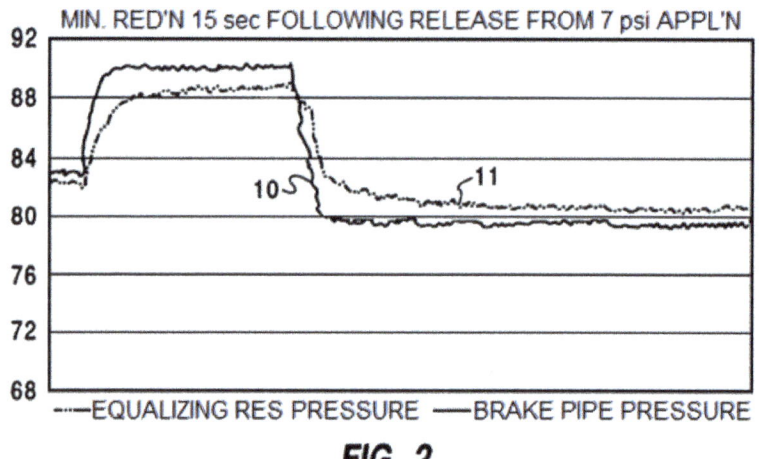

FIG. 2

Figure 254 - Adaptive Air Brake (9). With Correction for Second Order Transient Effects. FIG 2: Chart of locomotive ER and BP pressure recorded on a 98-car train.

(input through pressure control unit **107**), and also to monitor the brake pipe **101** to determine whether it is currently undergoing a change in air pressure, as would be the case during the recharge of the brake pipe subsequent to the release of a previous brake application.

As the pressure within the equalising reservoir **105** drops in response to the engineman's pressure reduction, the charging rate of the air flowing into the brake pipe will decrease as the reducing pressure within the equalising reservoir and the slowly-increasing pressure within the brake pipe approach each other. When the flow rate into the brake pipe becomes negligible (effectively equal to zero), indicating that brake pipe **101** pressure is the same as that within the equalising reservoir **105**, the value of the pressure within the equalising reservoir at that time is stored. Using this stored value, the requested equalising reservoir pressure change is modified (increased) by a differential amount equal to the difference between the maximum pressure of the equalising reservoir and the stored equalising reservoir pressure value.

In operational tests of the adaptive braking system described above, it was found that additional sources of error exist. In particular, the use of the air line charging flow rate to signal the equivalent pressure in air line **201** introduces inaccuracies caused by hysteresis and frictional effects in relay valve **117**, especially at very low pressure differentials. In addition, second-order transient effects in the propagation of air pressure along the length of air line **201** are significant. That is, the actual pressure experienced at the respective car braking mechanisms during charging of the air line lags the pressure measured at the locomotive brake pipe due to the time required for the so-called 'pressure gradient' to be equalised across the entire length of the train.

Since the magnitude of the pressure correction generated by the adaptive braking system of FIG. 1 is based on the measurement of pressure in the locomotive brake pipe, the effect of the above-mentioned pressure gradient is to cause the system to underestimate the magnitude of the pressure shortfall at the braking mechanisms distributed over the length of the train, and an additional correction is therefore required.

The amount of additional correction was determined experimentally by executing consecutive pressure reductions in the adaptive braking system of FIG 1 and measuring the brake cylinder pressure and brake shoe movement of the brake mechanisms on the respective train cars as a function of time to determine the decay characteristics of the transient

SECOND ORDER CORRECTION SAMPLE LOOK-UP TABLE

PREVIOUS APPLICATION	TENTATIVE CORRECTION Δ
4 - 7.5	1
8 - 11.5	2
12 - 15.5	4
16 - 19.5	6
20 - 23.5	6
24 - 27.5	6
28+	6

FIG. 3

Figure 255 - Adaptive Air Brake (10). With Correction for Second Order Transient Effects. FIG 3: Tabular presentation of the results of tests to determine the magnitude of the pressure shortfall at car braking mechanisms distributed over the length of the train and obtain a second-order correction value to overcome the pressure gradient effect.

effects. That is, the brakes were first applied and released (sometimes herein referred to as a 'first application', not to be confused with the term First-Service [or 'Minimum Reduction'] application as frequently used in the industry), and then after a selected time interval, applied again. The response at the train cars was then measured and recorded. Various combinations of pressure reductions and time intervals between braking operations were used.

FIG 2 shows the locomotive equalising reservoir and brake pipe pressure recorded in one test on a 98-car train. The test sequence described above was followed and chart recordings were made of the brake pipe and equalising reservoir pressures on the lead locomotive. Manual recordings were also made of the final brake cylinder pressure 24 cars behind the lead unit (not shown). The chart starts with a 7-psi minimum reduction application in effect (equalising reservoir pressure = 83 psi and brake pipe pressure = 82 psi).

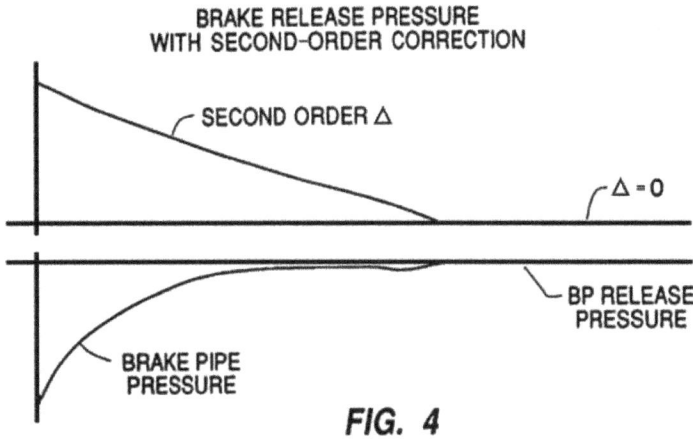

Figure 256: Adaptive Air Brake (11). With Correction for Second Order Transient Effects. FIG 4: Graphical illustration of how—as the magnitude of the pressure deficiency in the BP decreases over time (that is, as the BP charges up to its original pressure)—the size of the necessary second-order correction also decreases over time.

A brake release is then made, as indicated on the chart by the equalising reservoir pressure **11** and brake pipe pressure **10** starting to rise towards the release pressure, which is 90 psi. After a period of 15 seconds, a minimum (7 psi) application is made, and the equalising reservoir pressure **11** and brake pipe pressure **10** start to drop. (Note that at the point of application, the pressure **10** in the brake pipe had not recovered to 90 psi.) The equalising reservoir pressure **11** drops to 80 psi and brake pipe pressure **10** drops to 81 psi.

The results of numerous iterations of the above test procedure for various combinations of timing and pressure reductions demonstrated that the amount of the second-order correction required to overcome the pressure gradient effect is proportional to the magnitude of the pressure deficiency (due to incomplete charging of the air line) which prevails at the time that an application of the brakes is initiated. Thus, for any brake application, the deeper the previous pressure reduction was, the greater the 'gradient effect' and the required second-order correction will be. For example, if the release from a 10-psi reduction is followed immediately by a second reduction, the necessary second-order correction was determined to be 2.0 psi, while a 20-psi first reduction requires a correction of 6.0 psi.

In addition, since the magnitude of the pressure deficiency in the air line decreases over time—that is, as the air line charges up to its original pressure—the size of the necessary second-order correction also decreases over time. This characteristic is illustrated graphically in FIG 4 (see Figure 256 on pg 272).

Referring now to FIG 5, this illustrates an improved adaptive braking system in accordance with the present invention. To facilitate an understanding of the improved system, those elements which correspond to the prior art system of FIG 1 are designated by the same reference numerals.

In the improved system according to the invention, charging flow rate transducer **115** has been eliminated. Brake pipe flow rate transducer **141** and temperature transducer **143** pertain to other aspects of the prior art patent, and are also not shown. In operation, in accordance with the method of the present invention, a comparison of the pressure values sensed by brake pipe pressure transducer **131** and equalisation reservoir transducer **133** is used to determine the magnitude of the necessary first-order correction in lieu of utilising a

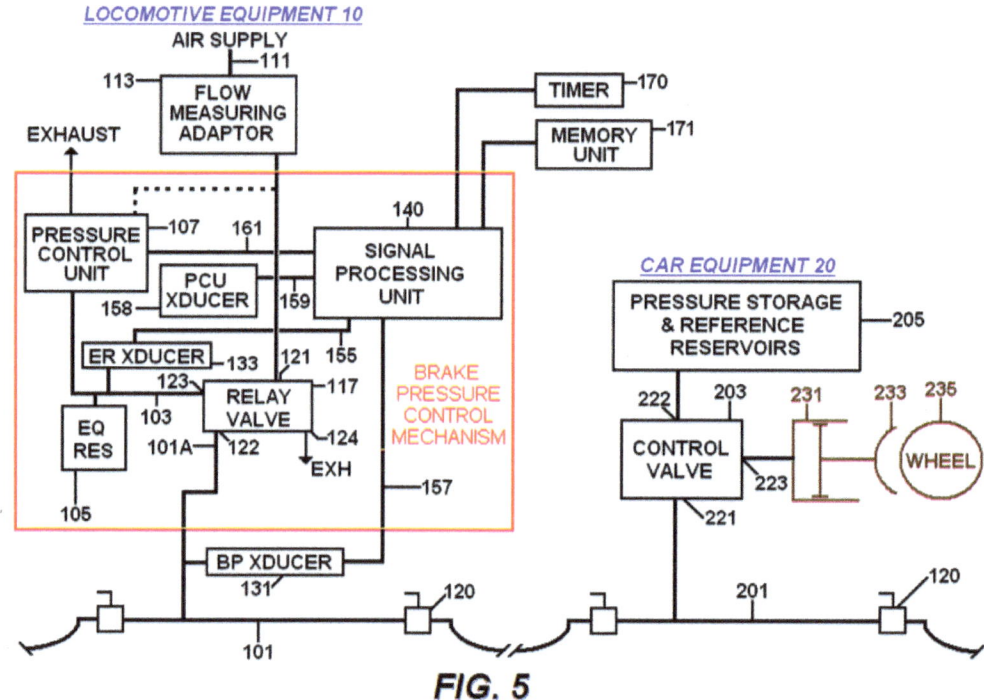

Figure 257: Adaptive Air Brake (12). With Correction for Second Order Transient Effects. FIG 5: A diagrammatic illustration of the improved adaptive air brake control system for a freight train.

flow rate transducer in the manner disclosed in US Pat. No. 4 859 000. That is, at the time when an application is initiated, signal processing unit **140** compares the respective pressure readings from transducers **131** and **133** and determines whether a difference exists. (Such a difference would occur where brake pipe **101**/train air line **201** have not fully charged following a previous application of the brakes.) If in fact a difference exists, the amount of the difference is then stored and is added to the equalising reservoir pressure change requested by the engineman by operation of control unit **107**. By determining the magnitude of the pressure correction in this manner rather than by use of a flow rate transducer, it was found that frictional and hysteresis errors are eliminated, and a more accurate first-order correction is effected.

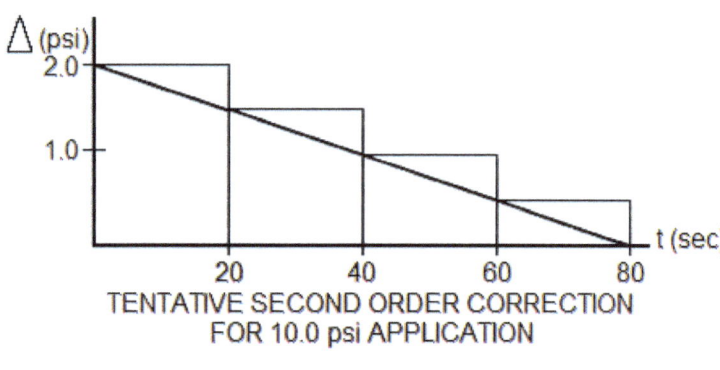

Figure 258: Adaptive Air Brake (13). With Correction for Second Order Transient Effects. FIG 6: The tentative second-order correction is decremented by 0.5 psi every 20 seconds.

With further reference to FIG 5, signal processing unit **140** has coupled to it a timer **170** and a memory **171**, and is also coupled to pressure control unit **107** to receive output signals representative of the magnitude and timing of pressure reductions and restorations as requested by the engineman. Memory **171** has stored the experimentally-determined look-up tables described above. In operation, each time the brakes are applied and released, signal processing unit **140** receives a signal from pressure control unit **107** indicating the amount of the total reduction (in psi), which is stored.

Signal processing unit **140** then queries the look-up table stored in memory **171** and reads out a tentative second-order correction factor based on the magnitude of the stored pressure reduction. As noted above, however, as brake pipe **101** and air line **201** are recharged following a pressure reduction, the amount of the first-order pressure differential decreases, and the amount of the second-order correction decreases accordingly, as illustrated in FIG 4. To account for this factor, timer **170** causes the magnitude of the tentative second-order correction read from memory **171** to be decremented by the processing unit **140** by a predetermined amount at fixed intervals, until all transient effects in brake pipe **101** and air line **201** have decayed to zero, at which time all adaptive braking action is disabled. That is, the brake pipe air line is fully-charged and no correction is necessary.

Based on the data determined in accordance with the procedures described above, the tentative second-order correction is decremented by 0.5 psi every 20 seconds, as illustrated in FIG 6.

It should be noted in this regard, however, that the magnitude of the required second-order correction values as stored in the look-up table, as well as the selection of the frequency and magnitude of the pressure decrement are dictated only by the response of the air line for the particular system, and will vary from train to train, dependent upon its length, car-type, air temperature and other factors. Moreover, for a particular system, other combinations of decrement and frequency may be used so long as they conform to the characteristic response of the system as depicted in FIG 6. (For example, for the system tested as described above, a decrement of 0.25 psi at 10-second intervals could be used.)

Upon initiation of a brake application prior to the complete charging of brake pipe **101** and air line **201**, the amount of the decremented second-order correction is added by signal processing unit **140** to the pressure reduction requested by the engineman and the first-order correction determined as described above, in order to effect an actual differential pressure reduction in air line **201** that will result in a braking force that accurately accords with that requested by the engineman.

The sequence of steps that are executed by signal processing unit **140** to implement the first- and second-order adaptive braking control system per the invention are illustrated in the flow diagram of FIG 7.

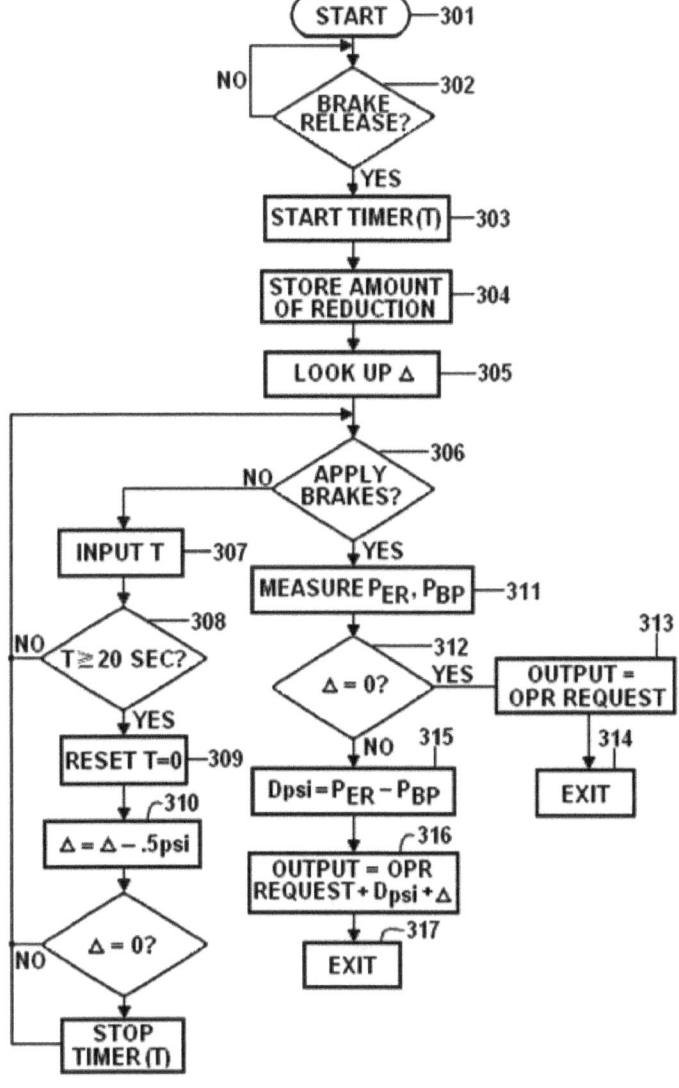

FIG. 7

Figure 259: Adaptive Air Brake (14). With Correction for Second Order Transient Effects. FIG 7: The sequence of steps executed by the signal processing unit to implement the first- and second-order adaptive braking control system.

For this diagram, it is assumed that the train is initially engaged in a braking application at <START> step 301. When the engineman initiates a <BRAKE RELEASE> (step 302) by operating pressure control valve 107, signal processing unit 140 starts timer 170 (step 303) and at the same time receives from pressure control unit 107 a signal representing the magnitude of the just-released pressure reduction that is stored (step 304). Now, signal processing unit 140 queries memory 171 and reads out a tentative second-order correction Δ corresponding to the stored value (step 305).

At step 306 <APPLY BRAKES?> a determination is made whether a subsequent braking application has been initiated. If a second application follows immediately upon a previous application, the output from step 306 is 'YES', and signal processing unit 140 then acquires pressure readings from equalisation reservoir 105 (PER) and brake pipe 101 (PBP) at step 311 <MEASURE PER, PBP>. If the second-order correction Δ = 0 (step 312), then brake pipe 101 and air line 201 have fully recharged and no correction is necessary, and the output from the control system is equal to the pressure reduction requested by the engineman (step 313).

If at the time the brakes are applied, however, the Δ is not 0 (step 312) then the recharging of the brake pipe has not reached steady state. Signal processing unit 140, therefore, determines the pressure differential Dpsi as the difference between PER and PBP (step 315), and the output of the control system is calculated by signal processing unit 140 as the sum of the pressure reduction requested by the engineman plus the first-order pressure correction Dpsi, plus the second-order correction Δ (step 316).

In the event that a period of more than 20 seconds elapses between the release of a first pressure reduction and the initiation of a second reduction, the amount of the second-order correction Δ necessary to counteract the transient pressure gradient in air line 201 is decremented as noted previously. This operation is accomplished by steps 307 through 310 as follows:

After a release of the brakes and the determination of a tentative second-order correction Δ from the look-up table in memory 171 (step 305), if a second application has not been initiated (step 306), signal processing unit 140 receives the output signal T (step 307) from timer 170 and determines whether 20 seconds have elapsed since the start of the timer in step 303. At the point when T = 20 sec., timer 170 is reset to zero (step 309) and signal processing unit 140 reduces the value of the second-order correction determined in step 305, by 0.5 psi (step 310). At this point, if the brakes still have not been applied, steps 307 through 310 are repeated until either the system reaches steady state (that is, the air line is fully recharged and no correction is necessary, step 313), or the brakes are applied and the decremented value of the second-order correction Δ is used to calculate the output reduction in steps 315–317 so that the correct braking force as requested by the engineman is applied to the wheels of the cars (20).

As noted previously, in practice, the magnitude of the actual correction value necessary to compensate for second-order transient effects is dictated by the response of the air line for the system, and thus will vary from train to train, dependent upon its length and other factors. Therefore, according to a preferred embodiment of the invention, signal processing unit 140 is adapted to monitor and automatically adjust to these train characteristics by modifying the look-up table based on actual train response during operation.

Whenever the brakes are released from a heavy application such as Full Service, the rate-of-change of the brake pipe pressure PBP is indicative of the train length and air brake system characteristics, such as effective air line volume. This effect is used to modify the adaptive braking feature to reflect these train characteristics. That is, the number of seconds required for the brake pipe pressure PBP to rise 20 psi from the Full-Service application pressure is used to select a train-specific look-up table and decay rate for the adaptive braking functions. (Alternatively, the decay in the charging flow rate can be used to select these parameters.) This function is repeated with each release from a Full-Service application during train operation. Thus, the system automatically adjusts to changing length as cars are added to or removed from the train consist. The steps executed by signal processing unit **140** to effect the adjustment referred to above are shown in the flow chart in FIG 8.

When the system is initially turned on, a default correction table is installed in memory unit **171** (step **401**). This table represents the characteristics of a 'nominal train', such as shown in FIG 3, and remains in effect until a release from a Full-Service (25 psi) application is made. Such a release is always part of a mandatory terminal brake test.

When the system detects a Full-Service application followed by a release of the brakes (steps **402**, **403**), a timing sequence is initiated by signal processing unit **140**, and at the same time brake pipe pressure PBP is read and stored (steps **404**, **405**). When the brake pipe pressure PBP has risen 20 psi, signal processing unit **140** again reads the timer (step **407**). If at this point the elapsed time since the initial reading is less than or equal to 8 seconds (that is, the brake pipe pressure PBP has risen 20 pounds in less than 8 seconds, indicating a relatively short train), signal processing unit **140** causes the correction look-up table stored in memory unit **170** to be changed to the table set forth in FIG 9a, in which the correction values are relatively smaller (step **409**).

If, on the other hand, the time required for brake pipe pressure PBP to rise 20 psi is greater than 8 seconds and less than or equal to 18 seconds (step **410**), no further adjustment of the correction look-up table is required, and the modification sequence is terminated (step **411**). However, if more than 18 seconds is required for the brake pipe pressure PBP to recover by 20 psi—indicating a relatively long train—signal processing unit **140** causes the input tables stored in memory unit **171** to be modified to conform to the tables set forth in FIG 9b, in which the correction values are relatively higher (step **412**). Now, the adjustment cycle is terminated.

It is apparent that numerous variations of the above process are possible. For example, instead of 8 and 18 seconds as the break points, 7 and 15 seconds could

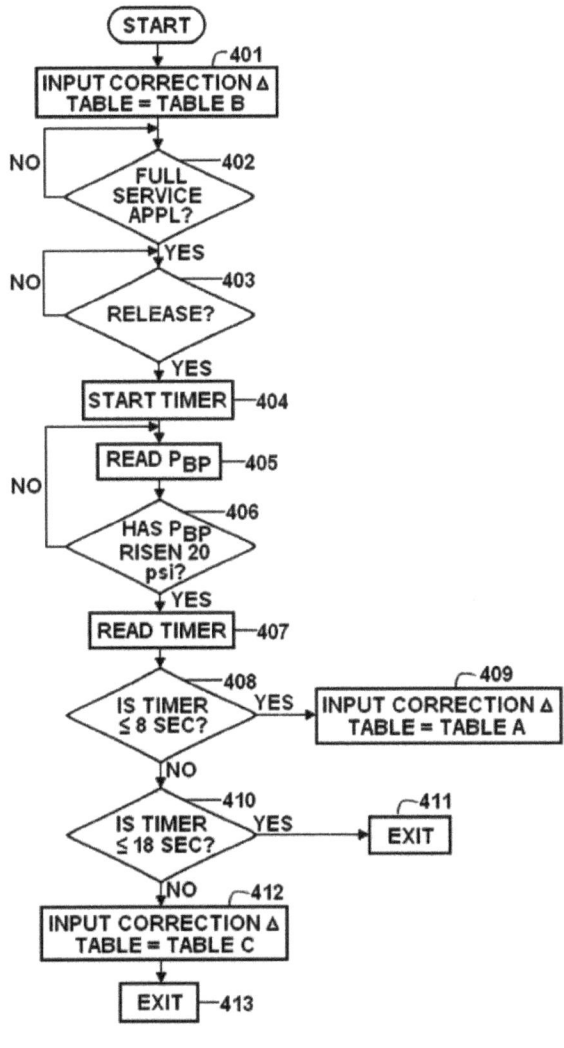

FIG. 8

Figure 260: Adaptive Air Brake (15). With Correction for Second Order Transient Effects. FIG 8: The steps executed by the signal processing unit to automatically adjust the system to changing length as cars are added to or removed from the train consist.

PREVIOUS REDUCTION	PSI
4 - 7.5	0.5
8 - 11.5	1.5
12 - 15.5	3.5
16 - 19.5	5.5
20 - 23.5	5.5
24 - 27.5	5.5
28+	5.5

DECREMENT VALUE = 0.5 PSI
DECREMENT INTERVAL = 15.0 SEC

FIG. 9a

Figure 261: Adaptive Air Brake (16). With Correction for Second Order Transient Effects. FIG 9a: When the system detects a Full-Service application followed by a release of the brakes, a timing sequence is initiated, and the BP pressure value is stored. When it has risen 20 psi, and if the elapsed time has been less than or is equal to 8 seconds, the signal processing unit changes the correction lookup table stored in the memory unit to that with the relatively smaller correction values depicted in this Figure.

PREVIOUS REDUCTION	PSI
4 - 7.5	1.0
8 - 11.5	2.5
12 - 15.5	4.5
16 - 19.5	6.5
20 - 23.5	6.5
24 - 27.5	6.5
28+	6.5

DECREMENT VALUE = 0.5 PSI

FIG. 9b

be used. Additionally, more than two break points could be used, with a correspondingly larger number of alternative correction tables, to achieve even greater accuracy in the adjustment. Finally, in lieu of using brake pipe pressure as the input variable, the charging flow rate in the brake pipe could be used, with—for example—a break point of 50 cubic feet per minute in place of step **408** as indicated in FIG 8.

Figure 262: Adaptive Air Brake (17). With Correction for Second Order Transient Effects. FIG 9b: If, on the other hand, the time required for brake pipe pressure PBP to rise 20 psi is greater than 8 seconds and less than or equal to 18 seconds (step 410), no further adjustment of the correction lookup table is required, and the modification sequence is terminated (step 411). However, if more than 18 seconds is required for the brake pipe pressure PBP to recover by 20 psi—indicating a relatively long train—signal processing unit 140 causes the input tables stored in memory unit 171 to be modified to conform to the tables set forth in FIG 9b, in which the correction values are relatively higher (step 412). Now, the adjustment cycle is terminated.

Figure 263: 'Trol-toon'. Harris Controls

GLOSSARY

CMOS

Complementary Metal–Oxide–Semiconductor is a technology for constructing integrated circuits and is a form of MOSFET (metal–oxide–semiconductor field-effect transistor) semiconductor. CMOS technology is used in microprocessors, microcontrollers, static RAM, and other digital logic circuits.

EPROM

Erasable Programmable Read-Only Memory. A PROM is a type of ROM that is typically programmed once and can't be changed after that. This type of memory is often used in hardware that has a dedicated purpose that will not change (e.g. firmware and RFID[109] chips). EPROM memory is similar except that it can be erased and reprogrammed. EPROM memory chips have an optical window on them that must be exposed to UV light, which will erase the memory and allow them to be reprogrammed. With this method, the memory must be completely erased before any new information is written. Since the UV light method is somewhat inconvenient, Electrically Erasable Programmable Read-Only Memory (EEPROM) has since been developed.

FSK

Frequency-Shift Keying is a frequency modulation scheme in which digital information is transmitted through discrete frequency changes of a carrier signal. The technology is used for communication systems such as telemetry, weather balloon radiosondes, caller ID, garage door openers, and low frequency radio transmission in the VLF and ELF bands. The simplest FSK is binary FSK (BFSK). BFSK uses a pair of discrete frequencies to transmit binary (0s and 1s) information. With this scheme, the '1' is called the mark frequency and the '0' is called the space frequency.

GMD

General Motors Diesel were a railway diesel locomotive manufacturer located in London, Ontario, Canada, and created in 1949 as the Canadian subsidiary of the Electro-Motive Division of General Motors (EMD) in the United States. In 1969, GMD was reorganised as the Diesel Division of General Motors of Canada Ltd. The plant was re-purposed to include manufacture of other diesel-powered General Motors vehicles such as buses. Following the US-Canada Free Trade Agreement in 1989, all of EMD's locomotives were built at the London facility. In 2005, the new owners of EMD, Progress Rail Services, renamed the Canadian subsidiary 'Electro-Motive Canada'. The plant was closed in 2012, with EMD's production remaining in LaGrange, IL and Muncie, IN.

Hall Effect

This refers to the production of a voltage difference (the Hall voltage) across an electrical conductor, transverse to an electric current in the conductor and to an applied magnetic field perpendicular to the current and was discovered by Edwin Hall in 1879.

---

[109] Scroll to RFID.

Hysteresis

From the Greek term meaning 'a coming short, a deficiency'. Refers to the dependence of the state of a system on its history (systems, organisms and fields that have memory). Generally, hysteresis means a time-lag between input and output in a system upon a change in direction. Hysteresis occurs in ferromagnetic and ferroelectric materials, as well as in the deformation of rubber bands and shape-memory alloys and many other natural phenomena. It can be a dynamic lag between an input and an output that disappears if the input is varied more slowly (known as rate-dependent hysteresis). In control systems, hysteresis can be used to filter signals so that the output reacts less rapidly than it otherwise would by taking recent system history into account.

Megacycles

More correctly expressed as 'megacycles per second', the term was replaced by 'hertz' from around 1960 (thus, 'Mc' became 'MHz').

Membrane switch

A printed circuit 'momentary switch' device in which at least one contact is on, or made of, a flexible underlying substance. Membrane switches require pressure to open and close a circuit, and their circuitry is most often screen-printed using conductive inks: typically made of silver, carbon, and/or graphite. They are a 'user interface' device (also 'operator' or 'man-machine' interface), as are display-based touchscreens, and mechanical switches such as push-button, toggle, rocker, and slide switches, the ultimate purpose being to enable an operator to communicate with a piece of equipment, instrument, or machinery.

PLC

Programmable Logic Controllers are microprocessor-based industrial digital computers having no keyboard, mouse or monitor, and built specifically to withstand harsh industrial environments. They include a programmable memory that stores instructions and implements functions that include sequencing, timing, logic, arithmetic, and counting. They originated in the US in the late 1960s in the automotive industry and were designed to replace relay logic systems. Before then, control logic for manufacturing was mainly via relays, cam timers, drum sequencers, and dedicated closed-loop controllers. PLCs are used for the control of manufacturing processes (such as assembly lines) or robotic devices, or any activity that requires high reliability, ease-of-programming and process fault diagnosis. Often tasked with controlling and monitoring numerous sensors and actuators, they also differ from regular computer systems in their extensive I/O (input/output) arrangements. Once programmed, a PLC will perform a sequence of events triggered by stimuli referred to as Inputs. It receives these stimuli through delayed actions such as counted occurrences or time delays.

Power assembly (locomotive)

This term refers to an Electro-Motive Diesel (EMD) prime mover subassembly designed for easy removal and replacement. As is typical with many heavy-duty internal combustion engines used in industrial applications, the design of EMD engines permits the cylinder liners, pistons, piston rings and connecting rods to be replaced at overhaul without removing the entire engine. The term has also become generic and is often used to refer to similar assemblies used in non-EMD engines where 'power pack' may be a more correct term. Power assemblies are large and heavy and overhead lifting equipment is required for a change-out. An EMD power assembly consists of the cylinder head assembly (including valves, springs, keepers, etc. less the fuel system components), cylinder liner, piston and piston rings, piston carrier, and connecting rod.

PWM

Pulse **W**idth **M**odulation is a technique made practical by modern electronic power switches. Pulse Width Modulation of a signal or power source is a method of transmitting information on a series of pulses and is an efficient way of providing intermediate amounts of electrical power between the settings *fully-on* and *fully-off*. The data being transmitted is encoded on the width of these pulses to either convey information over a communications channel or control the amount of power being sent to a load. PWM works well with digital controls that, because of their 'on/off' nature, can easily set the duty cycle.

RF

A **R**adio **F**requency signal is a broadcast wireless electromagnetic gesture used as a form of communication. RF propagation occurs at the speed of light and does not need a medium like air in order to travel. It became the early transmission solution for the remote control of distributed locomotives.

RFID

Radio-**F**requency **ID**entification uses electromagnetic fields to automatically identify and track tags attached to objects. An RFID tag consists of a tiny radio transponder: a radio receiver and transmitter. When activated by an electromagnetic interrogation pulse from a nearby RFID reader device, the tag transmits digital data—usually an identifying inventory number—back to the reader. There are two types. *Passive* tags are powered by energy from the RFID reader's interrogating radio waves. *Active* tags are powered by a battery and thus can be read at a greater range from the RFID reader. Unlike a barcode, the tag doesn't need to be within the line-of-sight of the reader, so it may be embedded in the tracked object.

SCADA

Supervisory **C**ontrol **A**nd **D**ata **A**cquisition is a system that allows an operator at a master facility to monitor and control processes that are distributed among various remote sites. As the name indicates, it is not a full control system but rather focuses on the supervisory level, being basically a software package that is positioned on top of hardware to which it is interfaced, generally via PLCs or other commercial hardware modules. The development of SCADA can be traced back to the early 1900s with the advent of telemetry, which involves the collection of data by sensing real-time conditions, and its wireless transmission to another location. The monitoring of remote conditions became possible with the convergence of electricity, telegraph, telephone, and wireless communication technology. Throughout the 1900s, more industries, such as gas, electric, and water utilities as well as mineral processing companies and chemical industries, have installed telemetry systems to monitor processes at remote sites and, later, Distributed Control Systems (DCS) to control these processes. DCS eventually evolved into SCADA.

Semiconductor

A semiconductor is a substance—usually a solid chemical element or compound—that can conduct electricity under some conditions, but not others. This characteristic provides a good medium for the control of electrical current. Semiconductor devices are components that exploit the electronic properties of these materials. The 'transistor' is perhaps the most well-known such device.

SSR

A **S**olid-**S**tate **R**elay is an electronic switching device that switches on or off when an external voltage (AC or DC) is applied across its control terminals. SSRs consist of a sensor which responds to an appropriate input (control signal), a solid-state electronic switching device which switches power to the load circuitry, and a coupling mechanism to enable the control signal to activate this switch without mechanical parts. The relay may be designed to switch either AC or DC loads. It serves the same function as an electromechanical relay but has no moving parts and therefore results in a longer operational lifetime. Packaged SSRs use power semiconductor devices such as thyristors and transistors, to switch currents up to around a hundred amperes and have fast switching speeds compared to electromechanical relays, and no physical contacts to wear out. Users of SSRs must consider an SSR's inability to withstand a large momentary overload the way an electromechanical relay can, as well as their higher 'On' resistance.

UDE

An **UnD**esired **E**mergency brake application is one that is uncommanded and unexpected when it occurs. UDEs are thought to originate from several causes: predominantly, individual car control valves that are inherently unstable and train slack action events that coincide with operator-initiated air brake activity.

Vellum

This was originally prepared animal skin or 'membrane' (or parchment), typically used as a material for writing on. Modern *paper vellum* is made of synthetic plant material and derives its name from its usage and quality similarities. In modern times (but prior to CAD), paper vellum has been used for a variety of purposes including tracing, technical drawings, plans and blueprints.

INDEX

26-C Automatic brake valve, 28, 180–196
28L-AV (locomotive air brake schedule, India), 130

A

A38 flow adapter, 98
AAR Conference, 111
ABDW (air brake operating valve), 186, 189, 256
Abydos (railroad location, Western Australia), 194
AC6000CW (locomotive), 202
Adaptive Air Brake Control System, 104, 256–269
    modified (improved), 269–278
adaptive air brake (prior system), 271
Aerotron, 134
Ahmed, Anwar, 137, 140, 143, 150, 155, *167*, *168*, *172*
Air Brake Control Console, 11, 23
air brake leakage test, 88
air braking, 3, 5, 9, 99, 40, 104, 134, 139
Airdrie, Alberta, Canada, 102
Albert Canyon, British Columbia, Canada, 45, 46
Alcatel, 24
Alco locomotives, 6, 50, 96, 97, 111, 130, 154, 160, *174*, 179, 183, *186*, 191, *195*
Alexandria, Virginia, USA, 14, 28, 29, 30
Algeria, 40, 98
Altoona, Pennsylvania, USA, 105
Alyth Diesel Shop, Calgary, Canada, 44, 45, 46, 47, 58, 60–61, 62, 65, 98
    testing at, *50*, *51*, *103*
Ambassador (car, India), 128
Andimeshk, Islamic Republic of Iran, 90, 97, 105
Angus shops, Montréal, 46, 50, 52, 56
antennae (radio), 35–36, 62, 122, 123, 144
Appleyard, Al, 57–58
applications engineering, 87–88, 98, 104, 105
Arlington Hill, Nebraska, USA, 37
Ashad (rental car driver), 87, 90, 92, 94
Asheville, North Carolina, USA, 10, 11, 13, 29
Asheville & Spartenburg Rail Road, 14
Assam, India, 104, 126, 137, 155, 157–158, 160

associated technology (LOCOTROL), 206–278
    Adaptive Air Brake, 256–278
    locomotive remote control, 207–226
    LOCOTROL I, 227–232
    LOCOTROL II and III, 233–248
    SOAR, 249–255
astronauts, 69
Atchison Topeka & Santa Fe (railroad), 35
Athens, Greece, 94–95
Atlanta, Georgia, USA, 14, 30, 32
Australia
    Milt Deno timeline, 96–97, 98, 99–102, 105, 106–107
    see also BHP Iron Ore (BHPIO); Hamersley Iron (iron ore mining company, Western Australia); Mount Newman Mining Company; Queensland, Australia; Rio Tinto (iron ore mining company, Western Australia, Australia); Robe River Iron Associates (RRIA); Western Australia
Automatic brake (train air brake system), 11, 102, 181, 184, 185, 187, 189, 259

B

backshop, defined, 50
back-to-back testing (LOCOTROL testing), *51*, 98, *103*, 104, *131*, 137, 138–139
Badarpur (BPB), Assam, India, 156, 157–158, 159
balancing speed, 191
Baltimore, Ohio, USA, 45, 46
Baltimore & Ohio Railroad, 45, 48
Bangladesh, 127–128, 157, 160, 162, 164
banker engines, 5, 106–107, 200, *see also* pushers
Barber, Wayne, 54, 67, 104, 141, 155, 157
BC Rail, 37–38, 76, 104
Beaver Hill, British Columbia, Canada, 110, 111, 119
Beavermouth, British Columbia, Canada, 45, 47, 111
Beijing, China, 101
Bellevue House, Kingston, Ontario, Canada, 81
Bendix Corporation, 11
Berwick Forge & Fabrication, 32
BHP Iron Ore (BHPIO), 40, 97, 101–102, 106, 179, 181, 201

Bihar, India, 126
Bing (railroad location, Western Australia), 195
Birmingham, South Carolina, USA, 12, 14, 29, 30
Bitter Creek, Wyoming, USA, 36
Black Mesa & Lake Powell Railroad, 81, 96
block diagrams
    Adaptive Air Brake Control System, 263–277
    LOCOTROL II, 182
    Multiple Consist Control, 21
    Tower Control, 76
    train operation, 23
Blue Ridge Mountain, North Carolina, USA, 13, 29
Bluefield, West Virginia, USA, 29
BM&LP (railroad) *see* Black Mesa & Lake Powell Railroad
BNSF (railroad), 38, 74, 203, *see also* Burlington Northern Santa Fe (railroad)
Bokaro Steel City, India, 137, 139, 145, 150, 152, 163
Bondamunda railyard, Rourkela, India, 104, 130, 133, *136*, 137, 143, 145, 150, 151, 153, 163, 164, *170*, *173*
Booth, Fred, 59, 60, 61–62, 64, 66, 70–72, 73, 77
Boucher, John, 156, 256
BOXN (rail car, India), 139, *169*
Brahmaputra Ashok Hotel, Guwahati, India, 156, 160
Brahmaputra River, India, 137, 156, 160, 162, *174*
Brake Control Centre, 11, *28*, 29
brake pipes, 9, 11, 29–30, 123, 130, 183–185, 257–269, 272–278
    continuity test, 181
    inspection and maintenance, 193
    leakage, 54, 88–89, 123, 124, 134
    LOCOTROL I, 227
    optimum brake pipe support, 115, 117
brake pipes (continued)
    parameter monitoring *see* Adaptive Air Brake Control System
    pressure/pressure reduction, 161, 184–185, 189–190, 192, 194, 196, 198, 249
    segmented systems, 99
    self-sealing connection, 106
    twin systems, 88, 89, 138
braking
    air braking, 3, 5, 9, 40, 99, 104, 134, 139
    ECP braking, 5, 6, 7, 35, 66, 124, 249, 256
    E-P braking, 34
    vacuum braking, 40, 104, 126, 130, 141, 156, 158

Brazil, 40, 108, 109
break-in-two, 11, 53, 119, 120, 122
Brisbane, Queensland, Australia, 96, 125, 180
British Columbia, 49, *55*, 70, 72–77, 110–111, *see also* BC Rail, *and names of specific places*
Britt, Ray, 32
broad-gauge, 88, 154, 157, *see also* wide-gauge
broken couplers, 111, 117, 198
broken knuckles, 53, 56, 111, 115–116, 117, 118, 119–120
Brook (Sir Isaac) School, 43
Brooklyn, New York City, USA, 25
Brosnan, Dennis William (Bill), 9, 12–13, 14, 26, *26*, 32
buff (coupler force), 9, 114, 116–117, 184
Bull's Gap, Tennessee, USA, 10, 11, 13
Burlington Northern (railroad), 37, 105
Burlington Northern Santa Fe (railroad), 35, 203, *see also* BNSF (railroad)
Burrard Inlet, British Columbia, Canada, 61, 73
Butte, Montana, USA, 35, 36

C

C&NW (railroad) *see* Chicago & North Western (railroad)
C40-8 locomotives, 36, 37, 38, *see also* Dash 8 (locomotive)
C42-8 locomotives, 37
Calcutta, India *see* Kolkata (Calcutta), India
Calgary, Alberta, Canada, 44–47, 50, *50–51*, 52, 54, 56, 58, 60, 61, 62, 64, 65, 83, 98, 102–103, 112
Canada
    Milt Deno timeline, 95–96, 97, 98–99, 102–103, 104, 106, 108
    see also names of specific places
Canadian Institute of Guided Ground Transport, 78
Canadian National Railway (CN Rail), 74, 78
Canadian Pacific Consulting Services, 78, 80, 82, 96, 97
Canadian Pacific Railway, 3, 40
    arctic oil and gas transport project, 78–79
    Assistant Chief Mechanical Officer, 44
    Assistant Chief of Motive Power & Rollingstock, 6, 110
    Chief Mechanical Officer, 43, 47, 77, 78
    Chief Mechanical Officer, Special Assistant to, 82, 85, 96
    coal train derailment, 83–84
    coal trains, 49–57, 61–67, 70–77, 83–84, 95, 110–124

Director of Mechanical Engineering, 77
General Locomotive Foreman, 58, 60, 62
LOCOTROL testing and implementation (summary), 40, 95–99, 102–103, 108, 256
LOCOTROL train operations, 110–124
Milt Deno CPR career summary, 42–44, 85, 95–99, 102–103, 256
Pacific Region, 44, 49–50, 56, 60, 64, 83, 102
Regional Civil Engineer, 64
Regional Purchasing Agent, 64
remote-controlled train operations testing, 49–57
wheelslip testing, 44–49
Canadian Railway Hall of Fame, 6, 109
Canadian Transport Commission, 83, 84
can-style radio antenna, 35
Cape Canaveral, Florida, USA, 24, 80
Cape Lambert, Western Australia, 105, 106, 107, 200
car dumpers, 65, 66, 74
Carbarton, Idaho, USA, 106
Casablanca, Morocco, 98
Century-Series locomotives, 6, 50
CGW (railroad) see Chicago Great Western (CGW) (railroad)
Chakradupur, India, 155
Charlotte, North Carolina, USA, 28, 29, 30
Chase, British Columbia, Canada, 111
Chattanooga, Tennessee, USA, 10, 11, 13, 29, 30, 32, 34
Chicago, Illinois, USA, 36, 38
Chicago & North Western (C&NW) (railroad), 37–39, 104, 106
Chicago Great Western (CGW) (railroad), 37
Chichester Range, Western Australia, 106, 181, 191–192, 200, *202*, *203*
Chief Electrical Engineer, GM, 47–48
Chief Mechanical Officer, CPR, 43, 47, 77, 78, 82
Special Assistant to, 82, 85, 96
China, 101
Chittaranjan Locomotive Works, India, 130, 154
Cincinnati, Ohio, USA, 30, 32
Class WDM-2 locomotives, 130–132, 133, 135, *165–166*
Class-9E locomotives, 198, 200
Class-34 locomotives, 198, 200
Class-59 locomotives, 105
Class-60 locomotives, 86
Cleveland, Ohio, USA, 24, 25, 26

Cleveland Electric Company, 32
Clovis, New Mexico, USA, 203
CMOS (Complementary Metal–Oxide–Semiconductor), 17, 279
CN Rail see Canadian National Railway (CN Rail)
coal trains
Canada, 49–57, 61–67, 70–77, 83–84, 95, 110–124
derailments, 83–84
Multiple Consist Control, 18
Queensland, 4, 124
USA, 10–14, 34–35, 37–39, 50, 106, 110–111, 197–198
Coal-train Coordinator, 65–66
Cocoa Beach, Florida, USA, 67, 69
Colombia River, British Columbia, Canada, 111
Columbia-Geneva Steel Division, US Steel, 11
Commonwealth Edison, 38
Communication Interrupt (LOCOTROL), 139, 192
Complementary Metal–Oxide–Semiconductor (CMOS), 279
Connaught Tunnel, British Columbia, Canada, 47, 52, 53–54, 62, 63, 72, 77, 123
Conrail, 105
Continental Hotel, 86
'Continuity' condition, 21, see also 'No-Continuity' condition
CONTROL (LOCOTROL operating state), 134, 184, 186
Control Console, 14, 19, 21, 22–24, *31*, *32*, *34*, *74*, *97*, *100*, *101*, *104*, 107, 122, *165–166*, *173*, 183, *183*, 187–189, 227, *237*
control valves, 13, 30, 88, 161, 189, 256, 258–259, 266, 270, 282
Controls & Composition Division (Harris Corp), 6, 26, 125
conventional train, 10, 19–20, 119, 122, 124
Coon (railroad location, Western Australia), 193
Coonarrie Creek, Western Australia, 193
Corneil, Ray, 79, 80–81, 82, 96
Corporation of Delta, British Columbia, Canada, 74
Cottbus, Germany, 109
Cowra (railroad location, Western Australia), 191
Cox, Lewis, 6, *71*, *72*, 140, *150*, 151, 157, *248*
CP Consulting see Canadian Pacific Consulting Services
CP Rail see Canadian Pacific Railway
CPR see Canadian Pacific Railway
Cranbrook, British Columbia, Canada, 61

Crane, L Stanley, 10, 12, 13, 26, 30, *30*
cresting grade territory, 117, 121–122
Cuba, 96
curve resistance, 116
cuttings, 122–123

D

Dalian, China, 101
Dallas, Texas, USA, 26, 203
Damchara, Assam, India, 159
Danaluk, Danny, 70
Daroud, Islamic Republic of Iran, 90
Dash 7 (locomotive), 191, *195*
Dash 8 (locomotive), 36, 37, 38, 97, 101–102, 106, 107, *108*, 180, *185*, 187, 194, 200–201, *see also* C40-8 locomotives
Davies (former Alyth Diesel Shop master mechanic), 59–60
Davies, Glen, 107, 201
Davies & Metcalfe, 104, 164
    P85 air braking system, 99
Dayton, Ohio, USA, 25
DD35 (locomotive), 35–36
DD35A (locomotive), 35
DDA40X (locomotive), 36
Delaruelle, Dale, 17, *31*, 32, 34, 54, *57*, 61, 67, *71*, 72–73, *72*, 74, 95, 249
    patents, 256, *see also* Adaptive Air Brake Control System
Delta, British Columbia, Canada, 74
demonstration trains, 197–205
Denius, Homer, 24
Deno, Bob, 76
Deno, Carlysle Frank, *42*
Deno, Ethel, 42–43
Deno, Milt
    CP Rail career summary, 42–44, 77, 85, 95–99, 102–103, 256
    death, 6
    family background, 42–43
    Harris Corporation, 84–87
    honours and awards, 6, 109
    illness, 104, 134–135
    LOCOTROL timeline, 95–109
    patents, 256, *see also* Adaptive Air Brake Control System
    at Queens University, 80–82
    retirement, 108
Denver, Colorado, USA, 107
Departure signal, 180, 188–189
Deutsch Bundesbahn, 108, 109
Didsbury, Alberta, Canada, 102
diesel inspectors, 44, 47, 49, 52, 57, 59, 63–66, 70–74, 75, 76, 77, 95, 256
Digitair, 183, 189, 194
Digital Telephone Systems, 26
Director of Mechanical Engineering, 77
dispatchers, 46, 58, 113, 181, 188, 193
distributed power (DP), 5–7, 10, 14, 17, 35, 36, 38–39, 44, 65, 79, 96, 97, 105, 107, 110, 113, 116, 119, 121, 179, 185, 187, 191–192, 198, 201, 227, 249
distributor valve, 88
Donald, British Columbia, Canada, 118
double-stacked container train, 203
DP (distributed power) *see* distributed power (DP)
Dracon Industries, 26
draft (coupler force), 9, 105, 114–115, 119, 121, 159, 184, 192
Drake St Shops, Vancouver, 44, *51*, 54
drawbar force, 114–116, 118, 119–120, 122, 192
Drumheller, George W, 24
Drumheller & North, 24
Drummond, Ken, 59, 60, 62
dual control stands, 29, 130, 165
Duluth, Winnipeg & Pacific Railroad, 37
Dum Dum International Airport, Kolkata, India, 157
Dumerta, India, 142, 143, 144, *150*, 151, 163
Duncan, Charlie, 46

E

ECP braking, 5, 6, 7, 35, 66, 124, 249, 256
Ektadar, Ali, 86, 88, 93
Electro-Motive Division (EMD) *see* EMD (GM Electro-Motive Division)
electronically controlled pneumatic braking *see* ECP braking
Ellis, Harvey, 17, *31*, 34
EMD *see* General Motors, Electro-Motive Division
Empire Mine, Michigan, USA, 37

Engine Run, 20
engine service (locomotive operations), 14, 119, 125, 249
E-P braking, 34
EPIC brake, 102, 107
EPROM, 152, 279
equalising reservoir (ER), 29, 131, 134, 135, 138, 163, *165, 166*, 181, 184, 190, 257, 258–266, 272–274
Ericsson group, 24
Erie, Pennsylvania, USA, 106, 107
Escanaba, Michigan, USA, 37, 38
Estrada De Ferro Vitória a Minas (railroad, Brazil), 108
Eugene, Oregon, USA, 106
Evans-Deakin Industries, 96, 97, 125, 180

F

F7B (locomotive), 11, 13, 37
Fairbanks-Morse (locomotive), 13, 37
Farinon Corporation, 26
'father of LOCOTROL', 7, 14, see also Selby, Don
Federal Communications Commission, 12, 19
feed valve, 22, 24, 122, 124, 156, 181, 196, 227–228
Ferrell, Dave, 79
Field, British Columbia, Canada, 44, 52, 64, 83, 95, 109
firecracker (radio antenna), 35–36
Fisher, Dick, 10–11, 12, 14, 28–30, 32
Flat Creek, British Columbia, Canada, 83
Flett, Bill, 44, 47, 49
'floating node' see node
Florida, USA, 6, 13, 24–25, 40, 41, 50, 61, 66, 67–73, 79–81, 85, 97, 109, 256
Florida East Coast Railroad, 111, 113
flow meter, 54, 258, 260
FM Half-Duplex carrier, 19, 22
Ford, Bacon & Davis, 74
formula (locating in-train power groups), 113
Fortescue River, Western Australia, flood plain, 191, *203*
Foster Yeoman Ltd (Mendip Rail), 105
Fraine, Jack, 56, 59, 61, 64, 66, 70, 71–72, 76, 77, 80
Franklin, Richard E, 11, 13
Fraser River, British Columbia, Canada, 74
Freeman, Gordon, 143
freewheeling, 46, 181
freight trains, 3, 4, 9, 14, 110, 112, 126, 138, 257, 260, 262, 274

Fremont, Nebraska, USA, 37
frequency shift keying (FSK), 18, 19, 21, 279
Function Selector Unit, 11, 29, 30

G

Galion, Ohio, USA, 13, 17, 24
Galloway, British Columbia, Canada, 122
Ganges River, 127, 128, 160, 162
Garden (railroad location, Western Australia), 192
Gas Arctic Northwest Project Study Group, 78–79
General Electric (GE), 26, 41, 95, 99, 106, 107, 108, 125, 164
　Edison Award, 133
　　locomotives, 38, 97–98, 101–102, 105–106, 107, 108, 155, 180, 183, 187, *195*, 198, 200–202, see also Dash 8 (locomotive)
General Electric (GE) Harris, 26, 99, 105, 106, 107, 108, 125
General Electric (GE) Transportation, 5, 7, 26
General Locomotive Foreman, 58, 60, 62
General Manager, Harris Corporation Controls Division, 84
General Manager, Radiation Corporation, 69, 70
General Managers, Indian Railways, 154, 156–157, 160
General Motors Chief Electrical Engineer, 47–48
General Motors Diesel (GMD), 44, 47, 48, 279
General Motors Electro-Motive Division, 10, 13, 29, 39, 48–49, 86, 96, 279
　　locomotives, *12*, 31, 36, 37, 46, 86, 105, 198, *237*, 280
General Motors of Canada Ltd, Diesel Division, 279
General Railway Signal Company, 9, 10
Germany, 40, 108, 109
Gheddis, Jim, 102
Gidgy (railroad location, Western Australia), 181, 188, 190
Gilbert, Tom, 11, 12–13, 26
Glacier, British Columbia, Canada, 45, 46, 83, 123, 124
Gladstone, Queensland, Australia, 99
Glass Hotel, Singapore, 127
GMD see General Motors Diesel (GMD)
Goetz, Lou, 84, 85
Goff, Q A Jnr, 197
Golden, British Columbia, Canada, 62, 63, 65, 70, 72, 111
Goldstein, Glenn, 34
Goldsworthy Mining Ltd, 195

Goodwin-Alco locomotives, 130
Gottbehuet, Ron, 17, 32, 34, 74
Green, Barry, 101, 125
Green, Eugene, 106
Greenville, South Carolina, USA, 12
Griffith, British Columbia, Canada, 119
GT26CW (locomotive), 86
Guwahati, India, 104, 126, 137, 138, 139, 141, 150–151, 152, 154, 155–156, 159, 160, 161, *174–175*, *177*

H

Haflong Hill, Assam, India, 158, *175–176*
Haldia, India, 126
Hall Effect, 18, 279
Hamersley Iron, 107
Harangajao, Assam, India, 158
Harris Aerospace, 31
Harris Controls, 5, 6, 25–26, 32, 35, 41, 44, *73*, 79, 85–86, 95, 96, 97, 106, 107, 111, 125, 140, 179, 256
Harris Corporation, 25–26, 40, 41, 82, 84, 111, 125, 163, 164, 197, 249
    Controls & Composition Division, 6, 26, 84, 125
    Railroad Product Line, 6, 26, 40, 107, 249
Harris-Intertype Corporation, 25–26
Hart, Charlie, 34
Hatia, India, 139, 145, 150, 152, 153, *172*
Hawker Siddeley Canada, 50
Hayward, Harold, 47–48, 70, 82, 95
head-end (front of train), 5, 9, 34, 37–38, 52, 72, 83, 105, 107, 110, 115–119, 121, 123, 144, 145, 180, 183, 188–190, 196, 200, 206
Heinz, Howard G, 10–11
helpers, 5, 9–10, 37, 98, 111, *see also* banker engines; pushers
Heneka, Steve, 104
Herndon, Don, 107, 108
Hesta (railroad location, Western Australia), 191, *203*
Hewett, M Worth, 14, 34
Hill, L A (Al), 55–56, 62, 64, 65, 76, 80
hind-end (rear of train), 5, 9, 111, 113, 114, 121, 122
Hinkle, Oregon, USA, 36–37, 106
Hi-Rail vehicles, 181, 193
Hoagie, Henry, 45–49
Honeywell International Inc, 11
Hong Kong, 163, 164
Hooghly River, India, 128, 162
Hooley, Vic, *52*, 53, 57
hosebag (air hose), 30, 88, 193
Host of America Motel, 67, 69
Hot Bearing/Hot Wheel detectors, 193
Hot Box Detectors, 197
Hotel Brahmaputra, Guwahati, India, 139, 160
Hotel Radhika, Rourkela, India, *129*, 133, 139, 152, 163, *167*
Howrah railway station, Kolkata, India, 126, 128, 129, 153, 155, 163
hump-controller, *101*, 121, 189
Huntington, Oregon, USA, 36
Hyatt Regency Hotel, Tehran, Iran, 86, 92
hysteresis, 270, 272, 274, 280

I

Iaeger, West Virginia, USA, 197
IDAC Wheel Slip Control System, 37, 48–49
Iden, Mike, 38–39
IDILB-08 (UPRR train), 203, *204–205*
Income Tax Clearance, 163
Independent (locomotive brake), 10, 11, 47, 101, 107, 118, 124, 130, 183–184, 185, 187, 189
Independent Control, 14, 21, 120, 121, 145, 181, *185*, 188, 190, 192, 227
Independent Motoring, 14, 34, 181, *185*, 192
India, 3, 40, 88, 101, 104, 125–165, *165–178*
    Milt Deno timeline, 101, 104
    see also names of specific places
Indian Railways, 3, 88, 101, 104, 125–165, *165–178*
    General Managers, 154, 156–157, 160
    Research, Design & Standards Organisation (RDSO), 144, 152, 156, 157, *175*
    Senior Locomotive Inspector, 137, 143
Indian River, Florida, USA, 67
Indian standard-gauge, 132, *see also* wide-gauge
Industrial Electric Company, 28
Institution of Mechanical Engineers, Railway Division
    Railway Engineering Journal, 110
Integrated Function Control, 107
International Railway Journal, 125
Intertype Corporation, 25, *see also* Harris-Intertype Corporation
Intracoastal Waterway, Florida, USA, 67

Iran, 40–41, 86–95, 164
    Milt Deno timeline, 97, 105
Iran Railways, 41, 85–95, 97, 105, 138
Iron Ore Company of Canada, 96
Isfahan, Iran, 90, 93
Islamic Republic of Iran *see* Iran
Ispat General Hospital, Rourkela, India, 104, *136*, *167*

J

Jacksonville, Florida, USA, 13
Japan, 97
    coal contract, 49, 52, 56, 62, 73
Jatinga, Assam, India, 158
Jilalan, Queensland, Australia, 99
John Sevier railroad yard, Knoxville, Tennessee, USA, 11, *13*, 27, *27*
Johnson, Don, 102, *248*
Johnson, Tom, 54, 57, 65, 77
judgement (train handling), 192, 195, 249, 256, 260

K

Kaiser Resources, 73
Kaiser Steel, 35
Kalgan (railroad location, Western Australia), 181, 186–187
Kamloops, British Columbia, Canada, 43, 54, 57, 63, 65, 121
Kansas City Southern (KCS), 33–34, 98
Karaj, Iran, 89, 90, 93
Karampada, India, 142, 143, 151, 152, *168*, *169*
Karratha, Western Australia, 105, 106
Khouribga, Morocco, 98
Kimball, Richard, 34, 38
kinetic energy, 194, 200
Kingston, Ontario, Canada, 78, 79, 81–82
Kiriburu, India, 142, 143, *168–171*
Kitsilano Secondary School, Vancouver, 43
Kline, Kerry, 125
Knersvlak (railroad location, South Africa), 200
Knorr-Bremse, 6, 99, 108, 201
knuckles *see* broken knuckles
Kolkata (Calcutta), India, 104, 126, 127–129, 134, 140, 150, 152, 154, 161–162, 163, 164
Kootenay River, British Columbia, Canada, 111
Kruk, Tony, 71

Kwality Inn, Ranchi, India, 145, 150, 152

L

Lake Louise, Alberta, Canada, 47
Langdon (railroad sub-division), British Columbia, Canada, 46, 47, 52
Leachman, Rob, 35, 36
Lead (LOCOTROL locomotive), 6, 14, 17, 18–24, 28, 29, 30, *31–32*, 35, 37, 38, 50, 52, 53–56, 62, 63, 66, 70, 72, 91, 99, 101, 105, 110, 111, 113, 115–123, 125, 130, 137–139, 144–145, 183–188, 190–192, 196, 197, 198, 200, 206, 227, *see also* Master (LOCOTROL locomotive)
leakage (brake pipe), 54, 88–89, 123, 124, 134
Leanchoil, British Columbia, Canada, 118
Leffingwell, Ralph, 17, 61, 67
Lethbridge, Alberta, Canada, 55, 56, 62, 95
LINK (LOCOTROL operating state), 134
Live Oak, Perry & Gulf Railroad, 11
LMD *see* Lumding, Assam, India
load-out, 73, 106
Locomotive Crew Foreman, 40, *see also* road foremen of engines (RFE)
locomotive engineers, 5, 6, 9, 11, 14, 19, 21, 23, 24, 29, 43, 73, 105, 117, 122, 125, 181, 184, 197, 227, 249, 256, 260
Locomotive Remote Control system, 206–216
LOCOTROL 102, 37
LOCOTROL 105, *32*, *50*, 95, 96, 97–98, *99*, *102*, 201, *222*, *225*
LOCOTROL 105SS, 6, 85, 179, 206
LOCOTROL I, *12*, 40, 206, 207–232
    operating processes, 227–232
    override, 227
    previous names, 206
LOCOTROL IDP, 97
LOCOTROL II, 6, 35, 37–38, 40, 95, 97–102, *101*, *102*, 104, 105–106, 179, *182–183*, 183, 196, 201, 233–241, 245–248
    commonalities with LOCOTROL III, 233–239
    handling a LOCOTROL train, 179–196
    hardware configuration, 241
    in India, 104, 125–178
    operating configuration, 240
    support equipment and services, 248
    technical data and specifications, 245–247
LOCOTROL III, 5, 6, 32, 36, 38, 97, 102–109, *102*, *103*, *104*, 179, 198, 200–201, 233–239, 242–248
    commonalities with LOCOTROL III, 233–239

hardware configuration, 244
operating configuration, 242–243
support equipment and services, 248
technical data and specifications, 245–247
LOCOTROL Operator's Manual, 126
LOCOTROL XA, 7, 179, 201
Loehne, Bob, 197, 198
logic cabinets, 17, 19, 22–23
London, England, 162
London, Ontario, Canada, 279
Long Beach, California, USA, 203
Loop 4 (railroad location, South Africa), 200
Loop 5 (railroad location, South Africa), 200
Lord shock-mounts, 29
Los Angeles, California, USA, 36, 203
Lost Springs Hill, Wyoming, USA, 37
Louisville & Nashville Railroad, 9, 14
'Low BP' safety feature, 161
Lucknow, India, 104
Lumding, Assam, India, 156–157, 158, 159

M

Macdonald, Sir John, 81
Maibong, Assam, India, 157
Manager of Systems Engineering, New York Air Brake, 141
Manchester, England, 164
Marion, North Carolina, USA, 13, 27, 28
Marion Machine, 28
master (LOCOTROL locomotive), 10, 11, 13, 27, 28, 32–33, 60, 70–71, 72, 74, 198, 206, *see also* Lead (LOCOTROL locomotive)
master mechanics, 28, 41, 42–43, 47, 58, 60, 64, 70, 77, 95, 256
MCC (Multiple Consist Control), 17–24
McCagg, Les, 78
McGill University, Montréal, Quebec, Canada, 80, 96
McGregor's Motor Inn, Revelstoke, British Columbia, Canada, 62, 76
Medicine Hat, Alberta, Canada, 44, 57–58
megacycles, 280
Meghataburu, India, 142, 143, 168, 169
Melbourne, Florida, USA, 13, 24, 26, 27, 38, 41, 50, 61, 67, *68*, *71*, *72*, 79–80, 84, 97, 107, 109
membrane switch, 189, 280

Merehead Quarry, England, 105
Mesa-J mine site, RRIA, Western Australia, 106
metre-gauge railways, 40, 87, 104, 108, 141, 154, 156, 157
Mexicali Airlines, 79
Mexico, 40, 97–98, 105
mid-train locomotives, 10, 37, 110, 191, 197
Midwest Power Systems, 38
Mills, Walter, 44
Minister of Railways (India), 154, 164
Minister of Railways (Iran), 86, 89
Mitchell, Dean Scott, 34
Mitchell, Pop, 181
Mitsubishi & Company, 73
MLW, 50, 57, 111, 183, 195
MLW-Worthington, 130, 131
Mode Selector, *34*, *97*, 181, 188, 189, 191, 192, 196, *227*
Montréal, Quebec, Canada, 46, 47–48, 50, *52*, 57, 70, 77, 79, 80, 82, 85, 95, 96, 162
Moore, J G, 10, 28
Moore, W H, 32
Moose Jaw Electric Railway, 42
Moose Jaw, Saskatchewan, Canada, 42, 43, 44, 69
Moroccan National Railways, 98
Morocco, 40, 98, 99
motive power groups, 3, 119
Motorola, 13, 18, 29
Mount Macdonald Tunnel, British Columbia, Canada, 77, *103*, 123
Mount Newman Mining Company, 6, 40, 96, 98, *101*, 107, 125, 138, 143, 179, 201
LOCOTROL units, *99–101*
railroad, Newman to Port Hedland ore train run, 180–196
Railroad Operations department, 125
Mountain (railroad subdivision), British Columbia, Canada, 44, 46, 47, 49, 52, 53, 56, 61, 64, 83, 96
Mozambique Railways, 164
Mukherjee, G C, 126, 163
Multiple Consist Control (MCC), 17–24
Multiple Unit mode, 14, 20, 23, 87, 181
Munich, Germany, 108, 109
Muscroft, Geraldine, 43

N

Nampa, Idaho, USA, 37, 106

Neal, Andrew, 102, 125
NEC *see* North Electric Company (NEC)
NEFR *see* North East Frontier Railway (NEFR), India
Nelson, British Columbia, Canada, 44
Netaji Subhash Chandra Bose International Airport, Kolkata, India, 127
New Delhi, India, 101, 104, 137, 138, 140, 152, 154, 164
New Orleans, Louisiana, USA, 14, 28, 30
New York Air Brake (NYAB), 28, 33, 34, 54, 62, 63, 64, 66–67, 104, 141, 201
New Zealand Railways, 5, 137
Newman, Western Australia, 98, 101, 125, 180–181, 183, *185, 186*
Newman to Port Hedland, Western Australia, ore train run, 180–196
Niles, Ohio, USA, 25
'No-Continuity' condition, 20–22, 122, *see also* 'Continuity' condition
nodal point, 191
node, 114, 116–117, 184, 186, 191–192
non-synchronous operation, 21, 184
Norfolk & Western Railroad, USA, 29, 34, 197–198
Norris Yard, 29
North, Charles N, 24
North East Frontier Railway (NEFR), India, 125, 126, 141, 154, 156, 157, 160, 164
North Electric Company (NEC), 7, 13–17, 24–25, 26–27, 30, 111, 206
    establishment, 24
    LOCOTROL advertisement, *16*
    LOCOTROL product names, 206
    Paricode Supervisory System advertisement, *15*
North Electric Manufacturing Co, 24
North Electric Works, 24
North Platte, Nebraska, USA, 38
Northwest Territories, Canada, 78
Notch Hill, British Columbia, Canada, 47, *103*
NYAB *see* New York Air Brake (NYAB)
NZ *see* New Zealand

O

Oberoi Grand, hotel, Kolkata, India, 126, 128–129, *128, 129*, 155, 163
Odisha (Orissa), India, 126
Okanagan Lake, British Columbia, Canada, 77
Old Fort, North Carolina, USA, 13, 29
Oleigis, Frank, 58

Omaha, Nebraska, USA, 38
operations inspector, 136
Oregon, USA, 36
override, 122–123, 227

P

Pacific Coast, 111
Pacific Region, Canadian Pacific Railway, 44, 49–50, 56, 60, 64, 83, 102
Page, Arizona, USA, 81
Pannawonica, Western Australia, Australia, 105, 106, 200
paper vellum, 282
Parker, Charlie, 6, 47, 50, 52–53, *52*, 56, 57, 80, 98, 180
    comments on CP Rail operations, 110–124
patents, 7, 17, 24, 99, 133, 256
    Adaptive Air Brake System, 256–269
    Adaptive Air Brake System, Improved, 269–278
PCM (Pulse Code Modulation), 24
penalty brake, 11, 161
Penticton, British Columbia, Canada, 77
People's Republic of China *see* China
Peterson, Harvey, 52
Pike, Charlie, 78, 79, 80, 82, 83, 84
Pilbara Iron (mining railroad network, Western Australia), 107
Pilbara region, Western Australia, 5, 96, 107, 179, 180–196, *201–203*
Piper, Harry, 47, 64
PLC (Programmable Logic Controller), 280, 281
Pocatella, Idaho, USA, 106
Port Hedland, Western Australia, 6, 40, 98, 101, 125, 140, 152, 179–180
    LOCOTROL units, *100, 195*
    Newman to Port Hedland ore train run, 180–196
Port of Vancouver, British Columbia, Canada, 74
    Roberts Bank terminal, 49, *49, 55*, 60, 63, 64, 65, 66, 72, 73–74, 77, 102, 110, 111
Portsmouth, Ohio, USA, 197–198
Potomac Yard, Virginia, USA, 9, 14, 30
Powder River Basin (USA) coal trains, 35, 37, 38, 106
power assembly, 95, 280
power group, 3, 111, 113–120, 123, 198
power supply, 17–18, 19, 23, 24, 30–31, 87
PRD Electronics, 25
Primmer, G, 106

Prince George, British Columbia, Canada, 104
printed circuit (PC) boards, 18, 19, 27, 280
Probe (computer), 102
Program Manager, 85–86, 94
Programmable Logic Controller (PLC), 280, 281
Proviso, Utah, USA, 37
pulled drawbar, 116
Pulse Code Modulation (PCM), 24
Pulse Width Modulation (PWM), 281
pushers, 5, 9, 10, 106, 110, 113–121, 200–201, *see also* banker engines; helpers
PWM (Pulse Width Modulation), 281

Q

QNS&L *see* Québec, North Shore & Labrador Railroad (QNS&L)
Qom, Iran, 89, 90, 93, 97
Quarry 5 (railroad location, Western Australia, Australia), 191
Québec, North Shore & Labrador Railroad (QNS&L), 78, 82, 95, 96, 97, 104, 106
Quebec-Cartier Mining, 97
Queens University, Kingston, Ontario, Canada, 78, 79, 80, 96
Queensland, Australia, 4, 6, 40, 96, 99, 201
Queensland Rail, 6, 40, 96, 99, 124, 179
Quincy, Illinois, USA, 25

R

Rabat, Morocco, 98
Radhika Hotel *see* Hotel Radhika, Rourkela, India
Radiation Incorporated, 17, 24–25, 27–28, 29, 30, 32, 50, 52–54, 57, 61, 66–72, 95, 111, 249
    acquired by Harris Controls, 25, 26, 32
    acquires North Electric, 14
    General Manager, 69, 70
    LOCOTROL pamphlets, 206–226
radio antennae *see* antennae (radio)
radio contact, 73, 122
Radio Control System (RCS), 17, 27–29, 34, 35–36, 197
radio frequency (RF), 6, 281
radio-controlled systems, 13, 17, 18, 123, 197
Radio-Frequency Identification (RFID), 281
Rail India Technical and Economic Services, 164
railroad diamond, 195

railroad managers, 40, 101–102, 125
Railroad Operations Department, Mount Newman Mining, 125
Railroad Product Line (Harris Corporation), 6, 26, 40, 107, 249
Railway Digest article, 180
Railway Engineering article, 110
Rakshi, India, 152
Ranchi, India, *144*, 145, 150, 152, 153, *169*, *177*
RCL (Remote-Control Locomotive), 7
RCS (Radio Control System), 17, 27–29, *34*, 35–36, 197
RDSO *see* Indian Railways Research, Design & Standards Organisation
Red Deer, Alberta, Canada, 102
Regina, Saskatchewan, Canada, 42, 44, 85
Regional Civil Engineer, Canadian Pacific Rail, 64
Regional Purchasing Agent, Canadian Pacific Rail, 64
Relay Interface Cabinet, 19, 33, 87, *165–166*, *174*
Remote (LOCOTROL locomotive), 9, 14, 17, 18–24, 28, 30, *31–32*, 32, 37–39, 50–57, 62–63, 66, 70, 72–73, 74, 83, 86, 87–89, 91, 97, 99, 105, 106, 107, *108*, 110–124, 125, 130, 137–139, 145, 156–157, 172, 183–196, 197, 198, 200, 227–228, *see also* Slave (LOCOTROL locomotive)
Remote Multiple Uniter (RMU), 10–14, 17, 35
Remote-Control Locomotive (RCL), 7, 99, 183
Revelstoke, British Columbia, Canada, 44, 45, 46, 47, 49, 50, 57, 61–66, 69, 70, 72–73, 76–77, 83–84, 95, 116, 119, 123
Reynolds, Art, 44–50, 52, *52*, 53, 54, 56–60
RF (radio frequency), 6, 281
RF Communication (company), 25–26
RFE (road foremen of engines), 14, 34, 38, 40, *50*, *52*, 53, 63, 64–65, 66, 74, 76, 83, 106, 126, 197
RFID (Radio-Frequency IDentification), 281
Rio de Janeiro, Brazil, 108
Rio Tinto (iron ore mining company, Western Australia, Australia), 107
RITES Ltd (Rail India Technical and Economic Services), 164
RMU (Remote Multiple Uniter), 10–14, 17, 35
road foremen of engines (RFE), 14, 34, 38, 40, *50*, *52*, 53, 63, 64–65, 66, 74, 76, 83, 106, 126, 197
road knowledge, 192
Robe River Iron Associates (RRIA), 40, 105, 106–107, 200–201
    iron ore trains, 201, *201*, *202*

Roberts Bank (terminal, Port of Vancouver), 49, *49*, *55*, 60, 63, 64, 65, 66, 72, 73–74, 77, 102, 110, 111
Robot 1, 55, 95, *96*
Robot cars, 62, 63, 71, 83
Rochester, New York, USA, 9, 26
Rocky Mountains (Rockies), 102, 111
Roll-By marker, 179, 188
rolling terrain, 121
rollingstock, 78, 86, 141, 196, 200
rotary couplers, 63, 65, 66, 73
Rourkela, India, 104, 126, 129, *136*, 137, 139, 150–153, 154, 155, 161, 163, *167*, *173*
Routledge, Fred, 38, 41, *57*
Rudola, Devendra, 140, 163, 164
Ruff, D, 12
ruling grade, 29, 38, 110, 113
Ruzek, Bill, 45, 47

S

Saggiesberg, South Africa, 200
Saginaw, Michigan, USA, 42
SAIL (Steel Authority of India Ltd), 142, 143
Salt Lake City, Kolkata, India, 126
Salt Lake City, Utah, USA, 35, 36, 37, 106
Saluda Grade, North Carolina, USA, 10, 14
Sandhill (railroad location, Western Australia, Australia), 186
Satellite Beach, Florida, USA, 80, 85, 97
SCADA (Supervisory Control And Data Acquisition) system, 7, 17, 25, 27, 30, 281
Schefferville, Quebec, Canada, 82
Scientific Timesharing Systems Ltd, 126, 129, 134
SD40-2 (locomotive), 35, 37, 48–49
SD45 (locomotive), 35–37, *103*, 197
SD90AC (locomotive), 36
Sears, Dick, 69
Segmented Brake Pipe, 99
Selby, Don, 7, 10, 14, 17, 26, 30, *31*, 32, 74
Selkirk Mountains, British Columbia, Canada, 53, 110, 111, 123
semiconductors, 24, 26, 281, 282
    Complementary Metal–Oxide–Semiconductor (CMOS), 279
Senior Electrical Engineer, South East Railways, India, 144
Senior Locomotive Inspector, Indian Railways, 137, 143, 144, *167*
Sept-Iles, Quebec, Canada, 78, 82, 95–96
SER *see* South Eastern Railway (SER), India
Sergeants Bluff, Iowa, USA, 38
Sevier railroad yard, Knoxville, Tennessee, USA, 11, *13*, 27, *27*
Seybold Machine Company, 25
Shaw (railroad location, Western Australia), 191–192, *203*
Shaw, George, 24
Shawnee Junction, Wyoming, USA, 37
Shorai (interpreter), 87, 88, 89, 92
Shuswap (railroad subdivision), British Columbia, Canada, 46, 47, 52, 61, 64
Siding Two (railroad location, Western Australia), 106, 200, *202*
Simultaneous Operating Airbrake Repeater (SOAR), 249–255
Singapore, 126–127, 164
Singh, Malkit, 136–137, 144, 152, 163, *172*, *173*
Sioux City, Iowa, USA, 38
Sir Isaac Brook School, 43
Sishen-Saldanha railway, South Africa, 198, 199, *200*
Ski-Doos, 63, 64
slack (train dynamic state), 10, 115–122, 124, 159, 181, 183, 184, 186, 282
Slave (LOCOTROL locomotive), 10, 11, 12, 13, 14, 27, 28, 30, 32, 33, 60, 206, *see also* Remote (LOCOTROL locomotive)
slimkabel (smart cable), 198
Smith, Eugene (Gene), 6, 37, 40, 41, *71*, *72*, 99, *103*, 104, 125, 133, *133*, *138*, 150, *172*, 200, 249
    patents, 256, *see also* Adaptive Air Brake Control System
Smith, Sterling, 78
SOAR (Simultaneous Operating Airbrake Repeater), 249–255
solid coupler, 65
Solid-State Relay (SSR), 281–282
South Eastern Railway (SER), India, 125, 126, 130, 138, 145–151, 153, 154, 164, *167*, *169–170*, *172–173*
South Hedland, Western Australia, 195
South Morrill, Nebraska, USA, 37
South Thompson River, British Columbia, Canada, 121
Southard, Gary, 14, 17
Southern Railway (SR), USA, 7, 9–14, *12*, 17, 26–35, 38–39, 50, 53, 95, 110, 206

Assistant Chief Mechanical Officer, 10
President *see* Brosnan, Dennis William (Bill)
VP, Research & Development *see* Crane, L Stanley
Southern Railway Historical Association, 10
Space Radio, 9
Spartanburg, North Carolina, USA, 10
Sparwood, British Columbia, Canada, 60, 61, 64, 65, 72, 73, 77, 111–112
Spencer, North Carolina, USA, 9, 13, 29, 30
SR *see* Southern Railway (SR), USA
SSR (Solid-State Relay), 281–282
St John's Ravenscourt (school), 43
stall (train), 54, 120, 123
steam locomotives, 9, 154, 160, *169*, *174*, *177*
Steel Authority of India Ltd (SAIL), 142, 143
Steele, George C, 24
Stephen, British Columbia, Canada, 47
Stewart, Bill, 43–44
Stinson, Bill, 78, 84, 85
Stoney Creek, British Columbia, Canada, 53–54, 111
strain-gauge, 9
Strait of Georgia, British Columbia, Canada, *49*, 74
Super Spool Valves, 11, 29
Supervisory Control And Data Acquisition (SCADA) system, 7, 17, 25, 27, 30, 281
Sutherland, Saskatchewan, Canada, 42
Swannanoa Tunnel, North Carolina, USA, 13, 29
Swissvale, Pennsylvania, USA, 12
Syracuse, New York, USA, 104
system description
    Adaptive Air Brake Control System, 260–278
    Locomotive Remote Control system, 206–216
    LOCOTROL I, 217–232
    LOCOTROL II and III, 233–248

T

Tabriz, Iran, 94–95, 97
Takkinnen, Lawrence, 65
Taylor, Clyde, 197, 198
Taylor, H C (Henry Clay), 7, 14, 28, 29, 32, 34
Taylor, Ian, 125
Tehran, Iran, 85, 86–87, 89, 90, 92, 93–94, 97, 105
telephone equipment manufacturers, 24
Telephone Improvement Company, 24
Thompson River, British Columbia, Canada, *55*

South Thompson River, 121
timeline (Milt Deno), 95–109
TMU (Train Multiple Uniter) system, 9–10, 14
Tower Control, 7, 40, 74, *74–76*, 102, 105, 106
track profile, 111, 121, 137, 192, 201
    Mt Newman Mining Railroad, 181, 184, 186, *187*, 188, *188*, *190*, *191*, *193*, *194*, *195*
    North East Frontier Railway, India, *158*
track/train dynamics conference, 110
train configuration, 114–116, 179
train dispatchers *see* dispatchers
train dynamics, 110, 117, 124, 192
train handling, 89, 110–111, 113, 119, 121–124, 143, 153, 158–159, 198, 200, 249, 256
    Newman to Port Hedland ore train run, 180–196
train length, 115, 123–124, 156, 201, 276
Train Multiple Uniter (TMU) system, 9–10, 14
train separation, 10, 11, 111, 114, 115, 117, 118, 119, 121–122, 123, 124, 202
trainline, 19, 20, 21, 23–24
Trans-Canada Highway, 77
Transnet Freight Rail, 198, *199*
Trenton, Nova Scotia, Canada, 50
tunnels, 13, 36, 122–123
    Connaught Tunnel, British Columbia, Canada, 47, 52, 53–54, 62, 63, 72, 77, 123
    derailment, Flat Creek, British Columbia, Canada, 83
    India, 137, 143, 157, 158
    Iran, 89, 90, 93, 105
    loading tunnels, 72, 73
    Mount Macdonald Tunnel, British Columbia, Canada, 77, *103*, 123
    Swannanoa Tunnel, North Carolina, USA, 13, 29
    USA, 198
turbocharger, 45–46, 47
Turner (railroad location, Western Australia), 194
26-C Automatic brake valve, 28, 180–196
28L-AV (locomotive air brake schedule, India), 130
twin-pipe systems, 88, 89

U

undesired emergency brake application (UDE), 30, 282
union action, 14
Union Pacific Railroad (UPRR), 35, 36, 38, 106, 107, 203, *204–205*

Union Switch & Signal (US&S), 10, 11, 12–13, 17, 28
unit trains, 36, 49, 57, 65, 95, 105, 110–111, 113, 119, 153
    Newman to Port Hedland ore train run, 180–196
United Kingdom
    Milt Deno timeline, 105
United States of America
    Milt Deno timeline, 96, 97, 104, 105–106, 107–108
    see also names of specific places
United Utilities, 24
Universal (LOCOTROL model), 130
University Computing Company, 26
University of Edmonton, 78
University of Wisconsin, 42
UNLINK (LOCOTROL non-operational state), 196
unmanned ore trains, 107
unmanned remote distributed power, 14
unmanned trailing locomotive group, 106–107
UPRR (railroad) *see* Union Pacific Railroad (UPRR)
US Steel, Columbia-Geneva Steel Division, 11
US&S (Union Switch & Signal), 10, 11, 12–13, 17, 28

V
vacuum braking, 40, 104, 126, 130, 141, 156, 158
Vancouver, British Columbia, Canada, 43, 49, 53, 54, 55, 56, 57, 58, 61, 64, 65–66, 71, 72, 77, 80, 95, 104
    Drake St Shops, 44, *51*, 54
    LOCOTROL test personnel, *52*
    Port of Vancouver Roberts Bank terminal, 49, *49*, *55*, 60, 63, 64, 65, 66, 72, 73–74, 77, 102, 110, 111
Varanasi, India, 101
vellum, 18, 282
Veracruz, Mexico, 98, 105
Vermilion, Butch, 106
Vitória, Espírito Santo, Brazil, 108
Vorhees, George, 140
Vulcano, Angelo, 71

W
Wabco, 9, 10, 11, 13–14, 17, 27–30, 34, 35, 77, 107
WAG-5 (locomotive), 144
wage savings, 110
Walkabout Hotel, Newman, Western Australia, 125
Wall, Mike, 106
Wallin, Johnny, *51*
Ward, Jim, 180, 181
Wasatch Grade, 36
Watertown, New York, USA, 66, 67, 104
WDM-2 locomotives, 130–132, 133, 135, *165–166*
Weber Canyon, Utah, USA, 36
Weeli (railroad location, Western Australia), 187–188
Wells, T, 106
West Bengal, India, 126
Western Australia
    Newman to Port Hedland ore train run, 180–196
    Pilbara region, 5, 96, 107, 179, 180–196, *201–203*
    Port Hedland, 6, 40, 98, *100*, 101, 125, 140, 152, 179–180, *195*
    see also BHP Iron Ore (BHPIO); Mount Newman Mining Company; Robe River Iron Associates (RRIA)
Western Railroad Properties Inc (WRPI), 37, 38
Westinghouse Air Brake Company, 10, 52, 57, 65, 130
Westinghouse Australia, 187
wheelslip, 44–49, 111, 119, 144, 145, 190, 229
White Power, 33, *34*
wide-gauge, 104, 126, 130, 132, 138, 157, 169, *see also* broad-gauge
Wilkie, Saskatchewan, Canada, 42
Williamson, West Virginia, USA, 197–198
Windermere (railroad subdivision, British Columbia, Canada), 61–62, 70, 72
Winnipeg, Manitoba, Canada, 42, 43
wireline, 6, 206
WRPI *see* Western Railroad Properties Inc (WRPI)

Y
Yandee (railroad location, Western Australia), 194
Yandi mine, Western Australia, 202
Yardman, 196
YDM-4 (locomotive), 141, 156, *173*, *174*
Yermo, California, USA, 37
York Canyon, New Mexico, USA, 35
Yukon, Canada, 78
Yule River, Western Australia, 193

www.ingramcontent.com/pod-product-compliance
Lightning Source LLC
Chambersburg PA
CBHW061131010526
44107CB00068B/2908